Association for Women in Mathematics Series

Volume 33

Focusing on the groundbreaking work of women in mathematics past, present, and future, Springer's Association for Women in Mathematics Series presents the latest research and proceedings of conferences worldwide organized by the Association for Women in Mathematics (AWM). All works are peer-reviewed to meet the highest standards of scientific literature, while presenting topics at the cutting edge of pure and applied mathematics, as well as in the areas of mathematical education and history. Since its inception in 1971, The Association for Women in Mathematics has been a non-profit organization designed to help encourage women and girls to study and pursue active careers in mathematics and the mathematical sciences and to promote equal opportunity and equal treatment of women and girls in the mathematical sciences. Currently, the organization represents more than 3000 members and 200 institutions constituting a broad spectrum of the mathematical community in the United States and around the world.

Titles from this series are indexed by Scopus.

Alina Bucur • Wei Ho • Renate Scheidler
Editors

Research Directions in Number Theory

Women in Numbers V

Springer

ASSOCIATION FOR WOMEN IN MATHEMATICS

Editors
Alina Bucur
Department of Mathematics
University of California, San Diego
La Jolla, CA, USA

Wei Ho
Department of Mathematics
Princeton University
Princeton, NJ, USA

Renate Scheidler
Department of Mathematics and Statistics
University of Calgary
Calgary, AB, Canada

ISSN 2364-5733 ISSN 2364-5741 (electronic)
Association for Women in Mathematics Series
ISBN 978-3-031-51676-4 ISBN 978-3-031-51677-1 (eBook)
https://doi.org/10.1007/978-3-031-51677-1

Mathematics Subject Classification: 05C25, 14G50, 11G20

This Springer imprint is published by the registered company Springer Nature Switzerland AG
The registered company address is: Gewerbestrasse 11, 6330 Cham, Switzerland

Preface

This volume is a compilation of research and survey papers in number theory, written by members of the *Women in Numbers* (WIN) network and their colleagues, principally by the collaborative research groups formed for the fifth Women in Numbers workshop. *Women in Numbers* 5 (WIN5), organized by Alina Bucur (University of California San Diego), Wei Ho (Institute for Advanced Study, Princeton University, and University of Michigan), and Renate Scheidler (University of Calgary), was to take place on November 15–20, 2020 at the Banff International Research Station in Banff, Alberta, Canada. To our great regret, the event could not take place in-person on site at BIRS as planned due to the COVID-19 pandemic.

The WIN network (`womeninnumbertheory.org`) was created for the purpose of increasing the number of active female researchers in number theory. The WIN conference series began in 2008 as the main mechanism for effecting this goal. It introduced a novel research mentorship model where women at all career stages, from graduate students to senior members of the community, joined forces to work in focused research groups on cutting-edge projects designed and led by experienced researchers. This model has proven so successful that to date there are over 25 research networks for women in mathematics, each of which holds *Research Collaboration Conferences for Women* as well as other conferences, workshops, special sessions, and symposia. The Association for Women in Mathematics, funded by the National Science Foundation ADVANCE program, is supporting this highly effective research mentorship model (`awm-math.org/programs/advance-research-communities`).

The goals for WIN5 were to generate research in significant topics in number theory, to broaden the research programs of women working in number theory (especially pre-tenure), to train graduate students and postdocs in number theory by providing experience with collaborative research and the publication process, to extend and strengthen a research network of potential collaborators in number theory and related fields, to enable faculty at small colleges to participate actively in research activities including the mentorship of graduate students and postdocs, and to highlight research activities of women in number theory.

The typical working model for WIN conferences consists of three phases. Prior to the event, participants are organized into project groups by research interest and asked to acquire the necessary background for their project topics. At the workshop, the groups collaborate on their projects and generally achieve significant progress during that time. Following the conference, the groups continue their collaborations remotely, carrying on with their research and eventually writing up their results for publication. The workshop is arguably the highlight of the WIN experience as it offers everyone the opportunity to get to know the participants outside their group, network with women in their field, and immerse themselves in intense collaborative research. The WIN5 project groups were formed prior to the outbreak of the COVID-19 pandemic. Sadly, the pandemic precluded an in-person event, thus depriving participants of this enriching experience. By necessity, the format of WIN5 was adapted and changed to fit the online world. To their great credit, almost all the groups chose to nevertheless proceed with their collaborations. After consulting with group leaders and experienced organizers of online conferences, the WIN5 organizers elected to decline BIRS's kind offer to host an online version of WIN5 during the originally scheduled dates. Instead, we felt that a sequence of more spread-out events would be more beneficial in supporting the collaborations academically, professionally, and socially. On June 30, 2020, we began with a virtual kick-off meet-and-greet plus Q&A for all WIN5 participants. This was followed by two panels on professional life in the virtual world and on grant writing, respectively. We held two online mini workshops in December 2020 and August 2021, where WIN5 groups gave talks to report on their progress. And throughout the period of pandemic-induced social distancing and forced working-from-home, we offered a weekly Wednesday evening virtual Happy Hour where WIN5 participants could voluntarily drop in and chat, usually over a non-virtual beverage of their choice.

For this volume, the editors solicited contributions from the WIN5 collaboration groups and sought additional articles through the Women in Numbers Network mailing list and other platforms. The ten articles collected here span algebraic, geometric, and computational aspects of number theory, including topics in algebraic and algorithmic number theory, algebraic and arithmetic geometry, arithmetic dynamics, Diophantine equations, modular forms, and additive number theory. Several papers in this volume stem from collaborations between authors with different mathematical backgrounds, allowing the group to tackle a problem using multiple perspectives and tools. All submissions were sent to anonymous referees for rigorous peer review.

Acknowledgments

We are grateful to the National Science Foundation (grant DMS-2012061), the Clay Mathematics Institute, the Number Theory Foundation, and Elsevier Publishers for the support they offered us. We are indebted to BIRS for offering to host the originally planned workshop and subsequently offering to host a virtual version.

We owe a big thank you to all WIN5 group leaders for going ahead with their projects despite not having an actual conference, and to all WIN5 participants for staying productive within the confines of solely online collaboration and persevering through the COVID-19 pandemic. The careful and dedicated work of all our anonymous referees, crucial in assuring the quality of this publication, is much appreciated. Lastly, we thank all our authors for contributing to this volume, Springer for publishing these contributions, and our readers for their interest.

The WIN5 Organizers and Editors:

- Alina Bucur (University of California San Diego)
- Wei Ho (Institute for Advanced Study, Princeton University, and University of Michigan)
- Renate Scheidler (University of Calgary)

November 2023

Contents

From Fontaine–Mazur Conjecture to Analytic Pro-p-groups: A Survey

Ramla Abdellatif, Supriya Pisolkar, Marine Rougnant, and Lara Thomas

1 Introduction

The Fontaine–Mazur Conjecture is a core statement in modern arithmetic geometry. Several formulations of this conjecture were given since its original statement (as appeared in [10]), and various angles have been adopted by numerous authors to try to tackle it. To state the original Fontaine–Mazur Conjecture (FMC), we need to introduce some definitions and notations. Let K be a number field, $\overline{K}$ be a fixed separable closure of K and $G_K := \mathrm{Gal}(\overline{K}/K)$ be the corresponding absolute Galois group. Given a prime number p, a finite extension F of the field $\mathbb{Q}_p$ of p-adic numbers and a profinite group G, an *F-representation of G* is a finite-dimensional vector space over F equipped with a continuous and linear action of G. When $F = \mathbb{Q}_p$, we call it a *p-adic representation of G*. A p-adic representation ρ of G_K is called *geometric* if it ramifies only at a finite number of places of K and if, for each place v of K above p, the restriction of ρ to G_v is potentially semi-stable, as defined

R. Abdellatif (✉)
Laboratoire Amiénois de Mathématique Fondamentale et Appliquée (LAMFA)—Université de Picardie Jules Verne, Amiens, France
e-mail: ramla.abdellatif@math.cnrs.fr

S. Pisolkar
Indian Institute of Science, Education and Research (IISER), Pune, India
e-mail: supriya@iiserpune.ac.in

M. Rougnant
Laboratoire de Mathématiques de Besançon, UFR Sciences et techniques, Besançon, France
e-mail: marine.rougnant@univ-fcomte.fr

L. Thomas
Département de Mathématiques, Faculté des Sciences et Techniques de Saint-Etienne, Université Jean Monnet, 23, rue du Docteur Michelon, Saint-Etienne, France
e-mail: lthomas@math.cnrs.fr

A. Bucur et al. (eds.), *Research Directions in Number Theory*, Association for Women in Mathematics Series 33, https://doi.org/10.1007/978-3-031-51677-1_1

by Fontaine in [11, Section 1.8]. Finally, for any integer r, we let $\mathbb{Q}_p(r)$ denote the r-th Tate twist of $\mathbb{Q}_p$, as defined in [27, §3].

Conjecture 1.1 (Fontaine, Mazur) An irreducible p-adic representation of G_K is geometric if, and only if, it is isomorphic to a subquotient of an étale cohomology group with coefficients in $\mathbb{Q}_p(r)$, for some $r \in \mathbb{Z}$, of a (projective, smooth) algebraic variety over K.

In short, the Fontaine–Mazur Conjecture predicts that p-adic representations of global Galois groups that are potentially semi-stable at primes dividing p and unramified outside finitely many places all come from algebraic geometry.

In the recent years, substantial progress has been made using p-adic representations and deformation theory, allowing, for instance, Kisin [17] to prove the original conjecture for families of two-dimensional representations when $K = \mathbb{Q}$. In this special case, the conjecture asserts that potentially semi-stable representations with odd determinant come from modular forms.

Conjecture 1.2 (Fontaine–Mazur Conjecture for $n = 2$) Let $\rho : \mathrm{Gal}(\overline{\mathbb{Q}}/\mathbb{Q}) \to GL_2(\mathbb{Q}_p)$ be an odd, irreducible representation that is unramified outside finitely many primes and whose restriction to the decomposition group at p is potentially semi-stable with distinct Hodge–Tate weights. Then ρ is the twist of a Galois representation associated with a modular form of weight $k \geq 2$.

Kisin's proof relies on an intimate connection between modularity lifting theorems, the Breuil–Mézard conjecture [2], and Breuil's p-adic local Langlands correspondence for GL_2 [3, 4]. In the same direction (meaning using local–global compatibility results and deformation theory), but using different tools (such as completed cohomology), Emerton proved further cases of this conjecture for two-dimensional representations and $K = \mathbb{Q}$ (see [9, Theorem 7.1.1]).

In this chapter, we are interested in a different approach, which can be motivated as follows. First note that Conjecture 1.1 implies the following conjecture, which is elementary to state but completely out of reach for now (see [7] for instance).

Conjecture 1.3 (Weak Fontaine–Mazur Conjecture) Every unramified pro-p-extension of K whose Galois group is p-adic analytic is finite.

Now recall that in [12], Golod and Shafarevich proved the existence of a number field K and a prime p such that K admits an everywhere unramified infinite pro-p-extension L. (Note that they actually provide a way to obtain infinitely many such number fields.) Conjecture 1.3 then claims that the Galois group of such an extension L/K cannot be an infinite analytic pro-p-group (i.e., isomorphic to a closed subgroup of $GL_n(\mathbb{Z}_p)$ for some positive integer n). Hence the idea is that a counter-example to Conjecture 1.3 would produce an everywhere unramified Galois representation with an infinite image, which cannot "come from algebraic geometry" in the sense of Conjecture 1.1.

From Lazard's seminal work on p-adic analytic groups [19, III, 3.4.3], we know that any finite-dimensional p-adic analytic group contains a finite index open uniform subgroup. We can thus reformulate Conjecture 1.3 as follows.

Conjecture 1.4 (Uniform Fontaine–Mazur Conjecture (UFMC)) There is no number field K for which there exists an infinite everywhere unramified Galois pro-p-extension L such that $\mathrm{Gal}(L/K)$ is uniform.

A major advantage in considering uniform groups is that they have a simple characterisation in terms of filtration by their subgroups, which does not hold for arbitrary analytic pro-p-groups. (Recall that a finitely generated pro-p-group is analytic if, and only if, it contains a uniform subgroup of finite index.)

The first evidence in favour of Conjecture 1.4 was given by N. Boston in [5, 6], using purely group-theoretic methods based on the connections between powerful and uniform pro-p-groups instead of representation-theoretic tools. This chapter provides an introduction to these purely group-theoretic tools and methods, which are not so well known among (young) arithmetic geometers, and a review of some of the main results they bring about various conjectures related to Conjectures 1.1 and 1.4. We also present some related questions on which are based recent, current and future works of the authors. We think that this survey paper may be of interest to anyone looking for a different viewpoint on the Fontaine–Mazur Conjecture, which does not require advanced knowledge in p-adic representation theory and highlights the number-theoretic nature of the problem.

This chapter is organised as follows. Section 2 gathers the definitions and basic results we need about (uniform) pro-p-groups. Section 3 is devoted to study the proofs of the main result of [5], which is the following special case of Conjecture 1.4, and of its generalisation to cyclic extensions of degree prime to p, which is the main result of [6].

Theorem 1.5 (Boston) *Given a prime number p and a number field F, let K be a normal extension of F of prime degree $\ell \neq p$ and such that p does not divide $h(F)$, the class number of F. Then there is no infinite, everywhere unramified pro-p-extension L of K such that L/F is Galois and $\mathrm{Gal}(L/K)$ is uniform.*

The main ingredient of the group-theoretic methods used to prove these results is the cyclic action on $\mathrm{Gal}(L/K)$ of a generator σ of $\mathrm{Gal}(K/F)$. This works quite well when the action of σ on $\mathrm{Gal}(L/K)$ has no non-trivial fixed point. The study of such actions also appears in the work of Hajir and Maire [14], in their attempt to extend Boston's strategy to (tamely) ramified extensions L/K. The challenge here is to understand the behaviour of the fixed points introduced by the ramification and the resulting constraints on the arithmetic of L/K.

This work of Hajir and Maire motivates our interest in the two following Fontaine–Mazur-style conjectures, where we consider finitely and tamely ramified p-adic representations. These conjectures are known, respectively, as the *Tame Fontaine–Mazur Conjecture* (TFMC) [10, 5a] and the *Uniform Tame Fontaine–Mazur Conjecture* (UTFMC).

Conjecture 1.6 (Tame Fontaine–Mazur Conjecture) Let K be a number field and S be a finite set of places of K that are prime to p. Set $G_S := \mathrm{Gal}(K_S/K)$, where K_S denotes the maximal pro-p-extension of K that is unramified outside S. Then any p-adic representation of G_S has finite image.

The idea behind this statement is that the eigenvalues of a Frobenius element must become roots of unity under the action of a finitely tamely ramified p-adic representation. In this case, the image of such a representation is solvable, hence finite by class field theory. Class field theory also helps to prove that the conjecture holds for one-dimensional representations, as we show in Sect. 4.2. For higher dimensional representations, Conjecture 1.6 seems for now out of reach in general.

Conjecture 1.7 (Uniform Tame Fontaine–Mazur Conjecture) Let K be a number field and Γ be a uniform pro-p-group of dimension $d > 2$ (hence infinite). Then there does not exist a finitely and tamely ramified Galois extension L/K whose Galois group $\mathrm{Gal}(L/K)$ is isomorphic to Γ.

We elaborate on how these statements connect in Sect. 4 and on how the assumptions made matter. In Sect. 5, we expose some connected questions that are of interest to us, as well as recent results proven in this direction by some of the authors.

General Notations From now on, we fix a prime integer p. We let $\mathbb{Q}_p$ be the field of p-adic numbers and $\mathbb{Z}_p$ be its ring of integers. Since the ring $\mathbb{Z}_p$ is isomorphic to the projective limit $\varprojlim_{k\geq 1} \mathbb{Z}/p^k\mathbb{Z}$, there exists, for any integer $k \geq 1$, a natural ring isomorphism $\psi_k : \mathbb{Z}_p/p^k\mathbb{Z}_p \simeq \mathbb{Z}/p^k\mathbb{Z}$. These ring isomorphisms allow us to define *principal congruence subgroups* in this p-adic setting as follows. Given any integer $n \geq 2$, we consider the special linear group $\mathrm{SL}_n(\mathbb{Z}_p)$, which is a maximal open compact subgroup of $\mathrm{SL}_n(\mathbb{Q}_p)$. For any integer $k \geq 1$, we can then define the k-th-*principal congruence subgroup* of $\mathrm{SL}_n(\mathbb{Z}_p)$ as $\Gamma_{n,k} := \ker\left(\mathrm{SL}_n(\mathbb{Z}_p) \twoheadrightarrow \mathrm{SL}_n(\mathbb{Z}/p^k\mathbb{Z})\right)$. Note that we choose to use this $\Gamma_{n,k}$ notation instead of the classical $K_n(k)$ notation to highlight the connection with the Galois context.

Assume that we are given a group G. For any subsets X and Y of G and any positive integer n, we write X^n for the subgroup of G generated by $\{x^n, x \in X\}$ and $[X, Y]$ for the subgroup of G generated by $\{[x, y] := x^{-1}y^{-1}xy,\ (x, y) \in X \times Y\}$. As usual, for any subgroup H of G, we let $[G : H]$ denote the index of H in G.

Finally, if Γ is a topological space and if Ω is a subset of Γ, then $\overline{\Omega}$ denotes the closure of Ω in Γ. If Γ is moreover a topological group, a closed subgroup of Γ is called *topologically characteristic* when it is stable under all **continuous** group automorphisms of Γ.

2 Preliminaries on Uniform Pro-p-groups

Our main reference for this section is [8]. The goal here is to recollect all the information we need on uniform pro-p-groups to understand how they connect to the Fontaine–Mazur Conjecture via Boston's method. The reader who is already familiar with uniform pro-p-groups can skip this section and go directly to Sect. 3.

2.1 *Some Basic Definitions Related to Profinite Groups*

We start by recalling some basic definitions, coming from [8, Chapter 1], which are fundamental for our purpose. The first definition we make is exactly [8, Definition 1.1].

Definition 2.1 A *profinite group* is a compact Hausdorff topological group whose open subgroups form a basis of neighbourhoods of the identity.

According to [8, Proposition 1.3], this is equivalent to the usual definition of profinite groups as inverse limits of finite groups.

Definition 2.2 A profinite group G is *finitely generated* if it contains a finite subset X such that X topologically generates G, i.e., such that the subgroup of G generated by X is dense in G.

Given a profinite group, we define its Frattini subgroup, which is a key tool in the forthcoming study of uniform pro-p-groups, as follows (see [8, Definition 1.8]).

Definition 2.3 Let G be a profinite group. The *Frattini subgroup of* G is defined as $\Phi(G) := \bigcap M$, where the intersection runs over all maximal proper open subgroups of G.

2.2 *Pro-p-groups: Definition and Basic Properties*

Among all profinite groups, we will mainly focus on those arising as projective limits of p-groups. Such groups are called pro-p-groups and are usually defined as follows (see [8, Definition 1.10]).

Definition 2.4 A *pro-p-group* is a profinite group in which every open normal subgroup has finite index equal to a power of p.

By [8, Proposition 1.12], we know that a group is a pro-p-group if, and only if, it is isomorphic to an inverse limit of p-groups. (Recall that a *p-group* is just a finite group of p-power order.)

Finitely generated pro-p-groups are nicely characterised (among pro-p-groups) by the topology of their Frattini subgroups, as proven in [8, Proposition 1.14].

Proposition 2.5 *A pro-p-group G is finitely generated if, and only if, its Frattini subgroup $\Phi(G)$ is open in G.*

Also note that for pro-p-groups, we can fruitfully go beyond Frattini subgroups via the notion of lower p-series, defined as follows (see [8, Definition 1.15]).

Definition 2.6 The *lower p-series* of a pro-p-group G is the series $(P_i(G))_{i\geq 1}$ of topologically characteristic subgroups of G defined by $P_1(G)=G$ and: $\forall\, i\geq 1,\ P_{i+1}(G) := \overline{P_i(G)^p[P_i(G), G]}$.

The next key proposition shows that the Frattini subgroup of a pro-p-group G is actually encoded in the lower p-series of G [8, Proposition 1.13].

Proposition 2.7 *For any pro-p-group G, we have* $\Phi(G) = P_2(G) = \overline{G^p[G, G]}$.

If G is a finitely generated pro-p-group, then [8, Proposition 3.7] ensures that the quotient $G/\Phi(G)$ is a finite-dimensional $\mathbb{F}_p$-vector space. We can hence consider the following integer.

Definition 2.8 For any finitely generated pro-p-group G, we let $d(G)$ be the dimension of $G/\Phi(G)$ as $\mathbb{F}_p$-vector space:

$$d(G) := \dim_{\mathbb{F}_p} G/\Phi(G)\ .$$

According to [8, Proposition 3.7], this integer is also equal to the minimal cardinality of a (topological) generating set for G. We will see later that, in good cases, $d(G)$ is also equal to the dimension of G as a p-adic analytic group, provided that such a structure exists on G.

2.3 *The (Uniform) Power of Pro-p-groups*

Following [8, Definition 2.1], we introduce now the following notion of powerfulness for (pro-)p-groups.

Definition 2.9 We say that a p-group G is *powerful* when one of the following holds:

- Either p is odd, and G/G^p is an abelian group
  or
- $p = 2$, and G/G^4 is an abelian group.

We say that a pro-p-group G is *powerful* when the analogous alternative, with G^p and G^4, replaced, respectively by their closure $\overline{G^p}$ or $\overline{G^4}$ in G, holds.

According to [8, Corollary 3.3], powerful pro-p-groups can equivalently be defined as inverse limits of powerful p-groups in which all transition maps are surjective.

We are now ready to follow [8, Definition 4.1] and define our first key notion.

Definition 2.10 A *uniformly powerful pro-p-group G* (or *uniform pro-p-group*) is a powerful, finitely generated pro-p-group G such that: $\forall\ i \geq 1$, $[P_i(G)\colon P_{i+1}(G)] = [G\colon P_2(G)]$.

The assumptions in Definition 2.10 ensure in particular that $P_2(G) = \Phi(G)$ is open in G. Another exciting feature of uniform pro-p-groups is that they are automatically endowed with a structure of p-adic analytic group. The reader interested in the general framework of p-adic analytic groups should refer to Lazard's seminal paper [19], which remains to our knowledge the best reference so far on this topic. For now, we only need the following existence result (see [8, Theorem 8.32]).

Theorem 2.11 *A topological group has a structure of p-adic analytic group if, and only if, it contains an open subgroup which is a uniform pro-p-group.*

In particular, this theorem ensures that uniform pro-p-groups are always endowed with a structure of p-adic analytic group. According to [8, Theorem 8.36], their dimension as analytic groups is then equal to the integer $d(G)$ introduced in Definition 2.8.

2.4 A Key Example: Principal Congruence Subgroups that are Uniform Pro-*p*-groups

The goal of this subsection is to prove the following result [8, Theorem 5.2], where the principal congruence subgroups $\Gamma_{n,k}$ are those we defined at the end of Sect. 1.

Theorem 2.12 *Let $n \geq 2$ be an integer.*

(1) *If p is odd, then the first principal congruence subgroup $\Gamma_{n,1} \subset \mathrm{SL}_n(\mathbb{Z}_p)$ is a uniform pro-p-group of dimension $n^2 - 1$.*
(2) *If $p = 2$, then the second principal congruence subgroup $\Gamma_{n,2} \subset \mathrm{SL}_n(\mathbb{Z}_2)$ is a uniform pro-p-group of dimension $n^2 - 1$.*

Looking, for instance, at the case $n = 2$, we obtain that:

- For any odd prime p, $\Gamma_1 := \ker\big(\mathrm{SL}_2(\mathbb{Z}_p) \twoheadrightarrow \mathrm{SL}_2(\mathbb{F}_p)\big)$ is a uniform pro-p-group of dimension 3.
- For $p = 2$, $\Gamma_2 := \ker(\mathrm{SL}_2(\mathbb{Z}_2) \twoheadrightarrow \mathrm{SL}_2(\mathbb{Z}/4\mathbb{Z}))$ is a uniform pro-2-group of dimension 3.

Let us point out that for $n = 1$, $\Gamma_{1,m}$ is the trivial group for any integer $m \geq 1$ and any prime p. This explains why we assume $n \geq 2$ in Theorem 2.12.

Though the proof of this result is essentially given in [8, Theorem 5.2], we collect here the relevant facts and definitions it is based on to ease later reference. For convenience, once n is fixed, we write Γ_m instead of $\Gamma_{n,m}$ to avoid this cumbersome notation as often as possible. Said differently, we now fix an integer $n \geq 2$ as well as a prime p and we set:

$$\forall\, m \geq 1,\ \Gamma_m := \ker(\mathrm{SL}_n(\mathbb{Z}_p) \to \mathrm{SL}_n(\mathbb{Z}/p^m\mathbb{Z}))\,.$$

Lemma 2.13 *For every $m \geq 1$, the group Γ_m is a pro-p-group. More precisely, we have*

$$\Gamma_m \cong \varprojlim_{r\geq 1} \Gamma_m/\Gamma_{m+r},$$

with each Γ_m/Γ_{m+r} being a finite p-group. Moreover, Γ_m/Γ_{m+1} is isomorphic to the additive group $\mathfrak{sl}_n(\mathbb{F}_p)$ of matrices in $M_n(\mathbb{F}_p)$ with trace 0.

Proof Let m be a positive integer. The obvious projection map $\Gamma_m \to \varprojlim_{r\geq 1} \Gamma_m/\Gamma_{m+r}$ is injective since we have $\bigcap_{r\geq 1} \Gamma_{m+r} = \{I_n\}$. Conversely, any compatible sequence in $\varprojlim_{r\geq 1} \Gamma_m/\Gamma_{m+r}$ provides an element of $M_n(\mathbb{Z}_p)$ that is congruent to I_n modulo p^m. The continuity of the determinant map ensures that such an element must be of determinant 1; hence, we have the desired isomorphism.

We now show that each Γ_m/Γ_{m+r} is a (finite) p-group. By induction on r, it suffices to prove that Γ_m/Γ_{m+1} is a (finite) p-group. To do this, we consider the map $\varphi_m : \Gamma_m \to M_n(\mathbb{Z}_p)$ defined by $\varphi_m(x) := p^{-m}\,(x - I_n)$. One directly checks that $\varphi_m(\Gamma_{m+1})$ is contained in $M_n(p\mathbb{Z}_p)$, and thus we get a well-defined function $\overline{\varphi_m} : \Gamma_m/\Gamma_{m+1} \longrightarrow M_n(\mathbb{Z}_p)/M_n(p\mathbb{Z}_p) \simeq M_n(\mathbb{F}_p)$. We claim that $\overline{\varphi_m}$ is a group homomorphism for the usual additive group structure on $M_n(\mathbb{F}_p)$. Indeed, let $I_n + p^m a$ and $I_n + p^m b$ be elements of Γ_m. As $(I_n + p^m a)(I_n + p^m b) = I_n + p^m(a + b) + p^{2m}ab$, we have

$$\varphi_m\big((I_n + p^m a)(I_n + p^m b)\big) = \varphi_m(I_n + p^m(a+b) + p^{2m}ab) = a + b + p^m ab\,.$$

Since $a + b + p^m ab \equiv a + b \bmod p$, we obtain as expected that

$$\overline{\varphi_m}\big((I_n + p^m a)(I_n + p^m b)\big) = a + b \bmod p = \overline{\varphi_m}\big(I_n + p^m a\big) + \overline{\varphi_m}\big(I_n + p^m b\big)\,.$$

Now assume that $[I_n + p^m a] \in \Gamma_m/\Gamma_{m+1}$ lies in $\ker \overline{\varphi_m}$. This means that a is in $M_n(p\mathbb{Z}_p)$, hence that $I_n + p^m a$ belongs to Γ_{m+1}, i.e., that $[I_n + p^m a] \in \Gamma_m/\Gamma_{m+1}$ is trivial. All this proves that $\overline{\varphi_m}$ is injective. Since $M_n(\mathbb{F}_p)$ is a finite-dimensional $\mathbb{F}_p$-vector space, we obtain that $\Gamma_m/\Gamma_{m+1} \simeq \mathrm{Im}\,\overline{\varphi_m}$ is a finite p-group.

To conclude, we are left to check that the image of $\overline{\varphi_m}$ is isomorphic to $\mathfrak{sl}_n(\mathbb{F}_p)$. Given an element $I_n + p^m a$ of Γ_m, let $f_a(X) = \sum_{k=0}^{n} \alpha_k X^k \in \mathbb{Z}_p[X]$ be the characteristic polynomial of a. Then the characteristic polynomial of $-p^m a$ is equal to $f_{-p^m a}(X) = \sum_{k=0}^{n} \alpha_k (-p^m)^k X^k$. By evaluating it at 1, we get that $1 =$

$\det(I_n + p^m a) = f_{-p^m a}(1) = \sum_{k=0}^{n} \alpha_k(-p^m)^k$. Reducing this equality mod p^m shows that $\alpha_0 - 1 \in p^m\mathbb{Z}_p$, hence subtracting 1 from it and dividing the result by p^m give that

$$-\alpha_1 + \sum_{k=2}^{n} \alpha_k(-p^m)^{k-1} = 0 .$$

This implies in particular that $-\alpha_1 = \mathrm{tr}(a)$ is in $p^m\mathbb{Z}_p$, hence that $\mathrm{tr}(a) \equiv 0 \bmod p$, which proves that $\overline{\varphi_m}\,(I_n + p^m a)$ lies in $\mathfrak{sl}_n(\mathbb{F}_p)$, as required. Conversely, let us check that the image of $\overline{\varphi_m}$ is equal to $\mathfrak{sl}_n(\mathbb{F}_p)$. First recall that, since $\overline{\varphi_m}$ is injective, it is also a homomorphism of $\mathbb{F}_p$-vector spaces. We are hence left to prove that a basis of $\mathfrak{sl}_n(\mathbb{F}_p)$ is contained in $\mathrm{Im}\,\overline{\varphi_m}$. For $1 \le i, j \le n$, let e_{ij} be the matrix whose (i, j)-th entry is 1, and all other entries are 0. For $1 \le i \le n-1$, set $d_i := e_{ii} - e_{nn}$. Then an $\mathbb{F}_p$-basis for $\mathfrak{sl}_n(\mathbb{F}_p)$ is given by

$$\left\{e_{ij} \colon 1 \le i \ne j \le n\right\} \sqcup \{d_i \colon 1 \le i \le n-1\} .$$

As $\varphi_m(I_n + p^m e_{ij}) = e_{ij}$ for all $1 \le i \ne j \le n$ while $\varphi_m(I_n + p^m d_i) = d_i$ for all $1 \le i \le n-1$, we obtain that $\varphi_m(\Gamma_m/\Gamma_{m+1}) = \mathfrak{sl}_n(\mathbb{F}_p)$, as claimed. □

Until the end of this section, we essentially follow [8, Section 5.1], making the changes required to pass from GL_n to SL_n and filling in some missing details. First note that, for all $x \in \Gamma_m$ with $m \ge 1$, we have $x^p \in \Gamma_{m+1}$. Indeed, write $x = I_n + p^m a$ for some $a \in M_n(\mathbb{Z}_p)$. Since a and I_n commute, we have

$$x^p = (I_n + p^m a)^p = I_n + \sum_{k=1}^{p} \binom{p}{k} p^{mk} a^k .$$

As p divides $\binom{p}{k}$ for all $1 \le k \le p-1$, we obtain that $x^p \equiv I_n \bmod p^{m+1}$, as claimed.

Following [8, Lemma 5.1], we will now see that, unless $(p, m) = (2, 1)$, the p-th power map induces a surjection $\Gamma_m \to \Gamma_{m+1}$.

Lemma 2.14 *Let p be a prime and m be a positive integer. Assume that p is odd or that $m \ge 2$. Then every element of Γ_{m+1} is the p-th power of an element of Γ_m.*

Proof Let $a \in M_n(\mathbb{Z}_p)$ be such that $\det(I_n + p^{m+1}a) = 1$. The goal is to find some $b \in M_n(\mathbb{Z}_p)$ such that $\det(I_n + p^m b) = 1$ and $(I_n + p^m b)^p = I_n + p^{m+1}a$. We will find it by successive approximations, which means that we will produce a sequence $(x_k)_{k\ge 1}$ of elements of $M_n(\mathbb{Z}_p)$ such that

$$\forall\, k \ge 1,\ \det(I_n + p^m x_k) \equiv 1 \bmod p^{m+1+k} \quad \text{and}$$

$$(I_n + p^m x_k)^p \equiv I_n + p^m a \bmod p^{m+1+k} .$$

Then $b := \lim_{k\to\infty} x_k$ will be as expected. First, assume that $I_n + p^{m+1}a \equiv (I_n + p^m x_r)^p \bmod p^{m+r+1}$. Since we have $\det(I_n + p^{m+1}a) = 1$, we already obtain that $\det(I_n + p^m x_r)^p \equiv 1 \bmod p^{m+r+1}$. Now, as we know that $\det(I_n+p^m a) \in 1+p^m\mathbb{Z}_p$, it actually follows that $\det(I_n+p^m x_r) \equiv 1 \bmod p^{m+r+1}$. This shows that we do not need to check at each step whether the determinant congruence holds when constructing the sequence $(x_k)_{k\geq 1}$ as it will automatically be true.

Let us first construct x_1. Note that, as long as m is in the supposed range, we have

$$(I_n + p^m a)^p = I_n + p^{m+1}a + \sum_{k=2}^{p} \binom{p}{k} (p^m a)^k ,$$

which is congruent to $I_n + p^{m+1}a \bmod p^{m+2}$, so we can set $x_1 := a$.

Now suppose that there exists some integer $r \geq 1$ for which we built some $x_r \in M_n(\mathbb{Z}_p)$ satisfying $I_n + p^{m+1}a \equiv (I_n + p^m x_r)^p \bmod p^{m+r+1}$. Then there exists some $c \in M_n(\mathbb{Z}_p)$ such that

$$(I_n + p^m x_r)^p = I_n + p^{m+1}a + p^{m+r+1}c .$$

Note that expanding the left-hand side of this equality provides a $\mathbb{Z}$-linear combination of powers of x_r, which shows that both a and c can be expressed as a $\mathbb{Q}_p$-linear combination of powers of x_r. In particular, this implies that any two among a, c, and x_r commute. Let us thus set

$$z := (I_n + p^m x_r)^{-(p-1)}c \text{ and } x_{r+1} := x_r - p^r z .$$

Then x_{r+1} satisfies the desired congruence. Indeed, we have

$$\begin{aligned}
(I_n + p^m x_{r+1})^p &= ((I_n + p^m x_r) - p^{m+r} z)^p \\
&= \sum_{k=0}^{p} \binom{p}{k} (I_n + p^m x_r)^{p-k} (-p^{m+r} z)^k \\
&= (I_n + p^m x_r)^p + \sum_{k=1}^{p} (-1)^k \binom{p}{k} (I_n + p^m x_r)^{p-k} p^{(m+r)k} z^k \\
&= I_n + p^{m+1}a + p^{m+r+1}c \\
&\quad + \sum_{k=1}^{p} (-1)^k \binom{p}{k} (I_n + p^m x_r)^{p-k} p^{(m+r)k} z^k ,
\end{aligned}$$

which implies that

$$(I_n + p^m x_{r+1})^p \equiv I_n + p^{m+1}a + p^{m+r+1}c - p(I_n + p^m x_r)^{p-1}p^{m+r}z \mod p^{m+r+2},$$

i.e., $(I_n + p^m x_{r+1})^p \equiv I_n + p^{m+1}a \mod p^{m+r+2}$.

By construction, we have $x_{r+1} - x_r = -p^r z$, and thus $b := \lim_{r\to\infty} x_r$ is well defined, which ends the proof. □

The next two results come from [8, Theorem 5.2].

Corollary 2.15 *Let p be a prime integer.*

- *If p is odd, then $P_i(\Gamma_1) = \Gamma_i$ for all $i \geq 1$.*
- *If $p = 2$, then $P_i(\Gamma_2) = \Gamma_{i+1}$ for all $i \geq 1$.*

Proof Set $\varepsilon = 0$ if p is odd and $\varepsilon = 1$ if $p = 2$. We show by induction on $i \geq 1$ that $P_i(\Gamma_{1+\varepsilon}) = \Gamma_{i+\varepsilon}$. When $i = 1$, this follows from the definition of P_1, so we can suppose that $P_i(\Gamma_{1+\varepsilon}) = \Gamma_{i+\varepsilon}$ for some $i \geq 1$. Then Lemma 2.14 ensures that $P_i(\Gamma_{1+\varepsilon})^p = \Gamma_{i+\varepsilon}^p = \Gamma_{i+1+\varepsilon}$.

We now claim that $[P_i(\Gamma_{1+\varepsilon}), \Gamma_{1+\varepsilon}] = [\Gamma_{i+\varepsilon}, \Gamma_{1+\varepsilon}] \subseteq \Gamma_{i+1+\varepsilon}$. Indeed, let $I_n + p^{i+\varepsilon}a \in \Gamma_{i+\varepsilon}$ and $I_n + p^{1+\varepsilon}b \in \Gamma_{1+\varepsilon}$. First, we have

$$(I_n + p^{i+\varepsilon}a)^{-1} = I_n + \sum_{k=1}^{\infty}(-1)^k p^{(i+\varepsilon)k}a^k\,,$$

which is congruent to $I_n - p^{i+\varepsilon}a \mod p^{i+1+\varepsilon}$. This directly implies that

$$[I_n + p^{i+\varepsilon}a, I_n + p^{i+\varepsilon}b] \equiv I_n \mod p^{i+1+\varepsilon}$$

and hence that $[P_i(\Gamma_{1+\varepsilon}), \Gamma_{1+\varepsilon}] \subseteq \Gamma_{i+1+\varepsilon}$. Thus we have $P_i(\Gamma_{1+\varepsilon})^p[P_i(\Gamma_{1+\varepsilon}), \Gamma_{1+\varepsilon}] \subseteq \Gamma_{i+1+\varepsilon} = P_i(\Gamma_{1+\varepsilon})^p$, so the first inclusion must be an equality. As $\Gamma_{i+1+\varepsilon}$ is open (hence closed) in $\Gamma_{1+\varepsilon}$, it follows that we actually have $P_{i+1}(\Gamma_{1+\varepsilon}) = \Gamma_{i+1+\varepsilon}$, as expected. By induction on $i \geq 1$, this ends the proof. □

Theorem 2.16 *Let p be a prime integer.*

- *If p is odd, then Γ_1 is a uniform pro-p-group of dimension $n^2 - 1$.*
- *If $p = 2$, then Γ_2 is a uniform pro-2-group of dimension $n^2 - 1$.*

Proof As in the proof of Corollary 2.15, we give a uniform proof by setting $\varepsilon = 0$ if p is odd and $\varepsilon = 1$ if $p = 2$. According to Lemma 2.13, $\Gamma_{1+\varepsilon}$ is a pro-p-group; hence, Propositions 2.5 and 2.7 ensure that $\Gamma_{1+\varepsilon}$ is finitely generated if, and only if, $P_2(\Gamma_{1+\varepsilon})$ is open in $\Gamma_{1+\varepsilon}$. But we know from Corollary 2.15 that $P_2(\Gamma_{1+\varepsilon}) = \Gamma_{2+\varepsilon}$, which is open in $\Gamma_{1+\varepsilon}$, so the finite type condition is satisfied.

Now, recall that Lemma 2.14 gives $\Gamma_{1+\varepsilon}^p = \Gamma_{2+\varepsilon}$, while Lemma 2.13 shows that $\Gamma_{1+\varepsilon}/\Gamma_{2+\varepsilon} \cong \mathfrak{sl}_n(\mathbb{F}_p)$ is abelian: thus $\Gamma_{1+\varepsilon}$ is powerful. Similarly, Corollary 2.15 gives that $P_i(\Gamma_{1+\varepsilon}) = \Gamma_{i+\varepsilon}$, while Lemma 2.13 shows that $\Gamma_i/\Gamma_{i+1} \cong \mathfrak{sl}_n(\mathbb{F}_p)$; hence, the index condition from Definition 2.10 is satisfied by $P_i(\Gamma_{1+\varepsilon})$ for any $i \geq 1$. All this shows that $\Gamma_{1+\varepsilon}$ is uniform.

Finally, we compute the dimension using Lemma 2.13, Corollary 2.15 and Definition 2.10 as follows. From these statements, we have that $\Gamma_{1+\varepsilon}/\Phi(\Gamma_{1+\varepsilon}) = \Gamma_{1+\varepsilon}/P_2(\Gamma_{1+\varepsilon}) = \Gamma_{1+\varepsilon}/\Gamma_{2+\varepsilon} \cong \mathfrak{sl}_n(\mathbb{F}_p)$. Since $\dim_{\mathbb{F}_p} \mathfrak{sl}_n(\mathbb{F}_p) = n^2 - 1$, we are done. □

We end this section with a well-known result that will be used in Sect. 4. Its proof is as outlined in [20, Page 1271]. Let G be a finitely generated pro-p-group. For any $i \geq 1$, set $H^i(G) := H^i(G, \mathbb{F}_p)$, and let $d_p(G) := \dim_{\mathbb{F}_p} H^1(G)$ be the p-rank of G. Note that in this setting, we have $d(G) = d_p(G)$.

Theorem 2.17 *Any uniform group of dimension 1 or 2 has a quotient isomorphic to* $\mathbb{Z}_p$.

Proving this theorem requires the following result of Lazard [19, V, Proposition (2.5.7.1)].

Theorem 2.18 *Let G be a uniform group of positive dimension d. Then, for all $i \geq 1$, we have*

$$H^i(G) \cong \bigwedge^i (H^1(G)),$$

where the exterior product is induced by the cup product.

Proof of Theorem 2.17 Let G be a uniform pro-p-group of dimension $d \in \{1, 2\}$.

If $d = 1$, then G is isomorphic to $\mathbb{Z}_p$ so we are done.

If $d = 2$, then Theorem 2.18 implies that $\dim_{\mathbb{F}_p} H^2(G) = \dim_{\mathbb{F}_p} \bigwedge^2 H^1(G) = 1$, which means that $H^2(G)$ is actually isomorphic to $\mathbb{F}_p$. Now recall that $H^1(G) = H^1(G^{ab})$, where G^{ab} is a finitely generated $\mathbb{Z}_p$-module; hence, it can be written as

$$G^{ab} \simeq \mathbb{Z}_p^r \times \prod_{s=1}^{n} \mathbb{Z}/p^{i_s}\mathbb{Z} \tag{2.1}$$

for some nonnegative integers $r, n, i_1, \ldots, i_n$. Also recall that the short exact sequence

$$1 \longrightarrow \mathbb{F}_p \longrightarrow \mathbb{Q}_p/\mathbb{Z}_p \xrightarrow{\times p} \mathbb{Q}_p/\mathbb{Z}_p \longrightarrow 1$$

leads to the following long exact sequence of cohomology:

$$\begin{aligned} 0 \longrightarrow H^1(G^{ab}, \mathbb{F}_p) \longrightarrow H^1(G^{ab}, \mathbb{Q}_p/\mathbb{Z}_p) &\xrightarrow{\times p} H^1(G^{ab}, \mathbb{Q}_p/\mathbb{Z}_p) \\ \longrightarrow H^2(G^{ab}, \mathbb{F}_p) \longrightarrow \ldots \end{aligned}$$

If we let K and C denote, respectively, the kernel and the cokernel of $\times p$, and $d_p(K)$ and $d_p(C)$ denote their respective dimensions as $\mathbb{F}_p$-vector spaces, then what we said above directly shows that

$$d_p(C) \leq \dim_{\mathbb{F}_p}(H^2(G)) \text{ and } d_p(K) \geq d_p(G) .$$

This implies that $d_p(K) - d_p(C) \geq d_p(G) - \dim_{\mathbb{F}_p}(H^2(G)) = 1$. But (2.1) ensures that we also have $d_p(K) - d_p(C) = r = \mathrm{rk}_{\mathbb{Z}_p} G^{ab}$; thus, we obtain that $\mathrm{rk}_{\mathbb{Z}_p}(G^{ab}) \geq 1$, and G^{ab} surjects onto $\mathbb{Z}_p$. □

3 Boston's Proof of a Special Case of Fontaine–Mazur Conjecture

In this section, we study the proof of Theorem 1.5 as given by Boston in [5, 6]. It heavily relies on Lazard's extensive study of p-adic analytic Lie groups, as given in [19]. Recall in particular that Lazard defined the notion of p-saturated groups [19, IV.3.3.1] and used it to give an algebraic characterisation of p-adic analytic groups as topological groups containing a (topologically) finitely generated, open, p-saturated pro-p-group with an integer-valued filtration [19, III.3.2.2]. In [8], the notion of uniform pro-p-groups is used to transfer Lazard's work in a group-theoretic manner and to obtain an analogue characterisation of p-adic analytic groups for odd p, where uniform pro-p-groups play the same role as p-saturated pro-p-groups with integer-valued filtration do in Lazard's statement [8, Page 81]. Note that in [5], Boston used Lazard's terminology (*p-saturated with integer values*, as defined in [19, II.1.2.10 and III.3.3.1]) instead of the *uniform group* one. We will stick to the uniform group formulation in the sequel, but let us recall here the main statement of [5] (namely Theorem 1), as originally stated.

Theorem 3.1 (Boston) *Let K be a normal extension of prime degree $\ell \neq p$ of a number field F such that $p \nmid h(F)$, the class number of F. Then there is no infinite everywhere unramified Galois pro-p-extension L of K such that L is Galois over F and* $\mathrm{Gal}(L/K)$ *is p-saturated with integer values.*

This result gives some evidence for Conjecture 1.3 in a special case. Using the strength of fixed-point-free automorphisms, Boston generalised this result to the case of cyclic extensions L/K, where $[L : K]$ is not necessarily a prime, see [6, Theorem 1]. To do this, he introduced the class of *self-similar* groups (as defined below), which contains the class of uniform groups, and he showed that, under some condition $H(G, n)$ related to fixed-point-free automorphisms (which is conjectured to always hold), Theorem 3.1 carries over for finite cyclic extensions of degree coprime to p, with 'uniform' replaced by 'self-similar'. Let us define the new notions aforementioned.

Definition 3.2 A pro-p-group G is *self-similar* when there exists a filtration by open, characteristic subgroups $G = G_1 \supseteq G_2 \supseteq \cdots$ such that G_i/G_{i+1} is abelian for all $i \geq 1$ and $\bigcap G_i = \{1\}$, together with a family of group isomorphisms

$$\phi_i : G_i/G_{i+1} \to G_{i+1}/G_{i+2}$$

which commute with every continuous automorphism of G.

Note that the commutativity of all quotients G_i/G_{i+1} ensures that the second condition must only be checked for outer automorphisms of G. Also note that Definition 3.2 implies that a non-trivial self-similar group is always infinite, since $\lim_{i\to\infty} |G/G_i| = \lim_{i\to\infty} |G/G_2|^{i-1} = \infty$.

The next proposition justifies the relevance of this notion in Boston's work.

Proposition 3.3 *Any uniform group is a self-similar group.*

Proof Let G be a uniform pro-p-group. For any $i \geq 1$, set $G_i := P_i$, where P_i is as in Definition 2.6. Then $(G_i)_{i\geq 1}$ is a filtration as in Definition 3.2 and the map $[x \mapsto x^p]$ induces group isomorphisms $\varphi_i : G_{i-1}/G_i \to G_i/G_{i+1}$ that commute with any continuous automorphism of G; hence, G is indeed a self-similar group. □

We now make explicit the aforementioned condition $H(G, n)$ defined by Boston in [6, Definition 2].

Definition 3.4 Let G be a pro-p-group and n be a positive integer. We say that $H(G, n)$ *holds* when there is a function of n that is an upper bound for the derived length of every finite quotient of G that admits a fixed-point-free automorphism of order n.

We already mentioned above that $H(G, n)$ is conjectured to hold for any pro-p-group G and any integer $n \geq 1$. It is known for uniform groups thanks to [8, p. 52] and [26], as well as in few other cases (see Shalev's work [26] for a description of such cases). This means in particular that for a uniform group G, there is a (uniform) bound on the derived length of the quotients G/P_i for $i \geq 1$.

We can now state the generalisation of Theorem 3.1 as proven in [6, Theorem 1].

Theorem 3.5 (Boston) *Let F be a number field such that p does not divide the class number of F. Let K be a cyclic extension of F of degree $n \geq 2$ co-prime to p. Then there is no infinite, everywhere unramified, Galois pro-p-extension L of K such that L is Galois over F and* $\mathrm{Gal}(L/K)$ *is self-similar and satisfies the property $H(\mathrm{Gal}(L/K), n)$ as stated in Definition 3.4.*

Before proving this theorem, we recall a classical result of Schur and Zassenhaus [23, Chapter 4].

Theorem 3.6 (Schur–Zassenhaus) *Let $1 \to \Gamma \to \mathcal{G} \to \mathcal{G}/\Gamma \to 1$ be a short exact sequence of profinite groups, with Γ a finitely generated pro-p-group and $\mathcal{G}/\Gamma$ of finite order co-prime to p. Then $\mathcal{G}$ contains a subgroup Δ_0 isomorphic to* *the quotient* $\Delta := \mathcal{G}/\Gamma$*, and* Δ_0 *is unique up to conjugation in* $\mathcal{G}$*. In particular,* $\mathcal{G}$ *is isomorphic to a semi-direct product of* Δ *and* Γ*:* $\mathcal{G} = \Gamma \rtimes \Delta_0 \cong \Gamma \rtimes \Delta$*.*

Proof of Theorem 3.5 To make things more comfortable to the reader, we address here the case of uniform groups: let us mention that the general case of self-similar groups can be proven along the same lines, as done in [6]. Suppose by contradiction that there exists an infinite, everywhere unramified, Galois pro-p-extension L of K such that L/F is Galois and $G := \mathrm{Gal}(L/K)$ is uniform and satisfies condition $H(G,n)$, where $n := [K : F]$. By the Schur–Zassenhaus theorem, the following extension splits:

$$1 \longrightarrow G \to \mathrm{Gal}(L/F) \longrightarrow \mathrm{Gal}(K/F) \longrightarrow 1\ .$$

Let us pick an element σ of $\mathrm{Gal}(L/F)$ that lifts a generator of the cyclic group $\mathrm{Gal}(K/F)$ under the splitting above. As G is a normal subgroup of $\mathrm{Gal}(L/F)$, σ induces an action by conjugation on G. Now, since G is a uniform group, we can consider its filtration by the characteristic subgroups P_i given in Definition 2.6. Each P_i is preserved under the action of σ (as it is a characteristic subgroup of G); hence, we get an action by conjugation of σ on each quotient G/P_i.

Suppose that, for all $i \geq 1$, σ has no non-trivial fixed point in G/P_i. As G satisfies $H(G,n)$, there is a uniform bound on the derived length of each of the quotients G/P_i as i goes to ∞. Such a bound cannot exist since a repeated use of the finiteness of class number shows that the maximal unramified pro-p-extension of any fixed derived length must be a finite extension.

This shows that there exists a positive integer i such that the action of σ on G/P_i has a non-trivial fixed point. Assume that i is minimal for this property and let τ be a non-trivial fixed point of σ in G/P_i. By minimality of i, τ must map to the identity element in G/P_{i-1}, i.e., lives in P_{i-1}/P_i, since the action of σ is compatible with $P_{i-1} \supseteq P_i$ and $G/P_i \twoheadrightarrow G/P_{i-1}$. Now recall that Proposition 3.3 ensures that the isomorphism $\phi_{i-1} : P_{i-2}/P_{i-1} \to P_{i-1}/P_i$ is σ-equivariant; hence, $\phi_{i-1}^{-1}(\tau)$ should define a non-trivial fixed point for σ in P_{i-2}/P_{i-1}, which contradicts the minimality of i if $i \geq 3$.

If $i = 2$, then our assumption is that the action by conjugation of σ on G/P_2 has a non-trivial fixed point. By definition, we have $P_2 = \overline{G^p[G,G]}$; hence, G/P_2 is naturally an $\mathbb{F}_p$-vector space. Said differently, it defines a representation over a field of characteristic p of the group $\langle\sigma\rangle$ generated by σ. As this group is, by definition of σ, of cardinality n, which is co-prime to p, we can apply Maschke's theorem to decompose G/P_2 as a direct sum of irreducible representations of $\langle\sigma\rangle$ over $\mathbb{F}_p$. As τ is fixed under the action of σ, this decomposition can be written as $G/P_2 = \mathbb{F}_p\tau \oplus W$ for some representation W of $\langle\sigma\rangle$ over $\mathbb{F}_p$.

The direct sum decomposition ensures that W also defines a subgroup and a quotient of G/P_2. Hence, we can define an extension M of K such that $\mathrm{Gal}(M/K) = \langle\tau\rangle$: we just let M be the fixed field (in L) of W over K. The field extension M/K is then an abelian extension of prime degree p, and we have a tower of fields of the form $F \subset K \subset M$.

We claim that there exists a cyclic extension of degree p over F that is contained in M. Indeed, $\mathrm{Gal}(K/F) \simeq \langle\sigma\rangle$ acts on $\mathrm{Gal}(M/K)$ by conjugation. Since K/F is assumed to be a normal extension, $\mathrm{Gal}(M/K)$ must be a normal subgroup of $\mathrm{Gal}(M/F)$. We can hence apply the Schur–Zassenhaus theorem to write $\mathrm{Gal}(M/F)$ as a semi-direct product of the following form:

$$\mathrm{Gal}(M/F) \simeq \mathrm{Gal}(M/K) \rtimes \mathrm{Gal}(K/F) . \tag{3.1}$$

Since τ is fixed under σ, the action of σ on $\mathrm{Gal}(M/K) = \langle\tau\rangle$ is trivial; hence, $\mathrm{Gal}(K/F)$ is also normal in $\mathrm{Gal}(M/F)$, which turns (3.1) into a direct product of the two groups appearing on the right-hand side. As $\mathrm{Gal}(M/K)$ is of order p while $\mathrm{Gal}(K/F)$ is of order n, with n and p being co-prime, we can conclude that there exists a Galois extension N of F of degree p (hence cyclic) in M.

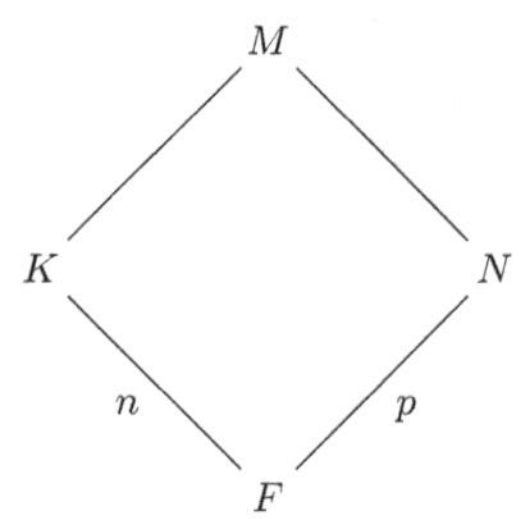

As M/K is unramified, so is N/F, which contradicts the fact that p is co-prime to $h(F)$. In all cases ($i \geq 3$ or $i = 2$), we have a contradiction: this proves that there exists no pro-p-extension L/K that is everywhere unramified and has for Galois group a uniform group. □

4 Some Results on the Tame Fontaine–Mazur Conjecture and Its Uniform Version

For this section, we will mostly refer to [13, 14, 22]. The goal here is to present some results attached to the Tame Fontaine–Mazur Conjecture (TFMC for short) and its Uniform version (TFMC-U for short), as well as related questions on which some of the authors are currently working.

4.1 Motivation and Background

We start by recalling some statements at the core of this section, namely the two versions of the Fontaine–Mazur Conjecture mentioned above, which will be proven to be equivalent to each other (see Sect. 4.3 below).

As explained in Sect. 1, TFMC is an analogue of Conjecture 1.3 for finitely and tamely ramified p-adic representations. Since tame representations are automatically potentially semi-stable (thanks to a theorem of Grothendieck [25, Appendix]), the Fontaine–Mazur Conjecture must imply (under some standard conjectures in algebraic geometry, see [18] for further details) that the following statement holds [10, Conjecture 5a].

Conjecture 4.1 (Tame Fontaine–Mazur Conjecture) Let p be a prime integer and K be a number field. Let S be a finite set of places of K that are all co-prime to p and let K_S be the maximal pro-p-extension of K that is unramified outside S. Set $G_S := \mathrm{Gal}(K_S/K)$. Then, for any positive integer n, any continuous Galois representation $\rho : G_S \to GL_n(\mathbb{Q}_p)$ has finite image.

In Sect. 4.2, we will see that TFMC for $n = 1$ comes from class field theory. When $n > 1$, this conjecture appears so far completely out of reach in general, though some preliminary evidence exists, as those given in [6, 13, 28]. When $K = \mathbb{Q}$ and $n = 2$, we already pointed out in Sect. 1 that the main contributions for now are due to Kisin [17] and Emerton [9], both using different methods from those developed in this chapter.

Recall that Theorem 2.11 claims that any finitely generated p-adic analytic group contains a uniform open subgroup. Moreover, Theorem 2.17 asserts that any uniform group of dimension 1 or 2 admits a quotient isomorphic to $\mathbb{Z}_p$. We can hence rephrase TFMC as follows.

Conjecture 4.2 (Tame Fontaine–Mazur Conjecture—uniform version for (K, d)) Let K be a number field and Γ be a uniform pro-p-group of dimension $d > 2$ (hence infinite). Then there is no finitely and tamely ramified Galois extension of K whose Galois group is isomorphic to Γ.

In the light of Conjectures 1.6 and 1.7 for $n = 2$, we would like to highlight the major contribution of Hajir and Maire in [14] before we go further. Among several key results, they proved an analogue of Theorem 1.5 for odd p, using purely group-theoretic and arithmetic methods such as the effect of a semi-simple cyclic action with fixed points on the group structure, the rigidity of uniform groups, the existence of Minkowski units, some arithmetic properties of Galois groups, etc.

To precise this, we need to introduce some notation and definitions, following [14, 1.2]. We keep the notations of Conjecture 4.1 and we let T be an auxiliary finite set of places of K such that $T \cap S = \emptyset$. Define K_S^T as the maximal pro-p-extension of K that is unramified outside S and in which any place in T splits completely, and let $G_S^T := \mathrm{Gal}(K_S^T/K)$ be the corresponding Galois group. Note that K_S^T is a subfield of K_S while G_S^T is a quotient of G_S and that $K_S^\emptyset = K_S$.

We can now state one of the main results of Hajir–Maire [14, Section 1.2, Theorem], which is also a special case of more general statements proven in [14, Section 2].

Theorem 4.3 (Hajir-Maire) *Let K/k be a quadratic extension with Galois group $\langle\sigma\rangle$. Assume that the odd prime p does not divide the class number of k. Suppose* *that for any finite set* Σ *of places of* k *all co-prime to* p*, there is no continuous Galois representation* $G_\Sigma(k) \twoheadrightarrow \Gamma_{2,1}$*. Then there exist infinitely many disjoint finite sets* S *and* T *of primes of* K *such that any place of* S *is prime to* p*,* $|S|$ *is arbitrarily large and*

(1) G_S^T *is infinite.*
(2) $G_S^T/\Phi(G_S^T)$ *has* $|S|$ *independent fixed points under the action of* σ*.*
(3) *There is no continuous Galois representation* $\rho : G_S^T \twoheadrightarrow \Gamma_{2,1}$ *with* $\left(K_S^T\right)^{\ker \rho}$ *Galois over* k*.*

Note that the last assertion is actually a strengthened notion of uniformness (called σ-uniformness) that gives further conditions on the action of σ on G_S^T, see [14, Definition 1.1]. Let us also mention that, although it is not obvious at first sight, this result is actually tightly connected to the Fontaine-Mazur Conjecture for uniform groups of constant type for order 2 automorphisms. The reader interested in more details on this topic should read [14, Corollary 2.6].

4.2 Class Field Theory Implies Tame Fontaine-Mazur Conjecture for $n = 1$

The goal of this section is to prove the tame Fontaine-Mazur Conjecture (Conjecture 1.6) for $n = 1$. This is certainly known to experts in the field, but beginners will find it useful to have a reference where this claim is fully proven; hence, we now give a detailed proof of the following statement. Note that it holds when $\mathbb{Q}_p$ is replaced by any p-adic field L, but we chose to stick to the case of p-adic representations to stay in the framework we used so far.

Theorem 4.4 (TFMC for $n = 1$**)** *Let* K *be a number field and* p *be a rational prime. Let* S *be a finite set of places of* K *in which no place has residue characteristic* p*. Let* K_S *denote the maximal pro-*p*-extension of* K *that is unramified outside* S *and let* $G_S := \mathrm{Gal}(K_S/K)$ *be its Galois group. Then every continuous group homomorphism*

$$\chi : G_S \to \mathrm{GL}_1(\mathbb{Q}_p) = \mathbb{Q}_p^\times$$

has finite image.

Proof Since G_S is a compact group (as it is a pro-p-group), its image under the continuous map χ must be a compact subgroup of $\mathbb{Q}_p^\times$; hence, it lands into $\mathbb{Z}_p^\times$. Also note that, since $\mathbb{Q}_p^\times$ is an abelian group, χ must factor through the abelianisation G_S^{ab} of G_S, which motivates the connection with class field theory. The latter provides indeed a canonical group homomorphism with dense image (called Artin reciprocity law)

$$\rho : \mathbb{A}_K^\times / K^\times \to \mathrm{Gal}(\overline{K}/K)^{\mathrm{ab}} ,$$

which becomes an isomorphism when $\mathbb{A}_K^\times / K^\times$ is replaced by its profinite completion. We are hence reduced to show that the composite group homomorphism $\tilde{\chi} := \chi \circ \rho : \mathbb{A}_K^\times / K^\times \to \mathbb{Z}_p^\times$ has finite image.

Saying that χ is unramified outside S means that χ is trivial on inertia groups for places outside S; hence, it also holds for $\tilde{\chi}$. Now let v be a place in S. We need to distinguish between the finite and Archimedean cases as follows.

If v is a finite place, then the inertia group at v corresponds to the subgroup of $\mathbb{A}_K^\times$ that is trivial at all places outside of v and equal to the subgroup $\mathcal{O}_{K_v}^\times$ of $K_v^\times$ at v. As all places in S are co-prime to p, the group of 1-units in $\mathcal{O}_{K_v}^\times$ is a pro-ℓ-group for some rational prime $\ell \neq p$. But the 1-units of $\mathbb{Z}_p^\times$ form a pro-p-group, and any continuous group homomorphism from a pro-ℓ-group into a pro-p-group is necessarily trivial: the image by $\tilde{\chi}$ of $\mathcal{O}_{K_v}^\times$ must hence be finite. Letting S^∞ denote the subset of finite places in S, we have proven that $\tilde{\chi}\left(\prod_{v \in S^\infty} \mathcal{O}_{K_v}^\times\right)$ is finite.

Let us now study what happens on $\prod_{v \in S^\infty} \varpi_v^{\mathbb{Z}}$, where ϖ_v denotes a uniformising element of $\mathcal{O}_v$. First note that the product is actually a restricted product, which means only finitely many places are such that ϖ_v has a non-trivial image in $\mathbb{Z}_p^\times$. Also recall that any element $x = (\varpi_v^{e_v})_{v \in S^\infty}$ of $\prod_{v \in S^\infty} \varpi_v^{\mathbb{Z}}$ defines a fractional ideal $\mathfrak{a}_x = \prod_v \mathfrak{p}_v^{e_v}$ of K, where $\mathfrak{p}_v$ denotes the prime ideal of $\mathcal{O}_K$ that corresponds to v. (This is well defined as only finitely many e_v can be non-zero.) Let h denote the class number of K: then $\mathfrak{a}_x^h$ is a non-zero principal ideal of $\mathcal{O}_K$ (by the definition of h), So $x \mod K^\times$ must have finite order dividing h. This shows that the quotient of $\prod_{v \in S^\infty} \varpi_v^{\mathbb{Z}}$ by the diagonal copy of $K^\times$ (in $\mathbb{A}_K^\times / K^\times$) has finite exponent dividing h; hence, its image by $\tilde{\chi}$ must satisfy the same condition in $\mathbb{Q}_p^\times$. As $\mathbb{Q}_p^\times$ contains finitely many elements of order dividing h, we can conclude that this image by $\tilde{\chi}$ is finite.

Finally, assume that v is an Archimedean place in S. Since $\mathbb{Z}_p^\times$ is a totally disconnected space, the identity component of $K_v^\times$ (which is either $\mathbb{R}_{>0}$ if v is real, or $\mathbb{C}^\times$ is v is complex) has trivial image under the continuous group homomorphism $\tilde{\chi}$. Letting S_∞ denote the set of Archimedean places in S, we are left to see that $\prod_{v \in S_\infty \text{ real}} \mathbb{R}^\times / \mathbb{R}_{>0}$ has finite image under $\tilde{\chi}$, which is straightforward as S and $\mathbb{R}^\times / \mathbb{R}_{>0} \simeq \mathbb{Z}/2\mathbb{Z}$ are finite sets, so the proof is complete. □

4.3 Equivalence of TFMC and TFMC-U

This subsection aims to prove that Conjectures 1.6 (TFMC) and 1.7 (TFMC-U) are equivalent. First note that we have the following nice result [8, Theorem 4.5], which makes unnecessary to assume that G is infinite in TFMC-U.

Proposition 4.5 *Any uniform pro-p-group is torsion-free.*

TFMC $\Longrightarrow$ TFMC-U Let G be a uniform pro-p-group such that there exists a finitely and tamely ramified Galois extension L/K with $\mathrm{Gal}(L/K) \cong G$. To prove that G must be finite, we will use the following key fact, proven in [8, Section 7.3].

Proposition 4.6 *Every pro-p-group of finite rank admits a faithful representation into $\mathrm{GL}_n(\mathbb{Q}_p)$ for some integer $n \geq 1$.*

Applied to G, it provides an integer $n \geq 1$ and a faithful representation $\iota \colon G \hookrightarrow \mathrm{GL}_n(\mathbb{Q}_p)$. Fixing an isomorphism ψ between $\mathrm{Gal}(L/K)$ and G as given in the statement, the composition of ι by ψ and by the natural projection map $\mathrm{Gal}(\overline{K}/K) \to G$ provides a Galois representation

$$\rho \colon \mathrm{Gal}(\overline{K}/K) \to \mathrm{GL}_n(\mathbb{Q}_p) .$$

Let S be the set of places in K that ramify in L. By assumption, S is a finite set and ρ factors through G_S. We are hence left to prove that any place in S is prime to p, as TFMC would then apply to ρ and give that $\mathrm{Im}(\rho) \simeq G$ is finite, as expected.

Recall that L/K is tamely ramified, which ensures that for any prime $\mathfrak{p}$ of K and any prime $\mathfrak{q}$ of L above $\mathfrak{p}$, the extension $L_{\mathfrak{q}}/K_{\mathfrak{p}}$ is tamely ramified. Moreover, there exists a unique intermediate Galois extension $K_{\mathfrak{p}} \hookrightarrow F \hookrightarrow L_{\mathfrak{q}}$ such that $L_{\mathfrak{q}}/F$ is totally ramified and $F/K_{\mathfrak{p}}$ is unramified. The 'tame quotient' of the decomposition group of $\mathfrak{q}$ above $\mathfrak{p}$ is then defined by $\mathrm{Gal}(L_{\mathfrak{q}}/F)$ and is in particular of (pro-)order co-prime to $\mathfrak{p}$. As G is a pro-p-group, all this implies that L/K must be unramified at places above p and hence that any place in S is prime to p, and we are done.

TFMC-U $\Longrightarrow$ TFMC Now, assume conversely that K is a number field and that S is a finite set of places of K that are all prime to p, and let $\rho : G_S \to \mathrm{GL}_n(\mathbb{Q}_p)$ be a continuous Galois representation. To use TFMC-U, we need to get a uniform pro-p-group out of this setting. First note that every maximal compact subgroup of $\mathrm{GL}_n(\mathbb{Q}_p)$ is conjugate to $\mathrm{GL}_n(\mathbb{Z}_p)$; hence, the compactness of G_S ensures that we can assume without loss of generality that $G := \mathrm{Im}(\rho)$ is a subgroup of $\mathrm{GL}_n(\mathbb{Z}_p)$.

From [19, (3.2.3.3)], we obtain that G is a p-adic analytic group, as a closed subgroup of the p-adic analytic group $\mathrm{GL}_n(\mathbb{Z}_p)$. Applying [8, Corollary 8.34] then shows that G contains an open normal uniform pro-p-subgroup Γ of finite index. The subfield extension of $K_S^{\ker \rho}/K$ on which Γ acts trivially then defines a finite Galois extension F/K (as Γ is normal of finite index in $G = \mathrm{Im}(\rho)$). Applying TFMC-U to F/K and Γ shows that Γ must be finite (and actually trivial because of Proposition 4.5), hence so does $G = \mathrm{Im}(\rho)$. Note that the ramification conditions of TFMC-U are satisfied since ρ is unramified outside S while the places in S are all prime to p, so we are done.

4.4 The Low-Dimensional Cases of TFMC-U ($d = 1$ and $d = 2$)

In this section, we will briefly see why we do not need to consider the cases $d \leq 2$ in Conjecture 1.7. First recall that any uniform pro-p-group of dimension 1 is isomorphic to $\mathbb{Z}_p$ and hence can be represented into $\mathrm{GL}_1(\mathbb{Z}_p) = \mathbb{Z}_p^\times$ using the map $p \mapsto 1 + p$. This boils down to TFMC-U for $n = 1$ and hence to TFMC for $n = 1$ (thanks to the content of Sect. 4.3), which was proven in Sect. 4.2 (see Theorem 4.4).

Now let Γ be a uniform pro-p-group of dimension 2. Then [8, Exercise 3.11] shows that Γ is *meta-procyclic*, which means that there exists an element γ in Γ such that $\overline{\langle\gamma\rangle}$ is a normal subgroup of Γ and the quotient $\Gamma/\overline{\langle\gamma\rangle}$ is pro-cyclic. Assuming the existence of a finitely and tamely ramified Galois extension L/K with $\mathrm{Gal}(L/K)$ isomorphic to Γ would then provide a $\Gamma/\overline{\langle\gamma\rangle} \simeq \mathbb{Z}_p$-extension of K that is finitely and tamely ramified. This contradicts TFMC-U for $d = 1$; hence, such an extension L/K cannot exist and TFMC-U is proven for $d = 2$.

4.5 Tame Ramification and Co-primality Matters

The goal of this section is to discuss why the two main assumptions in TFMC (tame ramification and co-primality with p of the places in S) cannot be dropped. We start by discussing the ramification assumption and show on a simple example why tame ramification matters in the statement of Conjecture 1.6. We produce a two-dimensional interesting representation $\rho\colon G_{\mathbb{Q},S} \to \mathrm{GL}_2(\mathbb{Q}_p)$ that is wildly ramified at p and whose image is infinite, and then we show that twisting by characters is not enough to get rid of the wild ramification property at p. This is based on the one-dimensional counterpart given by the p-adic cyclotomic character $\chi_p : G_{\mathbb{Q}} \to \mathbb{Z}_p^\times$, which is unramified outside p and wildly ramified at p (as the tame quotient of $G_{\mathbb{Q}_p}$ is prime to p while $\mathbb{Z}_p^\times$ is—up to a finite part—a pro-p-group). Note that the p-adic cyclotomic character is crystalline, hence de Rham, so it fits within the framework of Fontaine–Mazur's original conjecture.

Let $\mathcal{E}$ be a non-CM elliptic curve defined over $\mathbb{Q}$ and let p be an odd prime. Serre's open image theorem [24] ensures that the p-adic representation of $G_{\mathbb{Q}}$ attached to $\mathcal{E}$ has open image in $\mathrm{GL}_2(\mathbb{Z}_p)$ (and is actually surjective for almost all p). Moreover, this representation is unramified away from p and from the primes dividing the conductor of $\mathcal{E}$, which certainly form a finite set S. We are hence left to see that this representation is wildly ramified at p, but this comes from the fact that the determinant of this representation is the p-adic cyclotomic character, which is wildly ramified at p.

One can now wonder whether one can get rid of this wild ramification by simply twisting away the determinant. The answer is negative: indeed, let $\rho\colon G_{\mathbb{Q}} \to \mathrm{GL}_2(\mathbb{Z}_p)$ be the p-adic representation attached to $\mathcal{E}$, $\det\rho = \chi_p$ be its determinant and $\langle \cdot \rangle : \mathbb{Z}_p^\times \to 1 + p\mathbb{Z}_p$ denote the projection of $\mathbb{Z}_p^\times$ onto its pro-p part. As p is odd, every element of $1 + p\mathbb{Z}_p$ has a unique squareroot that is congruent to 1 mod p; hence, we get a character $\varepsilon : G_\mathbb{Q} \to 1 + p\mathbb{Z}_p$ whose square equals $\langle \chi_p \rangle$. If we set $\rho' := \rho \otimes \varepsilon^{-1}$, then $\det \rho'$ is of finite order, so the previous argument on the determinant cannot be used to show that ρ' is wildly ramified at p. Nevertheless, we can see that $\mathrm{Im}(\rho')$ cannot be finite (since it contains an open subgroup of $\mathrm{SL}_2(\mathbb{Z}_p)$); hence, TFMC predicts that ρ' must be wildly ramified at p. This last assertion can be proven without using TFMC. Indeed, first observe that if the image of the inertia contains an open subgroup of a unipotent subgroup, i.e., a conjugate of an open subgroup of

$$U = \left\{ \begin{pmatrix} 1 & u \\ 0 & 1 \end{pmatrix} : u \in \mathbb{Z}_p \right\},$$

then so does $\mathrm{Im}(\rho')$ since $\det(U) = \{1\}$. But U is a pro-p-group and the tame inertia quotient has pro-order prime-to-p; hence, ρ' cannot be tamely ramified at p and thus is wildly ramified at p.

Remark 4.7 Using the p-adic cyclotomic character χ_p, one can define infinitely many one-dimensional p-adic representations that provide infinite analytic extensions unramified outside p and wildly ramified at p. This is, for instance, the case of $\chi_p^{(p-1)k}$ for any positive integer k.

Let us now focus on the other main assumption, namely that the places in S must be co-prime to p. First recall that any $\mathbb{Z}_p$-extension must be unramified at primes with residue characteristic different from p. But $\mathbb{Z}_p$ is abelian; hence, the finiteness of the class number requires that $\mathbb{Z}_p$-extensions cannot be everywhere unramified. This shows that there is some ramification at places above p. This is why we need to impose the requirement claiming that S contains no place above p in Conjecture 1.6.

Following Fontaine and Mazur [10, page 44], one can also wonder whether the finiteness of S is automatic, or if it is an actual assumption, at least for semi-simple finite-dimensional p-adic representations. Ramakrishna answered this question under GRH when $n = 2$: in [22], he proved the existence of an explicit irreducible two-dimensional p-adic representation that ramifies at infinitely many primes but is potentially semi-stable at p. Also mention that Khare and Ramakrishna proved that the conditions S *is finite* and S *does not contain places above* p required in the statement of Conjecture 1.6 are actually independent (see [16]).

Since the finiteness of S is not automatic, even in low dimension, one can wonder whether this condition is necessary for Conjecture 1.6 to hold. This question has been addressed by Khare, Larsen and Ramakrishna in [15]. More precisely, recall that a p-adic representation is said *deeply ramified at a given prime* when it does not vanish on any of the corresponding higher ramification groups of finite index. With this definition in mind, one can ask whether there exists a p-adic representation that ramifies at infinitely many primes of a number field K but is not deeply ramified at p. The answer is yes, at least when $n = 2$ and $p > 5$, as proven by Khare–Larsen–Ramakrishna in [15, Main Theorem].

5 Some Future Directions

We end this chapter by giving some future directions of research: some are work in progress, and others are longer term projects. First recall that the results of Hajir–Maire we used to build Sect. 4, as well as more recent results obtained by Maire [21] in connection with these questions, require the prime p to be odd. A first natural question we plan to explore is hence the following one:

Question 5.1 To which extend do the previous results remain valid for $p = 2$?

Another interesting direction stems from the following setup. In [5, page 290], Boston (and Buell) pointed out an explicit example of a biquadratic field as being close to providing an actual counter-example to Conjecture 1.3. More precisely, they asked the following question.

Question 5.2 Set $K = \mathbb{Q}(\sqrt{-104}, \sqrt{229})$, $p = 3$ and let L be the maximal everywhere unramified pro-p-extension of K. Does $\mathrm{Gal}(L/K)$ have an infinite p-adic analytic pro-p-quotient?

Note that Boston's method cannot be used to test whether Conjecture 1.3 holds for this biquadratic field and this prime p, as we are not dealing with a cyclic extension. Moreover, we are in a case where all the non-trivial sub-extensions K/M involved are cyclic extensions with the class number divisible by 3, so this question does not fit within the framework of Boston's theorem (that requires the class number to be co-prime to p). In a recent work [1], two of us provided a negative answer to Question 5.2. Note that the methods developed in [1] open a door regarding the resolution of the following question, which extends Question 5.2 and is at the core of an ongoing project of the same authors.

Question 5.3 Given a biquadratic extension $K = \mathbb{Q}(\sqrt{a}, \sqrt{b})$, let L be the maximal everywhere unramified pro-3-extension of K. Does there exist a pair (a, b) such that $\mathrm{Gal}(L/K)$ has an infinite p-adic analytic pro-p-quotient?

Acknowledgments We want to thank the organisers of the workshop *Women in Numbers Europe 3* for giving us this opportunity to work on this project. We also warmly thank Jaclyn Lang for many stimulating discussions and for earlier work on this project. Finally, we are grateful to the referees for their careful reading of the manuscript and all their valuable questions, comments and suggestions, which helped us to upgrade the content of this chapter.

References

1. R. Abdellatif, S. Pisolkar, On a question of Nigel Boston about Fontaine-Mazur Conjecture, submitted (2022)
2. C. Breuil, A. Mézard, Multiplicités modulaires et représentations de $GL_2(\mathbb{Z}_p)$ et de $\mathrm{Gal}(\overline{\mathbb{Q}}_p/\mathbb{Q}_p)$ en $l = p$. Duke Math. J. **115**, 205–310 (2002), with an appendix by G. Henniart

3. C. Breuil, Sur quelques représentations modulaires et p-adiques de $GL_2(Q_p)$ I. Compositio Math. **138**, 165–188 (2003)
4. C. Breuil, Sur quelques représentations modulaires et p-adiques de $GL_2(Q_p)$ II. J. Inst. Math. Jussieu **2**, 1–36 (2003)
5. N. Boston, Some cases of the Fontaine-Mazur conjecture. J. Number Theory **42**(3), 285–291 (1992)
6. N. Boston, Some cases of the Fontaine-Mazur conjecture II. J. Number Theory **75**(2), 161–169 (1999)
7. N. Boston, Galois groups of tamely ramified p-extensions. J. de Théorie des Nombres de Bordeaux **19**, 59–70 (2007)
8. J.D. Dixon, J. D, M.P.F. du Sautoy, A. Mann, D. Segal, *Analytic pro-p groups*, 2nd edn. Cambridge Studies in Advanced Mathematics, vol. 61 (Cambridge University, Cambridge, 1999)
9. M. Emerton, Local-global compatibility in the p-adic Langlands programme for $\mathrm{GL}_2/\mathbb{Q}$, preprint (2011)
10. J.-M. Fontaine, B. Mazur, Geometric Galois representations, in *Elliptic Curves and Modular Forms*, ed. by J.H. Coates, S.T. Yau. Proceedings of a Conference in Hong-Kong, December 18–21, 1993; Ser. Number Theory, I (International Press, Cambridge, 1995)
11. J.-M. Fontaine, Représentations p-adiques semi-stables (Exposé III). Périodes p-adiques (Bures-sur-Yvette, 1988). *Astérisque* **No. 223** (1994), pp. 59–111
12. E. Golod, I. Shafarevich, On class field towers. Izv. Aka. Nauk SSSR **28**, 261–272 (1964)
13. F. Hajir, On the growth of p-class groups in p-class field towers. J. Algebra **188**(1), 256–271 (1997)
14. F. Hajir, C. Maire, Analytic Lie extensions of number fields with cyclic fixed points and tame ramification. J. Ramanujan Math. Soc. **37**, 63–85 (2022)
15. C. Khare, M. Larsen, R. Ramakrishna, Constructing semisimple p-adic Galois representations with prescribed properties. Am. J. Math. **127**(4), 709–734 (2005)
16. C. Khare, R. Ramakrishna, Finiteness of Selmer groups and deformation rings. Invent. Math. **154**(1), 179–198 (2003)
17. M. Kisin, The Fontaine-Mazur conjecture for GL_2. J. Am. Math. Soc. **22**(3), 641–690 (2009)
18. M. Kisin, S. Wortmann, A note on Artin motives. Math. Res. Lett. **10**(2–3), 375–389 (2003)
19. M. Lazard, Groupes analytiques$p$-adiques. Publ. Math. Inst. Hautes Études Sci. **26**, 389–603 (1965)
20. C. Maire, On the quotients of the maximal unramified 2-extension of a number field. Doc. Math. **23**, 1263–1290 (2018)
21. C. Maire, On Galois representations with large image, preprint (2021)
22. R. Ramakrishna, Infinitely ramified Galois representations. Ann. Math. **151**(2), 793–815 (2000)
23. L. Ribes, P. Zalesskii, *Profinite Groups*, 2nd edn. Ergebnisse der Mathematik und ihrer Grenzgebiete, vol. 40 (Springer, Berlin, 2010)
24. J.-P. Serre, Propriétés galoisiennes des points d'ordre fini des courbes elliptiques. Invent. Math. **15**, 259–331 (1972)
25. J.-P. Serre, J. Tate, Good reduction of abelian varieties. Ann. Math. **88**(2), 492–517 (1968)
26. A. Shalev, On almost fixed point free automorphisms. J. Algebra **157**, 271–282 (1993)
27. J. Tate, Relations between K_2 and Galois cohomology. Invent. Math. **36**, 257–274 (1976)
28. K. Wingberg, On the Fontaine-Mazur conjecture for CM-fields. Compositio Math. **131**(3), 341–354 (2002)

Orientations and Cycles in Supersingular Isogeny Graphs

Sarah Arpin, Mingjie Chen, Kristin E. Lauter, Renate Scheidler, Katherine E. Stange, and Ha T. N. Tran

2020 Mathematics Subject Classification Primary: 14G50, 94A60, 11G05, 14K04

Katherine E. Stange and Sarah Arpin were supported by NSF-CAREER CNS-1652238. Katherine E. Stange was also supported by Simons Fellowship 822143. Ha T. N. Tran was supported by the Natural Sciences and Engineering Research Council of Canada (NSERC) (funding RGPIN-2019-04209 and DGECR-2019-00428). R. Scheidler was supported by the Natural Sciences and Engineering Research Council of Canada (NSERC) (funding RGPIN-2019-04844). Mingjie Chen was supported by NSF grants DMS-1844206 and DMS-1802161.

S. Arpin
Mathematics Institute, Universiteit Leiden, Leiden, The Netherlands
e-mail: S.A.Arpin@math.leidenuniv.nl

M. Chen
Université Libre de Bruxelles, Bruxelles, Belgium
e-mail: mjchennn555@gmail.com

K. E. Lauter
Facebook AI Research, Meta, Seattle, WA, USA
e-mail: klauter@meta.com

R. Scheidler
Department of Mathematics and Statistics, University of Calgary, Calgary, AB, Canada
e-mail: rscheidl@ucalgary.ca

K. E. Stange (✉)
Department of Mathematics, University of Colorado, Boulder, CO, USA
e-mail: kstange@math.colorado.edu

H. T. N. Tran
Department of Mathematical and Physical Sciences, Concordia University of Edmonton, Edmonton, AB, Canada
e-mail: hatran1104@gmail.com

A. Bucur et al. (eds.), *Research Directions in Number Theory*, Association for Women in Mathematics Series 33, https://doi.org/10.1007/978-3-031-51677-1_2

1 Introduction

Fix primes $\ell < p$. The supersingular ℓ-isogeny graph $\mathcal{G}_\ell$ is the directed graph whose vertices are the $\overline{\mathbb{F}}_p$-isomorphism classes of supersingular elliptic curves and whose edges are isogenies, up to post-composition by an automorphism. The graph is $(\ell+1)$-regular, has around $p/12$ vertices and consists of one connected component. However, it has few obvious symmetries: Its structure more closely resembles that of a random regular graph in its statistical behaviour, including its graph spectrum. It is a Ramanujan graph.

The graph $\mathcal{G}_\ell$ was introduced into cryptography in 2006 in [15], where the hard problem of finding paths in the graph was proposed. In the cryptographic setting, p is always taken to be a large prime of cryptographic size and ℓ is a small prime like $\ell = 2$ or 3. A key exchange protocol called SIDH based on the path-finding problem in $\mathcal{G}_\ell$ was proposed in [21] and was considered for standardization in the NIST Post-Quantum Cryptography process (2017–2022). Although SIDH has been broken [11, 40, 47], so far there do not exist polynomial time algorithms (polynomial in $\log p$) to solve either the *path finding problem* (to find a path joining two given supersingular curves) or the *endomorphism ring problem* (to compute the endomorphism ring, either as a maximal order in a quaternion algebra or by giving a basis of endomorphisms) of a supersingular curve. These problems are closely related to one another [24, 57] and to other cryptographic protocols such as CSIDH [13] and SQISign [22].

The graph $\mathcal{G}_\ell$, like any superhero worth its salt, has an 'alter ego', which is as the graph of maximal orders in a quaternion algebra $B_{p,\infty}$ ramified at p and ∞. For each elliptic curve vertex, the endomorphism ring of the curve is such a maximal order. However, even viewed in this way, the graph is still its apparently disordered self. To reveal some familiar structure with which to navigate the graph, one can focus, one at a time, on the quadratic rings embedded in the quaternion algebra.

The aforementioned structure recalls to mind the analogous graph for ordinary elliptic curves, whose endomorphism rings are merely quadratic orders. One can form an ordinary ℓ-isogeny graph just as one does a supersingular one: Vertices are isomorphism classes of ordinary elliptic curves, and edges are ℓ-isogenies [26, 52]. In this case, the endomorphism rings of the curves in a connected component are all orders in the same quadratic field. Stratifying the graph according to the conductors of these orders, with larger orders at higher 'altitude', we see at the top a cycle, or 'rim', from which trees descend down the sides of the 'volcano' (see Fig. 1). This simple, organized picture is in contrast to the complexity of the supersingular graph.

The storyline of the present chapter is that the (finite) supersingular graph $\mathcal{G}_\ell$ can be understood as the result of superimposing infinitely many volcanoes, each obtained by focusing on a single quadratic field embedded inside the quaternion algebra $B_{p,\infty}$ and its intersection with the various maximal orders. The volcanoes in this case are not the volcanoes of the ordinary ℓ-isogeny graph, but rather the connected components of the *oriented supersingular ℓ-isogeny graph associated with a quadratic field K*, denoted by $\mathcal{G}_{K,\ell}$. This graph has recently been studied by Colò-Kohel [17] and Onuki [42] in the cryptographic context. The vertices of $\mathcal{G}_{K,\ell}$ are pairs (E, ι) consisting of an isomorphism class of supersingular curves, together

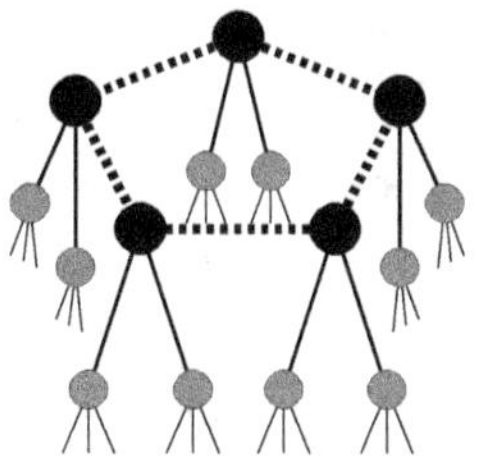

Fig. 1 An oriented 3-isogeny volcano. The rim is comprised of the five black vertices. These vertices are primitively oriented by a 3-fundamental order $\mathcal{O}$. The gray vertices at altitude 1 of the volcano are primitively oriented by the suborder of $\mathcal{O}$ of index 3. The five thick dotted edges along the rim indicate horizontal oriented 3-isogenies, and the remaining black edges are ascending/descending

with an embedding $\iota \colon K \to \mathrm{End}^0(E)$ of a quadratic field into the endomorphism algebra of the curve, called a *K-orientation*. Edges are again ℓ-isogenies. (For the precise definitions, see Sect. 2.)

Upon forgetting orientations, each of these volcanoes covers the supersingular ℓ-isogeny graph. In particular, each rim maps to a closed walk in the ℓ-isogeny graph.

We concern ourselves with a particular type of closed walk in $\mathcal{G}_\ell$: one with no backtracking and which is not a repeat of a smaller closed walk. We keep its direction (traversed 'forward' or 'backward'), but forget any choice of basepoint, and call this an *isogeny cycle* (Definition 3.1). The main result of this chapter is that each isogeny cycle is obtained, by forgetting orientations, from exactly one oriented volcano rim (up to conjugation of the rim; see Sect. 3.1 for definitions).

Theorem 1 (See the Precise Version in Theorem 3.2) *Fix distinct primes $\ell < p$. Let $r > 2$. The isogeny cycles of length r in $\mathcal{G}_\ell$ are in bijection with the rims of length r of the union of all oriented supersingular ℓ-isogeny volcanoes over $\overline{\mathbb{F}}_p$, up to conjugation of the orientations.*

The exact statement is discussed in Sect. 3.1, and a detailed example is given in Sect. 6. At the highest level, the bijection is given by a simple process. Given a rim, one can forget the orientation information to obtain an isogeny cycle in the supersingular graph. Conversely, given an isogeny cycle, one can compose the sequence of isogenies it comprises, to obtain an endomorphism. This endomorphism induces an orientation, placing the initial curve on a rim. The proof, however, spans around a third of the paper, in large part because of the existence of special cases and issues spawned by extra automorphisms and quadratic fields in which p ramifies. In particular, *the bijection is not canonical* due to these considerations. See also Remark 2.17, where we make some philosophical comments on the obstacles presented by extra automorphisms in the supersingular isogeny graph.

We use the bijection of Theorem 1 to count isogeny cycles of length r in $\mathcal{G}_\ell$. The essential ingredient is that the rims of K-oriented ℓ-isogeny volcanoes correspond to cycles of length r in the permutation on the oriented rim vertices arising from the action of an element of order r in the class group of an order of K.

Theorem 2 (See the Precise Versions in Theorem 7.1, Corollaries 7.3 and 7.7) *The number of isogeny cycles of length r in $\mathcal{G}_\ell$ is given by*

$$\frac{1}{r}\sum_{\mathcal{O}\in\mathcal{I}_r}\epsilon_{\mathcal{O},\ell}h_{\mathcal{O}}, \tag{1}$$

where $\mathcal{I}_r$ is a set of quadratic orders satisfying certain splitting properties for ℓ and p specified in Sect. 5, and $\epsilon_{\mathcal{O},r}$ is defined in Theorem 5.1 and satisfies $1 \leq \epsilon_{\mathcal{O},r} \leq 2$, with $\epsilon_{\mathcal{O},r} = 2$ whenever p is unramified in $\mathcal{O}$. The quantity (1) is bounded above by

$$\frac{2\pi e^{0.578}\log(4\ell)}{3}\ell^r(\log r + c) + O(\ell^{3r/4}\log r)$$

as $r \to \infty$, where the constant c is explicit, and the implied constant in the big O notation can be made explicit using Corollary 7.7. For $p \equiv 1 \pmod{12}$, the quantity (1) asymptotically approaches $\frac{\ell^r}{2r}$ as $r \to \infty$.

The asymptotic for the number of isogeny cycles is as expected for Ramanujan graphs (see Sect. 7.1), and we give a standard but self-contained proof for expander graphs. However, the class number formula and the explicit upper bound are consequences of the main bijection between rims and isogeny cycles. Figure 7 and Sect. 6 contain experimental data confirming our results.

Our strategy for counting isogeny cycles leads immediately to an algorithm to list all isogeny cycles of length r in $\mathcal{G}_\ell$. In particular, the discriminants of elements of $\mathcal{I}_r$ are bounded by ℓ^r. The j-invariants of the isogeny cycles are roots of the Hilbert class polynomials of the orders $\mathcal{O} \in \mathcal{I}_r$. The cycle in $\mathcal{G}_\ell$ can be determined by use of the modular polynomial Φ_ℓ. This generalizes a simple congruence condition check on p used in [15, Section 5.3.4] to rule out the existence of small cycles for the Charles–Goren–Lauter hash function. The exact value of $\epsilon_{\mathcal{O},r}$ in the ramified cases depends upon the theory of the ring class field of K and is given in terms of the class number and genus number (Sect. 4).

In the course of the proof, we need to clarify a point about the number of orientations of a curve. When p does not split in $\mathcal{O}$, Onuki [42] considered the set $\mathrm{SS}_{\mathcal{O}}^{\mathrm{pr}}$ of pairs (E, ι) of a supersingular isogeny graph together with a primitive $\mathcal{O}$-orientation (meaning a K-orientation ι such that $\iota(K) \cap \mathrm{End}(E) = \mathcal{O}$) and the set $\rho(\mathcal{E}\ell\ell(\mathcal{O}))$ of pairs $(E, \iota) \in \mathrm{SS}_{\mathcal{O}}^{\mathrm{pr}}$ obtained by reducing an elliptic curve over $\mathbb{C}$ with normalized complex multiplication by $\mathcal{O}$ (where normalized means that $[\alpha]^*\omega = \alpha\omega$ for the invariant differential ω; see Sect. 2.3.) Onuki showed that these two sets are not always equal; sometimes $\#\mathrm{SS}_{\mathcal{O}}^{\mathrm{pr}} = 2\#\rho(\mathcal{E}\ell\ell(\mathcal{O}))$. This raises the natural question of when this case occurs. In this chapter, we answer that it occurs if and only if p is inert in K. (In the ramified case, $\#\mathrm{SS}_{\mathcal{O}}^{\mathrm{pr}} = \#\rho(\mathcal{E}\ell\ell(\mathcal{O}))$ exactly.)

We prove this using the ring class field and the action of conjugation (Theorem 4.4) but note that the result is also proved using Deuring lifting in [25, Lemma 3.2].

We also explicitly find the in- and out-degrees of the vertices of all possible K-oriented ℓ-isogeny graphs. The K-oriented graph is $(\ell + 1)$-regular except at vertices corresponding to curves with extra K-automorphisms (i.e. automorphisms that preserve the orientation), where the out-degree may be slightly less than $\ell + 1$. This is made explicit in Remark 2.7 and Proposition 2.11. In connection with this, we provide a new direct proof of the volcano structure of the graph and the count of ascending, descending and horizontal edges even in exceptional cases, in Proposition 2.15.

The motivation for this chapter is prior work by the same authors, summarized in Sect. 8, in which the oriented volcano structure is used to navigate the supersingular ℓ-isogeny graph [5]. In particular, given an initial vertex in the supersingular ℓ-isogeny graph and given an endomorphism of that curve, one asks to navigate to a target curve (typically $j = 1728$). The endomorphism provides an orientation of the initial curve and hence a location in an oriented supersingular ℓ-isogeny volcano. Then, using methods of determining ascending and descending directions, one can navigate the oriented volcano. The target curve must be given an orientation on the same volcano (algorithms are provided in [5]). From both curves, one ascends to the rim and hopes to connect the paths. Having found a path on an oriented volcano, one takes the image in the original supersingular ℓ-isogeny graph.

The algorithms provided in [5] explicitly relied on heuristics about oriented volcanoes. They also implicitly hinted at the sort of bijection given in our first theorem above. In Sect. 9 of the present chapter, we revisit the explicit heuristics of [5] and provide some partial results. In particular, we do the following.

(1) We give an explicit bound on the distance from uniform of the distribution of endpoints of non-backtracking random walks in $\mathcal{G}_\ell$ from an initial set of vertices when $p \equiv 1 \pmod{12}$ (Proposition 9.1).
(2) We estimate the expected distance of any fixed vertex to a set of j-invariants of a given size (Corollary 9.2).
(3) We fix an oriented volcano and bound the depth at which any j-invariant first occurs (Sect. 9.2).
(4) We fix a j-invariant and consider all the possible orientations by a given field and the distribution of this collection of oriented curves on the various oriented volcanoes (Sect. 9.3). We are able to prove a result that differs slightly from the heuristic used in [5].

As an additional minor point of interest, Lemma 2.9 shows that the kernel of an endomorphism is fixed by an extra automorphism if and only if its field of definition lies in the field of definition of the automorphism. As a corollary, we can classify the behaviour of loops at $j = 0$ and $j = 1728$ in Corollary 2.13; this is a new method of proof, and slight generalization, of a known result [1, 36, 44].

The topic of closed walks[1] in the supersingular ℓ-isogeny graph was first studied in [15], where it was observed that closed walks give endomorphisms, and so the splitting behaviour of p in various extensions controlled the existence of small cycles. The endomorphisms created by closed walks were studied in the thesis of Kohel [33] and further in [7], with the goal of understanding when such cycles are independent and generate an endomorphism ring. These sources also discuss the computation of the trace of a cycle. The neighbourhoods of 0 and 1728 are particularly intricate, and their loops and cycles are discussed in [1, 35, 36, 44, 58]. Cycles in the SIKE graph which pass through the secret key are discussed in [43].

Further back in the literature, Gross [28] considered counting endomorphisms of a fixed degree by computing the traces of matrices derived from certain modular forms of weight 2 for the group $\Gamma_0(N)$. Thus Gross counts closed walks of a different sort and in particular allows backtracking (which we do not). As the degree increases, the proportion of the count due to backtracking increases as well. We provide a comparison at the end of our extended example in Sect. 6.

The study of K-orientations is in some sense as old as the study of optimal embeddings in quaternion algebras and has naturally appeared in the study of isogeny graphs, for example, in [37]. In the context of isogeny-based cryptography, the graph $\mathcal{G}_{K,\ell}$ was studied by Colò and Kohel in order to propose OSIDH, a new key exchange protocol [17] (see also [14]). Onuki considered the difference between $\mathrm{SS}_{\mathcal{O}}^{\mathrm{pr}}$ and $\rho(\mathcal{E}\ell\ell(\mathcal{O}))$ [42]. In cryptography, the class group action on oriented curves has been used constructively [16], and the hardness of the group action problem has been considered both classically and quantumly [12, 56]; a new variation called the $\mathcal{O}$-uber isogeny problem appears in [20] (see also [14]). The volcano structure provides some algorithmic approaches to hard problems, which includes the recent work of the present authors [5] and related work [56]. Recent work also asks about the number of orientable curves [34]; see also [37]. Finally, specific orientations underlie the line of work in [23] and [20].

Outline of the Chapter Section 2 contains the background on oriented and unoriented supersingular ℓ-isogeny graphs. In Sect. 3, we state and prove Theorem 1, and this section is the heart of the chapter. Section 4 settles the question of when $\rho(\mathcal{E}\ell\ell(\mathcal{O})) = \mathrm{SS}_{\mathcal{O}}^{\mathrm{pr}}$ and discusses the action of conjugation on $\mathrm{SS}_{\mathcal{O}}^{\mathrm{pr}}$. Section 5 specifies exactly the number of rims associated with a given quadratic order, in terms of class and genus numbers. Section 6 provides an extended example of the main bijection of Theorem 1. Section 7 counts the number of isogeny cycles in $\mathcal{G}_\ell$ asymptotically, as well as exactly in terms of class numbers, and gives an explicit upper bound. Section 8 gives an overview of our previous work and how it motivates the work in this chapter, as well as the heuristics it depends upon. Section 9 partially addresses some of these heuristic questions by use of methods of expander graphs.

[1] Often simply called cycles in the isogeny literature; we use the term *closed walk* to more accurately follow the graph theory literature.

2 Mathematical Preliminaries on Oriented Supersingular ℓ-Isogeny Graphs and Their Volcanoes

2.1 The Supersingular ℓ-Isogeny Graph $\mathcal{G}_\ell$

Let $\ell < p$ be distinct primes. We let $\mathcal{G}_\ell$ denote the supersingular ℓ-isogeny graph over $\overline{\mathbb{F}}_p$, defined as the directed graph whose vertices are isomorphism classes of supersingular elliptic curves over $\overline{\mathbb{F}}_p$ and whose directed edges are isogenies of degree ℓ up to post-composition by an automorphism. The number $\#\mathcal{G}_\ell$ of supersingular j-invariants over $\overline{\mathbb{F}}_p$ (see [50]) is

$$\#\mathcal{G}_\ell = \left\lfloor \frac{p}{12} \right\rfloor + \begin{cases} 0 & \text{if } p \equiv 1 \pmod{12}, \\ 1 & \text{if } p \equiv 5 \text{ or } 7 \pmod{12}, \\ 2 & \text{if } p \equiv 11 \pmod{12}. \end{cases} \tag{2}$$

Supersingular elliptic curves with j-invariants equal to 0 and 1728 have extra automorphisms, beyond the usual automorphisms $[\pm 1]$. Let $E_0 : y^2 = x^3 - 1$ and $E_{1728} : y^2 = x^3 - x$ with $j(E_0) = 0$, $j(E_{1728}) = 1728$. The curve $E_0/\overline{\mathbb{F}}_p$ is supersingular for any $p \equiv 2 \pmod 3$, and $E_{1728}/\overline{\mathbb{F}}_p$ is supersingular for any $p \equiv 3 \pmod 4$. For the sake of this discussion, we assume these curves are supersingular. We provide their explicit endomorphism rings below, referenced from [41, 51].

The automorphism group for E_{1728} is $\mathrm{Aut}(E_{1728}) = \{[\pm 1], [\pm i]\}$, where $[i](x, y) := (-x, iy)$ for $i \in \mathbb{F}_{p^2}$ with $i^2 = -1$. Write π_p for the p-power Frobenius endomorphism: $\pi_p(x, y) := (x^p, y^p)$. In $B_{p,\infty} := \mathrm{End}(E_{1728}) \otimes_{\mathbb{Z}} \mathbb{Q}$, we have

$$\mathrm{End}(E_{1728}) = \left\langle 1, [i], \frac{1 + \pi_p}{2}, \frac{[i] + [i]\pi_p}{2} \right\rangle. \tag{3}$$

The automorphism group for E_0 is $\mathrm{Aut}(E_0) = \{[\pm 1], [\pm \zeta_3], [\pm \zeta_3^2]\}$, where $[\zeta_3](x, y) := (\zeta_3 x, y)$ with ζ_3 a root of $T^2 + T + 1$ in $\mathbb{F}_{p^2}$. We let $[\sqrt{-3}] = 2[\zeta_3] + 1$. In $B_{p,\infty} := \mathrm{End}(E_0) \otimes_{\mathbb{Z}} \mathbb{Q}$, we have

$$\mathrm{End}(E_0) = \left\langle 1, \frac{-1 + [\sqrt{-3}]}{2}, \pi_p, \frac{3 + [\sqrt{-3}] + 3\pi_p + [\sqrt{-3}]\pi_p}{6} \right\rangle. \tag{4}$$

Lemma 2.1 *Let $E \in \{E_0, E_{1728}\}$ and let $\varphi \in \mathrm{Aut}(E)$ such that $\varphi \neq \pm[1]$. If $\overline{\varphi}\theta = \theta\varphi$ for $\theta \in \mathrm{End}(E)$, then $\deg\theta \geq p$.*

Proof Let $\eta = [i]$ or $\eta = [\sqrt{-3}]$ depending whether $E = E_{1728}$ or $E = E_0$, respectively. The condition $\overline{\varphi}\theta = \theta\varphi$ implies that θ must have the form $\theta = s\pi_p + t(\eta\pi_p)$ where $s, t \in \mathbb{Q}$.

For $E = E_{1728}$, using the $\mathbb{Z}$-basis for $\operatorname{End}(E_{1728})$ in (3), we have $\theta = m\pi_p + n([i]\pi_p)$ where $m, n \in \mathbb{Z}$. Thus $\deg\theta = p(m^2 + n^2) \geq p$.

For $E = E_0$, using the $\mathbb{Z}$-basis for $\operatorname{End}(E_0)$ in (4), we have $\theta = \frac{m}{2}\pi_p - \frac{n}{2}(\eta\pi_p)$ where $m, n \in \mathbb{Z}$ are of the same parity. Thus $\deg\theta = \frac{p}{4}(m^2+3n^2)$. Since $m^2+3n^2 \equiv 0 \pmod 4$, it follows that $\deg\theta \geq p$. □

The graph $\mathcal{G}_\ell$ has regular out-degree $\ell + 1$ (the number of subgroups of $E[\ell]$ of size ℓ) [46]. It is very nearly an undirected graph, obtained by identifying each ℓ-isogeny with its dual. The exception is that since isogenies are identified under post-composition with an automorphism, at curves with extra automorphisms, this identification may fail (see Fig. 3). Nevertheless, this issue occurs for edges adjacent to at most two vertices ($j = 0, 1728$), so one often considers it an $(\ell + 1)$-regular undirected graph for the purposes of spectral graph theory, without affecting the validity of the conclusions. Modulo this slight lie, or when $j = 0$ and $j = 1728$ are ordinary and hence absent from the graph (they are absent if and only if $p \equiv 1 \pmod{12}$), $\mathcal{G}_\ell$ is an $(\ell + 1)$-regular Ramanujan graph. In particular, if its adjacency matrix has eigenvalues $\ell + 1 = \lambda_1 \geq \lambda_2 \geq \cdots \geq \lambda_n$, then

$$\max_{i>1} |\lambda_i| \leq 2\sqrt{\ell}. \tag{5}$$

2.2 *The Oriented Supersingular ℓ-Isogeny Graph $\mathcal{G}_{K,\ell}$*

Let $\ell < p$ be distinct primes. Fix a supersingular elliptic curve $E/\overline{\mathbb{F}}_p$. Then $\operatorname{End}^0(E) \cong B_{p,\infty}$, where $B_{p,\infty}$ denotes the definite quaternion algebra ramified precisely at p and ∞ (unique up to isomorphism). Let K be an imaginary quadratic field in which p does not split. This condition on p gives the existence of an embedding of K into $B_{p,\infty}$. Let $\mathcal{O}$ denote an order in K and $h_\mathcal{O}$ the class number of this order. Let $\Delta_\mathcal{O}$ denote its discriminant. An order $\mathcal{O}$ of K is *ℓ-fundamental* if its conductor is relatively prime to ℓ.

Definition 2.2 (K-orientation) A *K-orientation* of E is an embedding $\iota\colon K \to \operatorname{End}^0(E)$.

Definition 2.3 ($\mathcal{O}$-orientation) A K-orientation $\iota\colon K \to \operatorname{End}^0(E)$ is an *$\mathcal{O}$-orientation* if $\iota(\mathcal{O}) \subseteq \operatorname{End}(E)$. Furthermore, we say that an $\mathcal{O}$-orientation is *$\mathcal{O}$-primitive* if $\iota(\mathcal{O}) = \operatorname{End}(E) \cap \iota(K)$.

Every K-orientation $\iota\colon K \to \operatorname{End}^0(E)$ is $\mathcal{O}$-primitive for a unique order $\mathcal{O}$ of K.

In the literature, for example, [37], one will find the closely related term *optimal embedding*, in the context of a maximal order $\mathfrak{O}$ of a quaternion algebra A, for an embedding $\iota\colon K \to A$ such that $\iota(K) \cap \mathfrak{O} = \mathcal{O}$. An optimal embedding corresponds to a primitive orientation via the Deuring correspondence. For the theory of optimal embeddings, see [55, Chapter 30].

Definition 2.4 **(K-oriented Elliptic Curve)** A *K-oriented elliptic curve* is a pair (E, ι), where E is an elliptic curve and ι is a K-orientation of E.

Isogenies of elliptic curves induce isogenies of oriented elliptic curves in the following sense: If $\phi\colon E \to E'$ is an isogeny and $\iota\colon K \to \mathrm{End}(E)$ is a K-orientation of E, then we define the unique induced orientation $\phi_*\iota$ on E' for any $\alpha \in K$ as follows:

$$(\phi_*\iota)(\alpha) := \frac{1}{[\deg \phi]}\phi \circ \iota(\alpha) \circ \widehat{\phi}. \tag{6}$$

Definition 2.5 **(K-oriented Isogeny)** If $\phi_*\iota = \iota'$, then we say that $\phi\colon (E, \iota) \to (E', \iota')$ is an isogeny of K-oriented curves, also known as a *K-oriented isogeny*.

Two K-oriented curves (E, ι), (E', ι') are *K-isomorphic* if there exists an isomorphism $\eta\colon E \to E'$ such that $\eta_*\iota = \iota'$. Two K-oriented isogenies $\phi\colon (E_1, \iota_1) \to (E_2, \iota_2)$ and $\phi'\colon (E_1', \iota_1') \to (E_2', \iota_2')$ are considered *equivalent* if there are K-oriented isomorphisms η and η' so that the following diagram commutes:

$$\begin{array}{ccc} (E_1, \iota_1) & \xrightarrow{\phi} & (E_2, \iota_2) \\ {\scriptstyle\eta}\downarrow & & \downarrow{\scriptstyle\eta'} \\ (E_1', \iota_1') & \xrightarrow{\phi'} & (E_2', \iota_2') \end{array} \tag{7}$$

We will frequently abuse notation by writing a representative (E, ι) or ϕ in place of its equivalence class.

An automorphism φ of E is a *K-automorphism* of (E, ι) if $\varphi_*\iota = \iota$. This occurs if and only if $\varphi \in \iota(K)$.

Definition 2.6 **($\mathcal{G}_{K,\ell}$)** The *ℓ-isogeny graph $\mathcal{G}_{K,\ell}$ of K-oriented supersingular elliptic curves over $\overline{\mathbb{F}}_p$* is the directed graph whose vertices (E, ι) are K-isomorphism classes of K-oriented supersingular elliptic curves over $\overline{\mathbb{F}}_p$ and whose edges are equivalence classes of K-isogenies of degree ℓ between the given oriented curves.

Remark 2.7 The only vertices of $\mathcal{G}_{K,\ell}$ with automorphism group larger than $\{[\pm 1]\}$ are (E, ι) where either (i) E has j-invariant 1728 and $\iota(i) \in \mathrm{End}(E)$ or (ii) E has j-invariant 0 and $\iota(\zeta_3) \in \mathrm{End}(E)$.

Proposition 2.8 *The directed edges of $\mathcal{G}_{K,\ell}$ can be put into equivalence classes by making an isogeny equivalent to its dual isogeny. These classes are of size* 1 *or* 2. *The only self-dual edges are loops. By identifying isogenies with their duals, this graph can be drawn undirected.*

Proof With reference to (7), if ϕ and ϕ' are equivalent, then their duals are equivalent. Therefore an equivalence class consists of one directed edge and its dual or a single self-dual edge. □

Although the number of ℓ-isogenies originating at a supersingular curve is always $\ell + 1$, some of these may be equivalent, so that in the graph $\mathcal{G}_{K,\ell}$, the out-degree may be strictly smaller. As usual, the culprits are the extra automorphisms, as in Remark 2.7. Before we state and prove the exact out-degree, we need a lemma.

Lemma 2.9 *Let E be a curve with extra automorphisms. Given a non-zero isogeny $\phi\colon E \to E'$, its kernel $\ker\phi$ is fixed by $\mathrm{Aut}(E) \setminus \{[\pm 1]\}$ if and only if $E \cong E'$. If we further assume that $\deg\phi < p$, then $\ker\phi$ is fixed by $\mathrm{Aut}(E) \setminus \{[\pm 1]\}$ if and only if $E \cong E'$ and ϕ (up to post-composition with an isomorphism $E' \to E$) is an element in $\mathbb{Q}(\mathrm{Aut}(E)) \cap \mathrm{End}(E)$.*

Proof Let φ be a non-integer automorphism of E. First, suppose $\ker\phi = \varphi\ker\phi$. Then $\ker\phi = \ker\phi\varphi$, so there exists a unique $\eta \in \mathrm{Aut}(E')$ such that $\phi\varphi = \eta\phi$ by [50, III.4.11]. Then, working in $\mathrm{End}(E) \otimes_{\mathbb{Z}} \mathbb{Q} \cong \mathrm{End}(E') \otimes_{\mathbb{Z}} \mathbb{Q}$, the element $\phi\varphi\frac{\widehat{\phi}}{[\deg(\phi)]} = \eta \in \mathrm{Aut}(E') \subset \mathrm{End}(E')$ has the same minimal polynomial as $\varphi \in \mathrm{Aut}(E) \subset \mathrm{End}(E)$. The existence of non-integer elements in $\mathrm{Aut}(E)$ and $\mathrm{Aut}(E')$ with the same minimal polynomial gives $\mathrm{Aut}(E) \cong \mathrm{Aut}(E')$. Since $\mathrm{Aut}(E) \cong \mathrm{Aut}(E') \supsetneq \{[\pm 1]\}$, the automorphism group determines the isomorphism class of the curve and we have $E \cong E'$.

Next, suppose $E \cong E'$ via an isomorphism $\lambda\colon E' \to E$. In particular, $\mathrm{Aut}(E) \cong \mathrm{Aut}(E') \supsetneq \{[\pm 1]\}$. Then, $\lambda\phi \in \mathrm{End}(E)$ and $\ker(\lambda\phi) = \ker\phi$. By abuse of notation, we replace $\lambda\phi$ with ϕ and may henceforth assume $\phi \in \mathrm{End}(E)$ in this case. The endomorphism ϕ has the same minimal polynomial as $\varphi^{-1}\phi\varphi$, and so up to some automorphism $\eta \in \mathrm{Aut}(E)$, we have $\varphi^{-1}\phi\varphi = \eta\phi$, and thus $\ker\phi = \ker\phi\varphi$.

Now, assume $\deg\phi < p$ and suppose $E \cong E'$ and $\ker\phi$ is fixed by a generator φ of $\mathrm{Aut}(E)$. Then $\phi = \varphi^k\phi\varphi$ for some integer $0 \le k < |\mathrm{Aut}(E)|$. By cancellation, we know $\phi \neq \pm\phi\varphi$ since $\varphi \neq [\pm 1]$, hence $\overline{\varphi}\phi = \phi\varphi$ or else $\varphi\phi = \phi\varphi$. For $\varphi = [\zeta_6]$ (a primitive sixth root of unity), we have $k \in \{1, 2, 4, 5\}$. If $k \in \{1, 5\}$, then either $\overline{[\zeta_6]}\phi = \phi[\zeta_6]$ or $[\zeta_6]\phi = \phi[\zeta_6]$. Otherwise, if $k \in \{2, 4\}$, then observe that $\phi = [\zeta_6]^k\phi[\zeta_6] = [\zeta_6]^k([\zeta_6]^k\phi[\zeta_6])[\zeta_6] = [\zeta_6]^{2k}\phi[\zeta_6]^2 = [\zeta_3]^k\phi[\zeta_3]$, so $\overline{[\zeta_3]}\phi = \phi[\zeta_3]$ or $[\zeta_3]\phi = \phi[\zeta_3]$. Since $\deg\phi < p$, Lemma 2.1 rules out the case that $\overline{\varphi}\phi = \phi\varphi$. We are left to conclude that ϕ commutes with an element φ of $\mathrm{Aut}(E) \setminus \{[\pm 1]\}$, which implies that $\phi \in \mathbb{Q}(\varphi) \subseteq \mathrm{End}^0(E)$.

On the other hand, if $\ker\phi \neq \ker\phi\varphi$, then $\phi \neq \pm\varphi^k\phi\varphi$ for any k, so ϕ does not commute with φ, and hence $\phi \notin \mathbb{Q}(\mathrm{Aut}(E))$. □

Definition 2.10 (Reduced Automorphism Group) Let $\mathrm{Aut}'(E) := \mathrm{Aut}(E)/[\pm 1]$ denote the reduced automorphism group of the elliptic curve E.

Proposition 2.11 *Let (E, ι) be an oriented curve in $\mathcal{G}_{K,\ell}$ where ι is an $\mathcal{O}$-orientation. Then (E, ι) has out-degree $\ell + 1$, except at the oriented curves with extra K-automorphisms (those of Remark 2.7). In that exceptional case, the out-degree is $(\ell + 1 - s_\ell)/|\mathrm{Aut}'(E)| + s_\ell$, where s_ℓ is the number of $\phi \in \mathrm{End}(E) \cap \iota(K)$ of degree ℓ up to post-composition by $\mathrm{Aut}(E)$. In particular, the out-degree at (E, ι), when (E, ι) has extra K-automorphisms, is*

$$\begin{cases} (\ell-1)/2+2 & \mathcal{O}=\mathbb{Z}[i] \textit{ and } \ell \textit{ splits in } \mathcal{O}(s_\ell=2) \\ (\ell+1)/2 & \mathcal{O}=\mathbb{Z}[i] \textit{ and } \ell \textit{ is inert in } \mathcal{O}(s_\ell=0) \\ 2 & \mathcal{O}=\mathbb{Z}[i] \textit{ and } \ell=2(s_\ell=1) \\ 2(\ell-1)/3+2 & \mathcal{O}=\mathbb{Z}[\zeta_3] \textit{ and } \ell \textit{ splits in } \mathcal{O}(s_\ell=2) \\ 2(\ell+1)/3 & \mathcal{O}=\mathbb{Z}[\zeta_3] \textit{ and } \ell \textit{ is inert in } \mathcal{O}(s_\ell=0) \\ 3 & \mathcal{O}=\mathbb{Z}[\zeta_3] \textit{ and } \ell=3(s_\ell=1). \end{cases}$$

Proof With reference to diagram (7), by taking $\eta \in \mathrm{Aut}(E)$, the two equivalent K-oriented isogenies ϕ and ϕ' from (E, ι) must satisfy $\iota' = \eta_* \iota = \iota$ and $\eta \ker \phi = \ker \phi'$. The first equality implies $\eta \in \iota(K)$. Hence (E, ι) has out-degree $(\ell + 1)$ if $\mathrm{Aut}(E) \cap \iota(K) = [\pm 1]$. If $\mathrm{Aut}(E) \cap \iota(K) \supsetneq [\pm 1]$, the out-degree may be less than $(\ell + 1)$, and we investigate this case below.

If $\mathrm{Aut}(E) \cap \iota(K)$ contains non-trivial automorphisms, then it contains all of them, and $\mathrm{End}(E) \cap \iota(K) = \iota(\mathcal{O}_K) \supseteq \mathrm{Aut}(E)$. The $\ell + 1$ outgoing ℓ-isogenies fall into equivalence classes according to whether their kernels are related by an extra automorphism. The kernel of an outgoing ℓ-isogeny ϕ from E is fixed by $\eta \in \mathrm{Aut}(E)$ if and only if $\phi \in \mathrm{End}(E) \cap \iota(K)$ (as in the proof of Lemma 2.9, we may assume that $\phi \in \mathrm{End}(E)$ by post-composing with an isomorphism; we have assumed throughout the paper that $\ell < p$). Let S_ℓ be the set of equivalence classes of elements in $\mathrm{End}(E) \cap \iota(K)$ of degree ℓ up to pre-composition by $\mathrm{Aut}(E)$, and put $s_\ell = \#_\ell$. Note that $\mathrm{End}(E) \cap \iota(K) = \iota(\mathcal{O}_K)$, so $s_\ell = 0$, 1 or 2, depending on whether ℓ is inert, ramified or split in $\mathcal{O}_K$. By the above, a degree ℓ-isogeny ϕ is in S_ℓ if and only if it is fixed by all $\eta \in \mathrm{Aut}(E)$. Otherwise its kernel has an orbit of size $|\mathrm{Aut}'(E)|$. Thus the number of out-isogenies is $(\ell+1-s_\ell)/|\mathrm{Aut}'(E)|+s_\ell$. □

Remark 2.12 As an immediate corollary, we obtain a new method of proof and generalization of a known result about the number of loops at a vertex [1, 36, 44]. Specifically, the previous works concerned only large p (the last two sentences of the next corollary).

Corollary 2.13 *Let $j \in \{0, 1728\}$ correspond to a supersingular curve over $\overline{\mathbb{F}}_p$. Let $K = \mathbb{Q}(\zeta_3), \mathbb{Q}(i)$ for $j = 0, 1728$, respectively. Let n represent the number of loops at the vertex of j-invariant j, in $\mathcal{G}_\ell$. Then*

$$n = 1 + \left(\frac{\Delta_K}{\ell}\right) + k|\mathrm{Aut}'(E)|,$$

for some $k \in \mathbb{Z}$. If $4\ell < p$, then the number of loops at $j = 1728$ is exactly $1 + \left(\frac{\Delta_K}{\ell}\right)$. If $3\ell < p$, then the number of loops at $j = 0$ is exactly $1 + \left(\frac{\Delta_K}{\ell}\right)$.

Proof Let $\mathcal{O}$ be $\mathbb{Z}[\zeta_3]$ or $\mathbb{Z}[i]$ when $j = 0$ or 1728, respectively. Whether ℓ splits or ramifies or is inert determines the number of elements of $\iota(\mathcal{O})$ of degree ℓ, up to automorphisms: two, one or zero, respectively. These elements are all loops on the vertex. All other loops fall into groups of size $|\mathrm{Aut}'(E)|$, as in the proof of Proposition 2.11. For the last two sentences of the statement, we revisit the endomorphism rings of the vertex (Sect. 2.1), to see that the smallest elements not in $\iota(\mathcal{O})$ have norm $p/4$ and $p/3$, respectively. □

It is possible to compute k explicitly for any given value of ℓ in terms of p, by reference to the endomorphism rings as described in the proof.

The graph $\mathcal{G}_{K,\ell}$ consists of connected components, each with a volcano structure as pictured in Fig. 1. The rim vertices of any given volcano are K-orientations of supersingular elliptic curves which are $\mathcal{O}$-primitive for some fixed ℓ-fundamental order $\mathcal{O}$ of K. The oriented supersingular elliptic curves that appear at depths below the rim are primitively oriented by orders whose index in $\mathcal{O}$ is given by increasing powers of ℓ: At depth 1 from the rim, the curves are primitively $(\mathbb{Z}+\ell\mathcal{O})$-oriented, and in general, at depth k, the curves are primitively $(\mathbb{Z}+\ell^k\mathcal{O})$-oriented. The edges of the volcano can thus be classified according to whether they increase, decrease or preserve the depth as we move from domain to codomain.

Definition 2.14 (Horizontal, Descending and Ascending Isogenies) Let $\phi : (E,\iota) \to (E',\iota')$ be a K-oriented degree ℓ isogeny. If ι is a primitive $\mathcal{O}$-orientation and ι' is a primitive $\mathcal{O}'$-orientation, then ϕ is one of the following three types of isogeny:

(1) ϕ is *horizontal* if $\mathcal{O} = \mathcal{O}'$.
(2) ϕ is *descending* if $\mathcal{O} \supsetneq \mathcal{O}'$.
(3) ϕ is *ascending* if $\mathcal{O} \subsetneq \mathcal{O}'$.

The fact that each component has a volcano structure is the content of the following proposition; versions of this appear in various places in the literature [17, 42].

Proposition 2.15 *Let (E,ι) be such that ι is $\mathcal{O}$-primitive. If $\mathcal{O}$ is ℓ-fundamental, then (E,ι) has no ascending ℓ-isogeny and $\left(\frac{\Delta_{\mathcal{O}}}{\ell}\right)+1$ horizontal ℓ-isogenies. Otherwise, (E,ι) has one ascending ℓ-isogeny and no horizontal isogenies. All other isogenies from (E,ι) are descending.*

Proof This follows from [5, Proposition 4.8]. Write $\mathcal{O} = \mathbb{Z}[\theta]$ satisfying the hypotheses of that proposition. If $\mathcal{O}$ is ℓ-fundamental, then the norm of θ is not divisible by ℓ. Then θ acts to permute $E[\ell]$ and therefore has two non-zero eigenvalues. By [5, Proposition 4.8], there are two horizontal isogenies if and only if these eigenvalues are distinct and defined over $\mathbb{F}_\ell$, i.e. ℓ splits. If they are not distinct but defined over $\mathbb{F}_\ell$, there is one horizontal isogeny, i.e. ℓ ramifies. Otherwise there are no horizontal isogenies. If $\mathcal{O}$ is not ℓ-fundamental, then the norm of θ is divisible by ℓ and there is a zero eigenvalue (but only one, since ℓ does not divide θ). By [5, Proposition 4.8], this gives exactly one ascending isogeny. All other isogenies are descending. □

Let $\mathcal{O}$ be an order in K. Every primitive $\mathcal{O}$-oriented curve (E,ι) is on a volcano in $\mathcal{G}_{K,\ell}$. If $\mathcal{O}$ is ℓ-fundamental, then (E,ι) is on the *rim* of this volcano. If ℓ divides the conductor of $\mathcal{O}$, then the power of ℓ dividing the discriminant of $\mathcal{O}$ is referred to as the *depth* of the curve on this volcano.

Proposition 2.16 *A connected component of $\mathcal{G}_{K,\ell}$, when identifying ℓ-isogenies with their duals, has at most one cycle, passing through the rim vertices.*

Proof The fact that there is at most one ascending isogeny from any vertex below the rim implies that there are no cycles below the rim. Restricting to the rim, we have a 2-regular graph, hence a union of cycle graphs. But since the graph is connected by assumption, it consists of at most one cycle. □

Therefore, we also use the term *rim* to refer to the undirected cycle graph that one obtains as a subgraph of $\mathcal{G}_{K,\ell}$ by restricting to the rim vertices and identifying ℓ-isogenies with their duals. We call a rim *directed* if we choose a direction of traversal (disallowing backtracking).

Remark 2.17 We pause briefly to justify the choice of equivalence on isogenies for $\mathcal{G}_{K,\ell}$ (which differs from the notion of equivalence in $\mathcal{G}_\ell$). In fact, when forming a graph whose vertices are isomorphism classes of any type of object, the graph is well defined if we choose the edges to be maps up to pre- and post-composition by isomorphism, as in (7). In other words, it is reasonable to require of our graph definition that, under any two sets of representatives for the vertices, and any set of object-wise isomorphisms between those sets of representatives, the resulting graphs should be canonically isomorphic. This requirement, when unravelled, is essentially the requirement than the maps be taken up to pre- and post-composition. Historically, the supersingular isogeny graph is not defined in this robust way, instead identifying isogenies as equivalent merely up to post-composition. This allows for the computational convenience of identifying kernels with isogenies and also allows for an identification between non-backtracking walks and cyclic isogenies. However, it results in certain non-canonical behaviours. These behaviours can be observed only in proximity to the curves with extra automorphisms (having j-invariant $j = 0$ and $j = 1728$), particularly the phenomenon described in the second paragraph of Sect. 3.2. The historical definition of the supersingular isogeny graph resulted in substantial difficulties arising from extra automorphisms in proving our main Theorem 3.2.

2.3 The Sets $\mathrm{SS}_{\mathcal{O}}^{pr}$ and $\rho(\mathcal{E}\ell\ell(\mathcal{O}))$ and the Class Group Action

Having stratified each volcano into depths where the orientations are all primitive with respect to a fixed order, it is natural to consider the complete set of curves with such orientations.

Definition 2.18 ($\mathrm{SS}_{\mathcal{O}}^{\mathrm{pr}}$) Let $\mathrm{SS}_{\mathcal{O}}^{\mathrm{pr}}$ denote the set of primitively $\mathcal{O}$-oriented isomorphism classes of supersingular elliptic curves, up to K-isomorphism.

Here we recall the following conditions for the set $\mathrm{SS}_{\mathcal{O}}^{\mathrm{pr}}$ to be non-empty.

Proposition 2.19 ([42, Proposition 3.2]) *The set $\mathrm{SS}_{\mathcal{O}}^{pr}$ is not empty if and only if p does not split in K and does not divide the conductor of $\mathcal{O}$.*

If $\mathcal{O}$ is ℓ-fundamental, then $\mathrm{SS}^{\mathrm{pr}}_{\mathcal{O}}$ is a union of volcano rims in $\mathcal{G}_{K,\ell}$.

For the proof of our bijection in Sect. 3.3, we invoke the Deuring lifting theorem as outlined in [42, Section 3.2]. For this, we will need a number of definitions and lemmas about CM elliptic curves over number fields.

There is a number field L' (an extension of the ring class field L of $\mathcal{O}$) and a prime ideal $\mathfrak{p}$ above p in $\mathcal{O}_{L'}$ such that every elliptic curve with complex multiplication by $\mathcal{O}$ has a representative over L' with good reduction at $\mathfrak{p}$ [50, Section VII.5]. Fix such a choice.

Definition 2.20 ($\mathcal{E}\ell\ell(\mathcal{O})$) Let $\mathcal{E}\ell\ell(\mathcal{O})$ denote the set of isomorphism classes of elliptic curves over L' with endomorphism ring isomorphic to $\mathcal{O}$ and good reduction at $\mathfrak{p}$.

By the theory of complex multiplication, $\#\mathcal{E}\ell\ell(\mathcal{O}) = h_{\mathcal{O}}$. Write $[\cdot]_E\colon \mathcal{O} \to \mathrm{End}(E)$ for the isomorphism which is normalized so that $([\alpha]_E)^*\omega_E = \alpha\omega_E$, where ω_E denotes the invariant differential of E [42, Section 2.3], [49, Prop. II.1.1]. There is a map given by reduction modulo $\mathfrak{p}$:

$$\rho\colon \mathcal{E}\ell\ell(\mathcal{O}) \to \mathrm{SS}^{\mathrm{pr}}_{\mathcal{O}}, \quad E \mapsto (\tilde{E}, \iota_E),$$

where $\iota_E\colon K \to \mathrm{End}^0(\tilde{E})$ is determined by its restriction $\iota_E\colon \mathcal{O} \to \mathrm{End}(\tilde{E})$ being the reduction modulo $\mathfrak{p}$ of the isomorphism $[\cdot]_E$ (in other words, $\iota_E(\alpha) = [\alpha]_E \pmod{\mathfrak{p}}$ for all $\alpha \in \mathcal{O}$). In particular, from the normalization, it holds that $\iota_E(\alpha)^*\omega_{\tilde{E}} = \alpha\omega_{\tilde{E}}$ by reduction modulo $\mathfrak{p}$.

The map ρ is injective on $\mathcal{E}\ell\ell(\mathcal{O})$, so that its image $\rho(\mathcal{E}\ell\ell(\mathcal{O}))$ is a subset of $\mathrm{SS}^{\mathrm{pr}}_{\mathcal{O}}$ of size $h_{\mathcal{O}}$. There is an action of the p-power Frobenius on $\mathrm{SS}^{\mathrm{pr}}_{\mathcal{O}}$, given by $\pi_p \cdot (E, \iota) \mapsto (E^{(p)}, \iota^{(p)})$, where $E^{(p)} = \pi_p(E)$ and $\iota^{(p)} = (\pi_p)_*\iota$.

Proposition 2.21 ([42, Proposition 3.3]) *For all $(E, \iota) \in \mathrm{SS}^{pr}_{\mathcal{O}}$, at least one of (E, ι), $(E^{(p)}, \iota^{(p)})$ belongs to $\rho(\mathcal{E}\ell\ell(\mathcal{O}))$.*

As suggested by the above, surjectivity of ρ may indeed fail: $\#\,\mathrm{SS}^{\mathrm{pr}}_{\mathcal{O}} \in \{h_{\mathcal{O}}, 2h_{\mathcal{O}}\}$. Theorem 4.4 describes when this failure occurs.

We define an action of ideals of $\mathcal{O}$ on oriented elliptic curves $(E, \iota) \in \mathrm{SS}^{\mathrm{pr}}_{\mathcal{O}}$.

Definition 2.22 (Ideal Action on Oriented Elliptic Curves) Let $(E, \iota) \in \mathrm{SS}^{\mathrm{pr}}_{\mathcal{O}}$ for some order $\mathcal{O}$ of K. Let $\mathfrak{a}$ be an integral ideal of $\mathcal{O}$ coprime to p. Define the intersection:

$$E[\iota(\mathfrak{a})] := \bigcap_{\alpha \in \mathfrak{a}} \ker(\iota(\alpha)).$$

This group defines an isogeny $\phi^{(E,\iota)}_{\mathfrak{a}} : E \to E/E[\iota(\mathfrak{a})]$. The action of $\mathfrak{a}$ on (E, ι) is defined as $\mathfrak{a} * (E, \iota) := (\phi^{(E,\iota)}_{\mathfrak{a}}(E), (\phi^{(E,\iota)}_{\mathfrak{a}})_*\iota)$.

When clear from context, we will drop the superscript and just write $\phi_{\mathfrak{a}}$. Note that to define the action of the ideal $\mathfrak{a}$, we require its norm to be coprime to p (see the proof of [42, Proposition 3.6]).

Theorem 2.23 ([42, Theorem 3.4]) *Assume that p does not split in K and p does not divide the conductor of $\mathcal{O}$. Then the ideal action of Definition 2.22 defines a free and transitive action of $Cl(\mathcal{O})$ on $\rho(\mathcal{E\ell\ell}(\mathcal{O}))$.*

This action is the 'same' as that of the CM theory, in the sense that $\rho(\sigma \cdot (E, \iota)) = \sigma \cdot (\rho((E, \iota)))$ for $\sigma \in \mathrm{Cl}(\mathcal{O})$ [42, Proof of Proposition 3.6].

Finally, we relate $\mathrm{SS}_{\mathcal{O}}^{\mathrm{pr}}$ to the volcanoes of $\mathcal{G}_{K,\ell}$.

Definition 2.24 ($\mathcal{O}$-cordillera) A collection of volcanoes in $\mathcal{G}_{K,\ell}$. The collection of volcanoes whose rims are primitively oriented by $\mathcal{O}$ is called the *$\mathcal{O}$-cordillera.*

The set of vertices at the rim of the $\mathcal{O}$-cordillera is exactly the set $\mathrm{SS}_{\mathcal{O}}^{\mathrm{pr}}$.

Next, we slightly strengthen results of Colò-Kohel [17] and Onuki [42] to give an action on oriented isogenies by a direct product of the class group with Frobenius.

Definition 2.25 (Action of Frobenius on Oriented Elliptic Curves) The two-element group $\langle \pi_p \rangle = \{1, \pi_p\}$ generated by the Frobenius automorphism π_p of $\mathbb{F}_{p^2}$ acts on $\mathcal{G}_{K,\ell}$ by

$$\pi_p \cdot (E, \iota) = (E^{(p)}, \iota^{(p)}), \quad \pi_p \cdot \varphi = \varphi^{(p)},$$

where $\iota^{(p)} := (\pi_p)_*(\iota)$.

For any isogeny φ, we have $\pi_p \circ \varphi(x, y) = \varphi^{(p)}(x^p, y^p) = \varphi^{(p)} \circ \pi_p(x, y)$. Hence, one has

$$\iota^{(p)}(\alpha) = (\pi_p)_*(\iota)(\alpha) = \frac{1}{p}\pi_p \circ \iota(\alpha) \circ \widehat{\pi_p} = \frac{1}{p}\iota(\alpha)^{(p)} \circ \pi_p \circ \widehat{\pi_p} = \iota(\alpha)^{(p)}.$$

Since $\varphi \mapsto \varphi^{(p)}$ gives an isomorphism $\mathrm{End}(E) \cong \mathrm{End}(E^{(p)})$, this yields an action on $\mathrm{SS}_{\mathcal{O}}^{\mathrm{pr}}$ by $\langle \pi_p \rangle$. In fact, it is an action on the graph $\mathcal{G}_{K,\ell}$, i.e. it preserves adjacency.

Proposition 2.26 *The actions described in Definitions 2.22 and 2.25 above commute and hence give an action of $Cl(\mathcal{O}) \times \langle \pi_p \rangle$ on $\mathrm{SS}_{\mathcal{O}}^{pr}$. This action is transitive and its point stabilizers are either all trivial or all $\langle \pi_p \rangle$. In particular, $\# \mathrm{SS}_{\mathcal{O}}^{pr} \in \{h_{\mathcal{O}}, 2h_{\mathcal{O}}\}$.*

Proof We have $\pi_p \cdot \varphi_{\mathfrak{a}} \cdot (E, \iota) = (\varphi_{\mathfrak{a}})^{(p)} \cdot \pi_p \cdot (E, \iota)$. To avoid confusion we momentarily use the more specific notation $\varphi_{\mathfrak{a}}^{(E,\iota)}$ to denote the isogeny $\varphi_{\mathfrak{a}}$ with domain (E, ι) and kernel $E[\iota(\mathfrak{a})]$. Then

$$\ker((\varphi_{\mathfrak{a}}^{(E,\iota)})^{(p)}) = \ker(\varphi_{\mathfrak{a}}^{(E,\iota)})^{(p)} = E[\iota(\mathfrak{a})]^{(p)} = \bigcap_{\theta \in \iota(\mathfrak{a})} \ker(\theta)^{(p)}$$

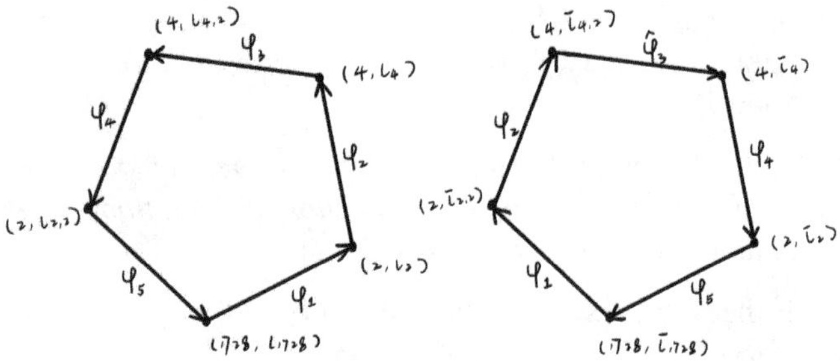

Fig. 2 Here $p = 31$, $\ell = 2$ and $K = \mathbb{Q}(\sqrt{-47})$. This figure shows the set $\mathrm{SS}_{\mathcal{O}_K}$ which consists of two cycles. We denote by (j, ι_j) the vertex with j-invariant j and orientation ι_j; if the j-invariant appears twice, the second orientation is denoted by $\iota_{j,2}$. The notation $\bar{\iota}_j$ denotes the conjugate orientation. Conjugation (Sect. 4) maps (j, ι_j) to $(j, \bar{\iota}_j)$ and $(j, \iota_{j,2})$ to $(j, \bar{\iota}_{j,2})$. On the other hand, π_p maps $(1728, \iota_{1728})$ to $(1728, \bar{\iota}_{1728})$, $(2, \iota_2)$ to $(2, \bar{\iota}_{2,2})$ and $(4, \iota_4)$ to $(4, \bar{\iota}_{4,2})$

$$= \bigcap_{\theta \in \iota(\mathfrak{a})} \ker(\theta^{(p)}) = \bigcap_{\theta \in \iota^{(p)}(\mathfrak{a})} \ker(\theta) = E^{(p)}[\iota^{(p)}(\mathfrak{a})]$$
$$= \ker((\varphi_{\mathfrak{a}}^{(E^{(p)}, \iota^{(p)})}).)$$

The calculation above implies that $(\varphi_{\mathfrak{a}}^{(E,\iota)})^{(p)} = \varphi_{\mathfrak{a}}^{(E^{(p)}, \iota^{(p)})}$. Thus,

$$\pi_p \cdot \mathfrak{a} \cdot (E, \iota) = \mathfrak{a} \cdot \pi_p \cdot (E, \iota).$$

So the action of $\mathrm{Cl}(\mathcal{O}) \times \langle \pi_p \rangle$ on $\mathrm{SS}_{\mathcal{O}}^{\mathrm{pr}}$ is well defined.

The restriction of this action to $\mathrm{Cl}(\mathcal{O})$ acts freely and transitively on a subset of $\mathrm{SS}_{\mathcal{O}}^{\mathrm{pr}}$ which contains at least one of (E, ι) or $(E^{(p)}, \iota^{(p)})$ [42, Theorem 3.4], from which the rest of the statement follows. Transitivity implies that the stabilizers are all of the same size. □

Remark 2.27 One might expect to see the dihedral group, not a direct product. We will see the dihedral group coming from class field theory in Sect. 4, where we consider the action of conjugation on $\mathrm{SS}_{\mathcal{O}}^{\mathrm{pr}}$. But in our definitions here, Frobenius acts on both curve and orientation. In other words, $\mathrm{SS}_{\mathcal{O}}^{\mathrm{pr}}$ has an action by a direct product and by a dihedral group, both extensions of the same action of $\mathrm{Cl}(\mathcal{O})$ by $\mathbb{Z}/2\mathbb{Z}$, but these are not necessarily the same. For an example of the action of Frobenius and the action of conjugation being different, see Fig. 2.

2.4 A Lemma of Algebraic Number Theory

We make explicit a small lemma from algebraic number theory that will be used in Sect. 7.

Lemma 2.28 *Let K be an imaginary quadratic field and let $\mathcal{O}$ be an order in K with conductor f. Assume that ℓ is a prime that does not divide f. Then ℓ splits (ramifies and is inert, respectively) in $\mathcal{O}$ if and only if ℓ splits (ramifies and is inert, respectively) in the maximal order $\mathcal{O}_K$.*

Proof By [18, Proposition 7.20], if $\mathfrak{a}$ is an $\mathcal{O}_K$-ideal above ℓ prime to f, then $\mathfrak{a}' = \mathfrak{a} \cap \mathcal{O}$ is an $\mathcal{O}$-ideal above ℓ with the same norm, and if $\mathfrak{a}'$ is an $\mathcal{O}$-ideal above ℓ prime to f, then $\mathfrak{a} = \mathfrak{a}'\mathcal{O}_K$ is an $\mathcal{O}_K$-ideal above ℓ with the same norm. Furthermore, these maps are inverse to one another, giving a bijection between ideals of $\mathcal{O}_K$ coprime to f, and invertible ideals of $\mathcal{O}$. Therefore the splitting, ramification and inertia properties of ℓ in $\mathcal{O}_K$ and in $\mathcal{O}$ are the same. □

3 Bijection Between Oriented Volcano Rims and Isogeny Cycles in the Supersingular ℓ-Isogeny Graph

3.1 Statements of Results

In this section, we give a bijection between 'isogeny cycles' (closed walks satisfying a non-backtracking and non-repeating condition; see below) on the supersingular ℓ-isogeny graph and rims (up to conjugation) of the volcanoes in all oriented supersingular ℓ-isogeny graphs. We first give our definition of isogeny cycles.

Definition 3.1 (Isogeny Cycle) An *isogeny cycle* is a closed walk, forgetting basepoint, in $\mathcal{G}_\ell$ containing no backtracking (no consecutive edges compose to multiplication-by-ℓ, Definition 3.5) which is not a power of another closed walk (i.e. not equal to another closed walk repeated more than once).

Isogeny cycles travelling the same vertices in opposite directions necessarily employ different directed edges and hence are considered different.

We need a few more items of terminology. The term *directed* refers to the notion of direction of traversal. Each rim corresponds to two directed rims, since there are two possible directions of traversal. The *conjugate* of a rim R is obtained by taking each vertex (E, ι) to $(E, \overline{\iota})$ and each edge $\phi\colon (E, \iota) \to (E', \iota')$ to $\phi\colon (E, \overline{\iota}) \to (E', \overline{\iota'})$. The conjugation of a rim is again a rim. Note that if $\phi_{\mathfrak{a}}^{(E,\iota)}\colon (E, \iota) \to (E', \iota')$, then $\phi_{\overline{\mathfrak{a}}}^{(E,\overline{\iota})}\colon (E, \overline{\iota}) \to (E', \overline{\iota'})$ since $E[\overline{\iota}(\overline{\mathfrak{a}})] = E[\iota(\mathfrak{a})]$. So in particular, the direction of a rim (E_0, E_1 etc.) is preserved by conjugation, but the ideal whose action traverses the rim in that direction is conjugated.

We now state our main theorem. For each integer $r > 2$, define two sets:

$$\mathcal{C}_r = \{ \text{ isogeny cycles in } \mathcal{G}_\ell \text{ of length } r\}.$$

$$\mathcal{R}_r = \left\{ \text{ directed rims } R \text{ of size } r \text{ in } \mathcal{G}_{K,\ell} : K \text{ is an imaginary quadratic field} \right\}.$$

Theorem 3.2 *Let $r > 2$. There is a bijection between $\mathcal{C}_r$ and $\mathcal{R}_r/\sim$, where $\sim$ denotes identifying any rim with its conjugate.*

Sketch of Proof Theorem 3.2 will be proved in Sect. 3.3. At the highest level, this bijection is easy to describe: Given a rim $R \in \mathcal{R}_r$, forgetting orientations gives a cycle $C \in \mathcal{C}_r$. Conversely, given a cycle $C \in \mathcal{C}_r$ passing through a curve E, composing its component isogenies in order, one obtains an endomorphism $\theta \in \mathrm{End}(E)$, which induces an orientation ι_θ. Lifting E to (E, ι_θ), the isogenies of the cycle will give a rim in $\mathcal{G}_{K,\ell}$, where K is the quadratic field generated by the minimal polynomial of θ. These two maps are inverse.

However, the devil is in the details, especially the devil whose name is *extra automorphisms*. In particular, if $\mathcal{G}_\ell$ has no curves with extra automorphisms, the proof is much briefer. In what follows, we have endeavoured to clearly mark areas where we address extra automorphisms, so that at a first read, they can be skipped. In particular, if $\mathcal{G}_\ell$ has no extra automorphisms, the reader can entirely skip Sect. 3.4, and the word 'safe' (Definition 3.13) can be ignored. Section 3.2 should still be read, however, because its definitions and statements will be used in the proof of Theorem 3.2.

3.2 Extra Automorphisms and 'Arbitrary Assignment'

Before we proceed, we must discuss a subtlety concerning extra automorphisms (see the related Remark 2.17). In particular, the composition of the isogenies along a walk in $\mathcal{G}_\ell$ is not well-defined, because the isogenies themselves are only defined up to post-composition by an automorphism. In the case of the automorphisms $[\pm 1]$, which commute with all the isogenies in the chain, this amounts to a single sign ambiguity. But extra automorphisms may wreak havoc: The endomorphism obtained by composing around a closed walk may not even have the same discriminant (and quadratic field) after a post-composition by an automorphism.

This would appear to lampoon any hope of associating endomorphisms nicely with closed walks. However, the remedy to this issue is the following observation. Consider two walks that differ by replacing a segment $j_1 \xrightarrow{\phi} 0 \xrightarrow{\psi} j_2$ with $j_1 \xrightarrow{[\zeta_3^k]\phi} 0 \xrightarrow{\psi[\zeta_3^{-k}]} j_2$ or $j_3 \xrightarrow{\phi} 1728 \xrightarrow{\psi} j_4$ with $j_3 \xrightarrow{[i]\phi} 1728 \xrightarrow{\psi[-i]} j_4$, where $[\zeta_3]$ and $[i]$ are extra automorphisms associated with primitive third and fourth roots of unity, respectively. In each case, the first arrow is the same directed edge of $\mathcal{G}_\ell$, but the second arrow differs as an edge of $\mathcal{G}_\ell$. The composition of the walk, however, is unchanged. In fact, there are typically three distinct walks $j_1 \to 0 \to j_2$ in $\mathcal{G}_\ell$ and three distinct resulting isogenies $j_1 \to j_2$, but there is no *canonical* identification between the elements of these two sets of size 3. One way to set a non-canonical identification is to consider the incoming isogeny $j_1 \to 0$ to be a particular fixed (but arbitrary) representative of its equivalence class (recall that the equivalence

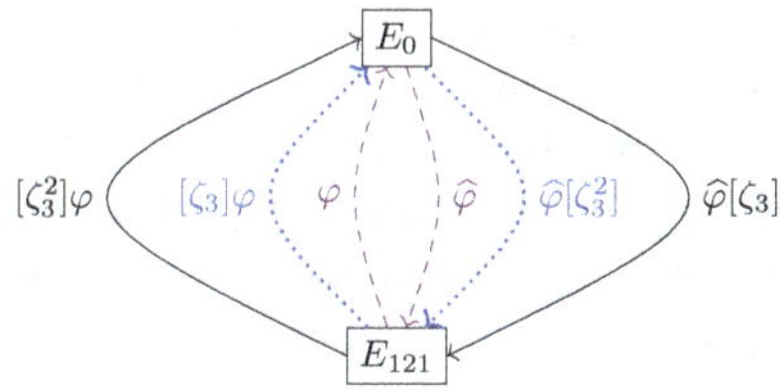

Fig. 3 This figure shows all the 2-isogenies between $j = 0$ and $j = 121$ for $p = 179$. Dual pairs are indicated by arrows matching in style. The isogenies entering $j = 0$ (upward in the picture) differ by post-composition by an automorphism: either $[\pm 1]$, $[\pm\zeta_3]$ or $[\pm\zeta_3^2]$. Therefore they are represented by a single directed edge in $\mathcal{G}_\ell$. The downward arrows from $j = 0$ to $j = 121$ are represented by distinct directed edges in $\mathcal{G}_\ell$. The two walks $\widehat{\varphi}[\zeta_3] \circ \varphi$ (up on dashed and down on solid) and $\widehat{\varphi} \circ [\zeta_3]\varphi$ (up on dotted and down on dashed) give the *same* endomorphism of $j = 121$, namely $\widehat{\varphi} \circ [\zeta_3] \circ \varphi$, despite being different walks in $\mathcal{G}_\ell$. In fact, the nine distinct closed walks of length 2 from 121 to 0 and back compose to only three distinct endomorphisms of $j = 121$, namely $\pm[2]$, $\pm\widehat{\varphi} \circ [\zeta_3] \circ \varphi$ or $\pm\widehat{\varphi} \circ [\zeta_3^2] \circ \varphi$. Finally, notice that *any* walk of length 2 from $j = 0$ to $j = 121$ and back is considered backtracking according to Definition 3.5. The same is not true for walks of length 2 from j=121 to $j = 0$ and back

class consists of all the isogenies with the same kernel, i.e. up to post-composition by an automorphism).

Definition 3.3 (Arbitrary Assignment) An *arbitrary assignment* is a choice of isogeny ϕ, up to sign, from every equivalence class of isogenies represented by a directed edge of $\mathcal{G}_\ell$.

There is no choice except when the codomain has extra automorphisms. **In what follows, we shall assume that such a fixed arbitrary assignment has been made for every arrow entering $j = 0$ or $j = 1728$.** This choice will affect the exact bijection we obtain in our main theorem, but not the fact that it is a bijection. Figure 3 is instructive as to the subtleties that may arise.

The following lemma is now immediate.

Lemma 3.4 *Assume we have made an arbitrary assignment for $\mathcal{G}_\ell$. Then any closed walk in $\mathcal{G}_\ell$ results in a unique endomorphism (up to sign) by composition of its component isogenies in order.*

The endomorphism obtained by composing the isogenies of a cycle in order will simply be called the *composition* of the cycle. The fact that composition is well defined allows us to define a notion of backtracking without ambiguity around curves with extra automorphisms.

Definition 3.5 (Backtracking Closed Walk) Given an arbitrary assignment on $\mathcal{G}_\ell$, two consecutive edges $\bullet \xrightarrow{\phi_0} \bullet \xrightarrow{\phi_1} \bullet$ are said to be *backtracking* if $\phi_1 \circ \phi_0 = [\ell]$, up to possible post-composition by an automorphism. A closed walk is considered to contain backtracking if any consecutive edges, including the last and first, are backtracking.

This definition of backtracking is the same definition as [7, Definition 4.3], except those authors leave out the composition of the last and first steps, which is more appropriate in their context. Note that this definition is not captured by the sequence of vertices alone. To wit, see Fig. 3 for an example of two length-2 walks, one of which is backtracking and one of which is not backtracking, but which are taken to each other by an isomorphism of the graph fixing vertices.

We also have a converse to Lemma 3.4.

Lemma 3.6 *Assume we have made an arbitrary assignment for $\mathcal{G}_\ell$. Consider an endomorphism $\theta \in \mathrm{End}(E)$ of ℓ-power degree. If θ is not divisible by $[\ell]$, then there is a unique closed walk in $\mathcal{G}_\ell$ with composition $\varphi\theta$, for some $\varphi \in \mathrm{Aut}(E)$. Furthermore, the walk has no backtracking, and φ is unique up to sign.*

Proof Since $[\ell] \nmid \theta$, the kernel $\ker(\theta) \cap E[\ell]$ has size ℓ. Define the isogeny $\phi_1 : E \to E'$ corresponding to this subgroup (using the arbitrary assignment if applicable). Then any isogeny cycle which composes to θ must factor through this isogeny; hence, this is the first step in the desired isogeny cycle: $\theta = \phi' \circ \phi_1$. Now take $\ker \phi' \cap E'[\ell]$ to find the next kernel and the map ϕ_2 so that $\theta = \phi'' \circ \phi_2 \circ \phi_1$. Continuing, we obtain a closed walk in the ℓ-isogeny graph composing to an isogeny with the same kernel as θ. Which is to say, $\pm\theta$, up to post-composition by an automorphism of E. But by construction, this walk is unique. It has no backtracking, since backtracking would produce a composition divisible by $[\ell]$. The same closed walk is obtained if we begin with θ or a post-composition of θ by an automorphism on E. Thus, by Lemma 3.4, the automorphism φ of the statement is unique up to sign. □

Finally, there is one more small lemma we will require, which shows that any closed walk composing to a power of an endomorphism is a power of a closed walk.

Lemma 3.7 *Suppose we have made an arbitrary assignment for $\mathcal{G}_\ell$ and let $\iota \colon K \to \mathrm{End}^0(E)$ be an $\mathcal{O}$-primitive orientation. Let $\alpha = u\beta^s$ for some $s > 1$, $\beta \in \mathcal{O}$ and some unit $u \in \mathcal{O}$. Assume that ℓ does not divide α and that a closed walk C of $\mathcal{G}_\ell$ composes to $\pm\iota(\alpha)$. Then $C = C_\beta^s$, where C_β is a closed walk of $\mathcal{G}_\ell$ that composes to $\pm\iota(\beta)$.*

Proof By Lemma 3.6, since ℓ does not divide β, there exists a unique closed walk C_β of ℓ-isogenies composing to $\varphi\iota(\beta)$ for some automorphism φ of E. If E has no extra automorphisms, then the proof is done.

If E has extra automorphisms, however, then we slightly modify C_β. Change the initial edge ϕ of C_β by replacing it with $\phi\varphi^{-1}$ (having the same domain and codomain). Call this modified closed walk C'_β. By the conventions of the arbitrary assignment (Definition 3.3), the composition around C'_β is $\varphi\iota(\beta)\varphi^{-1}$. The composition of $(C'_\beta)^s$ is $\varphi\iota(\beta)^s\varphi^{-1} = \varphi\iota(u^{-1})\iota(\alpha)\varphi^{-1}$. But then we have closed walks C and $(C'_\beta)^s$, composing to $\pm\iota(\alpha)$ and $\pm\varphi\iota(u^{-1})\iota(\alpha)\varphi^{-1}$, respectively. By the arbitrary assignment and Lemma 3.6, we can again modify the first edge of $(C'_\beta)^s$ to obtain C' which composes to $\pm\varphi\iota(u^{-1})\iota(\alpha)$. But then $C = C'$ by Lemma 3.6, and so $\varphi = \pm\iota(u) \in \iota(\mathcal{O})$. Therefore $\varphi \in \iota(\mathcal{O})$, in which case $(C'_\beta)^s$ composes to $\pm\iota(u^{-1})\iota(\alpha)$, and by Lemma 3.6, $C = (C'_\beta)^s$. □

3.3 *Proof of Theorem 3.2*

In order to prove Theorem 3.2, it is easier to first discuss rims and isogeny cycles with a basepoint (a marked point). Let E be a fixed vertex of $\mathcal{G}_\ell$. We define the following sets:

$$\mathcal{R}_{E,r} = \left\{ \begin{array}{l} \text{directed rims of size } r \text{ in } \mathcal{G}_{K,\ell} \\ \text{passing through } (E,\ \iota) \end{array} : \begin{array}{l} K \text{ is an imaginary quadratic field,} \\ \iota \text{ is a } K\text{-orientation on } E \end{array} \right\},$$

$$\mathcal{I}_{E,r} = \left\{ (\iota, \mathfrak{l}) : \begin{array}{c} \iota\colon K \to \mathrm{End}^0(E) \text{ a primitive } \mathcal{O}\text{-orientation,} \\ K \text{ is an imaginary quadratic field,} \\ \mathcal{O} \text{ is an } \ell\text{-fundamental order of } K, \\ (\ell) = \mathfrak{l}\bar{\mathfrak{l}} \text{ splits in } \mathcal{O}, \\ [\mathfrak{l}] \text{ has order } r \text{ in } \mathrm{Cl}(\mathcal{O}), \end{array} \right\}, \text{ and}$$

$$\mathcal{C}_{E,r} = \{\text{isogeny cycles of length } r \text{ in } \mathcal{G}_\ell \text{ starting and ending at } E\}.$$

Let $\sim$ denote conjugation on elements of $\mathcal{I}_{E,r}$, i.e. taking a pair $(\iota, \mathfrak{l})$ to $(\bar{\iota}, \bar{\mathfrak{l}})$. As described in Sect. 3.1, conjugation preserves the direction of the rim but conjugates the ideal whose action traverses the rim. We will continue to use the symbol $\sim$ to denote conjugation of a rim.

We assume a safe arbitrary assignment (Definition 3.13) has been made for $\mathcal{G}_\ell$. Depending upon this arbitrary assignment, we define the following maps:

$$\begin{array}{ccc} \mathcal{I}_{E,r}/\!\sim & \underset{\Psi'}{\overset{\Psi}{\rightleftarrows}} & \mathcal{R}_{E,r}/\!\sim \\ & {}_{\Theta}\nwarrow \quad \swarrow_{\Phi} & \\ & \mathcal{C}_{E,r} & \end{array} \tag{8}$$

The goal is to show that the diagram commutes, where every map is a bijection. Then Theorem 3.2 will follow by 'forgetting' the basepoint.

Disregarding conjugation for the moment, we first define maps

$$\mathcal{I}_{E,r} \underset{\Psi'}{\overset{\Psi}{\rightleftarrows}} \mathcal{R}_{E,r}\ .$$

In Lemma 3.8, we show that these are inverse bijections and that they provide inverse bijections even after a quotient by the conjugation relation. In Lemma 3.9 we will define Θ and show it is a bijection. Lemma 3.10 will define Φ and show it is a bijection. Finally, in Theorem 3.11, we show that the triangle in (8) commutes.

Definition of Ψ' Consider a directed rim $R \in \mathcal{R}_{E,r}$ passing through (E, ι). Denote by $\mathcal{O}$ the order such that ι is $\mathcal{O}$-primitive. Then $\mathcal{O}$ is ℓ-fundamental since (E, ι) is at the rim. Moreover, the $\mathcal{O}$-ideal (ℓ) splits and the $\mathcal{O}$-ideals above ℓ have order r in $\mathrm{Cl}(\mathcal{O})$ since (E, ι) belongs to a rim of size r. The direction along the rim corresponds to the action of one of the ideals above ℓ, which we denote $\mathfrak{l}$. Then Ψ' is defined as

$$\Psi' \colon \mathcal{R}_{E,r} \to \mathcal{I}_{E,r}, \quad R \mapsto (\iota, \mathfrak{l}).$$

Definition of Ψ On the other hand, consider $(\iota, \mathfrak{l}) \in \mathcal{I}_{E,r}$. The orientation ι on E gives rise to a directed rim of size r passing through (E, ι) by the consecutive action of $\mathfrak{l}$. Let Ψ send $(\iota, \mathfrak{l})$ to this directed rim in $\mathcal{R}_{E,r}$, i.e.

$$\Psi \colon \mathcal{I}_{E,r} \to \mathcal{R}_{E,r}, \quad (\iota, \mathfrak{l}) \mapsto R = ((E, \iota) \to_{\phi_{\mathfrak{l}}} \cdots).$$

Lemma 3.8 *The maps Ψ and Ψ' defined above are inverse bijections. They also provide bijections*

$$\mathcal{I}_{E,r}/\sim \;\underset{\Psi'}{\overset{\Psi}{\rightleftarrows}}\; \mathcal{R}_{E,r}/\sim ,$$

where $\sim$ denotes equivalence between conjugates.

Proof The fact that Ψ and Ψ' are inverse to each other follows from their definitions. It is also clear that conjugate rims correspond to conjugate orientations (while fixing E and K). □

Definition of Θ We now define a map Θ from $\mathcal{C}_{E,r}$ to $\mathcal{I}_{E,r}$. Let $C \in \mathcal{C}_{E,r}$. Starting at the vertex E, this cycle composes into an endomorphism $\pm\theta \in \mathrm{End}(E)$. This endomorphism $\pm\theta$ induces an orientation $\iota_\theta \colon K \to \mathrm{End}^0(E)$ on E (defined up to conjugation). Fixing a choice for ι_θ, let $\mathcal{O}$ be the order for which ι_θ is primitive. Then ℓ must split or ramify in $\mathcal{O}$ because $\theta \in \mathcal{O}$ has norm ℓ^r and is not divisible by ℓ (since $C \in \mathcal{C}_{E,r}$ has no backtracking). If $\phi_{\mathfrak{l}}^{(E,\iota_\theta)}$ is equal to the first isogeny of the cycle C in its indicated direction, for a prime $\mathfrak{l}$ above ℓ, then define $\Theta(C) := (\iota_\theta, \mathfrak{l})$.

Lemma 3.9 *The process described above gives a well-defined map $\Theta \colon \mathcal{C}_{E,r} \to \mathcal{I}_{E,r}/\sim$ (in particular, the ideal $\mathfrak{l}$ indicated in the definition above exists).*

Proof Let $C_{E,r} \in \mathcal{C}_{E,r}$ and let $\theta \in \mathrm{End}(E)$ be the composition of the edges in $C_{E,r}$ (unique up to sign). Let α denote the preimage of θ in K via ι_θ. The orientation ι_θ is a primitive $\mathcal{O}$-orientation for exactly one order $\mathbb{Z}[\alpha] \subseteq \mathcal{O} \subseteq \mathbb{Q}(\alpha)$. If $\mathbb{Z}[\alpha]$ has conductor divisible by ℓ, then α has discriminant and norm divisible by ℓ, hence trace divisible by ℓ, so it is itself divisible by ℓ. But this is impossible because the isogeny cycle $C_{E,r}$ has no backtracking. Hence $\mathbb{Z}[\alpha]$, and therefore $\mathcal{O}$, are ℓ-fundamental orders.

Note that α has norm ℓ^r. We consider the factorization of (ℓ) in $\mathcal{O}$. As mentioned in the definition of Θ, ℓ splits or ramifies, so $(\ell) = \mathfrak{l}\bar{\mathfrak{l}}$. Then we can factor (α), using its norm $(\alpha) = (\ell)^t \mathfrak{l}^s$ for some t, s (up to possibly replacing $\mathfrak{l}$ with its conjugate). However, $t = 0$ since ℓ does not divide α. Hence $s = r$, by comparing norms. That is, $(\alpha) = \mathfrak{l}^r$.

This shows that ℓ is represented by a class $[\mathfrak{l}]$ of order dividing r in $\mathrm{Cl}(\mathcal{O})$. We will show it is of order exactly r. If not, then $\mathfrak{l}^s = (\beta) \in \mathcal{O}$, so $\alpha = u\beta^s$ for some unit $u \in \mathcal{O}$. Then by Lemma 3.7, the cycle $C_{E,r}$ is a power of a smaller cycle, a contradiction to the definition of isogeny cycle.

Finally, since $(\alpha) = \mathfrak{l}^r$, the kernel $E[\mathfrak{l}]$ is contained in $\ker\theta$, and so $\phi_{\mathfrak{l}}$ defined on (E, ι_θ) is equal to the first isogeny of C in its direction of travel.

Note that our only arbitrary choice is that of ι_θ in place of its conjugate, but as we map into $\mathcal{I}_{E,r}/\sim$, the other choice results in the same image element, and the map is well defined. □

Definition of Φ Next we consider the map Φ. Given a directed rim $R \in \mathcal{R}_{E,r}/\sim$, we get a walk in $\mathcal{G}_\ell$ by forgetting orientations. If there are no extra automorphisms on any curves of the rim, this is well defined up to a sign on the isogenies. If there are extra automorphisms, we must use the process in Sect. 3.4 to safely forget orientations. The resulting cycle will be denoted $\Phi(R)$. Note that if R is conjugated, we obtain the same walk. This is clear when there are no extra automorphisms. When we use the process in Sect. 3.4, this follows from the observation that any valid set of representatives is again valid for the conjugate rim, in the proof of Proposition 3.14.

Lemma 3.10 *Let $R \in \mathcal{R}_{E,r}$. Then $\Phi(R) \in C_{E,r}$.*

Proof Let W be the walk obtained from R by forgetting the orientations. We need to show that W does not have backtracking and is not repeating smaller cycles. Let $(\iota, \mathfrak{l}) = \Psi'(R) \in \mathcal{I}_{E,r}/\sim$. First, we show that the action of $\mathfrak{l}$, which dictates the direction of R, cannot produce a backtracking path. Suppose the action of $\mathfrak{l}$ produces a backtracking path ϕ_1, ϕ_2. Then there is an endomorphism φ of E such that $\varphi[\ell] = \phi_2\phi_1$ has the same kernel as $\phi_{\mathfrak{l}^2}$, namely $E[\ell]$. This implies $\mathfrak{l}^2 = (\ell)$, so $r = 2$, a contradiction.

Now suppose that the walk W in $\mathcal{G}_\ell$ repeats a smaller closed walk $k > 1$ times. Let $\phi \in \mathrm{End}(E)$ be the composition of the smaller closed walk. Then $\phi = \phi_{\mathfrak{a}}$ for the ideal $\mathfrak{a} = \mathfrak{l}^{r/k}$, and $\phi^k = \phi_{\mathfrak{b}}$ where $\mathfrak{b} = \mathfrak{l}^r = (\beta)$ for some $\beta \in \mathcal{O}$. Hence as an element of $\mathrm{End}(E)$, $\phi^k = \iota(\beta)$, which implies that $\phi \in \iota(K)$. But then letting α be such that $\iota(\alpha) = \phi$, we have $\alpha^k = \beta$ and $\mathfrak{l}^{r/k} = (\alpha)$. This contradicts the rim being of size r. Thus, the walk W in $\mathcal{G}_\ell$ cannot repeat a smaller walk $k > 1$ times. □

Theorem 3.11 *Given a safe arbitrary assignment for $\mathcal{G}_\ell$, we define maps Φ and $\Phi' := \Psi \circ \Theta$ as above. Then the sets $C_{E,r}$ and $\mathcal{R}_{E,r}/\sim$ are in bijection under the inverse pair of maps Φ and Φ'.*

Proof We make a safe arbitrary assignment (Definition 3.13), to define the maps above. We will show that the two maps are inverse to each other.

Suppose $C \in \mathcal{C}_{E,r}$ is taken to $[R] \in \mathcal{R}_{E,r}/\sim$ by Φ'. If θ is the composition of C, then R is the rim passing through (E, ι_θ) ($\overline{R}$ is the rim passing through $(E, \overline{\iota_\theta})$, respectively). In particular, $\mathfrak{l}^r = (\alpha)$ where $\iota_\theta(\alpha) = \theta$. Therefore, composing the isogenies around R results in the endomorphism $\pm\theta$. By Lemma 3.6 (and using Proposition 3.14 in the case of extra automorphisms), the rim R maps to C when orientations are forgotten (i.e. under Φ).

Next, suppose $[R] \in \mathcal{R}_{E,r}/\sim$ is taken to $(\iota_\theta, \mathfrak{l}) \in \mathcal{I}_{E,r}$ by Ψ'. Let $C = \Phi(R)$. Then C composes to θ (using Proposition 3.14 in the case of extra automorphisms). Thus C maps to $(\iota_\theta, \mathfrak{l}) \in \mathcal{I}_{E,r}$ under Θ. □

The discussion so far relies on a choice of a basepoint E. We now prove the main theorem (Theorem 3.2) as a corollary to the basepointed version (Theorem 3.11).

Proof of Theorem 3.2 Make a safe arbitrary assignment for $\mathcal{G}_\ell$ and use it to define the maps of diagram (8) as described above.

There is a map $\mathcal{C}_{E,r} \to \mathcal{C}_r$ given by forgetting the basepoint of the isogeny cycle. There is also a map $\mathcal{R}_{E,r}/\sim \to \mathcal{R}_r/\sim$ given by forgetting the basepoint of the rim.

Given $[R] \in \mathcal{R}_r/\sim$, a basepoint can be chosen in r ways. For each such basepoint (E, ι), we obtain $R_{E,\iota} \in \mathcal{R}_{E,r}$, which we map to $C_{E,\iota} := \Phi(R_{E,\iota}) \in \mathcal{C}_{E,r}$ via Theorem 3.11. We must show that the resulting $C_{E,\iota}$ are all equal when basepoints are forgotten. But the map Φ to isogeny cycles is agnostic to the basepoint: This is clear for rims without extra automorphisms, as there are no choices to make, and is a consequence of Proposition 3.14 when there are extra automorphisms.

Similarly, a given $C \in \mathcal{C}_r$ corresponds to $C_i \in \mathcal{C}_{E_i,r}$ for E_i the i-th curve in the cycle. Each of these basepointed cycles gives a corresponding basepointed rim $R_i := \Phi'(C_i) \in \mathcal{R}_{E_i,r}$. We must show that the resulting R_i are all equal, when basepoints are forgotten. Define $(\iota_i, \mathfrak{l}_i) = \Theta(C_i) \in \mathcal{I}_{E,r}$. Define the composed endomorphism $\theta_i \in \mathrm{End}(E_i)$ for each E_i. Then by construction, $\iota_i = \iota_{\theta_i}$ and $\phi_i = \phi_{\mathfrak{l}_i}^{(E_i,\iota_i)}$. It suffices to show that all $(\iota_i, \mathfrak{l}_i)$ give rise to the same rim. The θ_i are related by $\theta_{i+1} = \phi_i\theta_i\phi_i^{-1}$. Thus, we obtain $\iota_{i+1} = (\phi_i)_*\iota_i$. In particular, the orientations ι_i are all associated with the same field K, and they are all primitive for the same order $\mathcal{O}$ of K, since ϕ_i is horizontal. Therefore (E_i, ι_i) and (E_{i+1}, ι_{i+1}) are elements of $\mathrm{SS}_{\mathcal{O}}^{\mathrm{pr}}$ connected by a horizontal isogeny ϕ_i of degree ℓ, i.e. they are on the same rim. As for direction, by construction $\phi_i = \phi_{\mathfrak{l}_i}^{E_i}$ for all i. Since ϕ_i all indicate the same direction, the $\mathfrak{l}_i$ all indicate the same direction. □

3.4 Mapping from $\mathcal{G}_{K,\ell}$ to $\mathcal{G}_\ell$

We now deal with the one remaining item pushed aside during the proof: intricacies introduced by extra automorphisms when defining a map from rims to closed walks by forgetting orientations. One can hope to map any walk in $\mathcal{G}_{K,\ell}$ to a walk in $\mathcal{G}_\ell$ by forgetting orientations. However, isogenies in the oriented isogeny graph are taken up to pre- and post-composition by automorphisms (see (7)), so when mapping down to $\mathcal{G}_\ell$, there is a choice to be taken if the domain has extra automorphisms. Even more to the point, if we obtain a closed walk from a rim, by making the choice arbitrarily, we may end up with a closed walk whose composition does not lie in $\iota(K)$. (Recall that in order to obtain a bijection, we hope to recover the rim from the isogeny cycle using this composition.)

In this section, we sidestep this issue, with the help of a special type of arbitrary assignment, by showing that it is nevertheless possible to accomplish something more limited: to define a map from rims to isogeny cycles such that the resulting isogeny cycle has a composition in $\iota(K)$. We emphasize that this is a non-issue (and this entire section can be skipped) if $\mathcal{G}_\ell$ has no curves with extra automorphisms. To be somewhat more precise, consider a rim R in $\mathcal{G}_{K,\ell}$. If we choose representatives of the vertices and edges (which are, a priori, only equivalence classes), then we can forget orientations to obtain a closed walk in $\mathcal{G}_\ell$. However, the choice of representatives must be made with care. First, the edges must belong to the arbitrary assignment: If not, then when composing around the rim before and after forgetting orientations, we may differ by a post-composition and hence may produce an endomorphism from the wrong quadratic field (which we will see momentarily in Example 3.12 below). Second, the choice of representatives must be consistent: When we traverse the entire rim, beginning at representative (E, ι), we must return to representative (E, ι) and not merely an isomorphic oriented curve. In the proof that follows, we spend the vast majority of our energy in showing that such a choice of representative exists and is unique.

Example 3.12 An example demonstrates some of the problems the proof will need to overcome. Consider the example $p = 11$ and $\ell = 3$ where $\mathcal{G}_3$ has two loops at $j = 1728$. Recall the endomorphism ring of the curve with j-invariant $j = 1728$ from Sect. 2.1. There are eight endomorphisms of degree $\ell = 3$, namely

$$\frac{\pm 1 \pm \pi_p}{2} \quad \text{and} \quad \frac{\pm[i] \pm [i]\pi_p}{2}.$$

The first group have trace ± 1 and the second group have trace 0. The endomorphisms

$$\pm\frac{1 + \pi_p}{2}, \quad \pm\frac{[i] + [i]\pi_p}{2}$$

have the same kernel, which we denote κ. This corresponds to one of the loops, and these four endomorphisms differ by post-composition by an automorphism (namely $\pm 1, \pm[i]$). The endomorphisms

$$\pm\frac{1 - \pi_p}{2}, \quad \pm\frac{[i] - [i]\pi_p}{2}$$

all have the same kernel, namely the image of κ under $[i]$. This corresponds to the second loop, and these four endomorphisms also differ from one another only by post-composition by an automorphism.

If, in the arbitrary assignment, we choose $\frac{1+\pi_p}{2}$ for the first loop and $\frac{1-\pi_p}{2}$ for the second, then these endomorphisms generate orientations by $\mathbb{Z}\left[\frac{1+\sqrt{-11}}{2}\right]$. The endomorphisms $\pm([i]+[i]\pi_p)/2$ of trace zero generate orientations by $\mathbb{Z}[\sqrt{-3}]$. The graph $\mathcal{G}_{\mathbb{Q}(\sqrt{-3}),\ell}$ has one rim in $\mathrm{SS}_{\mathbb{Z}[\sqrt{-3}]}$, namely the loop on the oriented curve (E,ι), where E has j-invariant 1728 and $\iota(\sqrt{-3}) = [\pm\sqrt{-3}]$ (case (i) of Remark 2.7). Under this arbitrary assignment, there is no place to map this rim in such a way that the composition of the resulting loop will match the order $\mathbb{Z}[\sqrt{-3}]$ as desired.

This motivates modifying Definition 3.3 so that the two loops must compose to endomorphisms from different fields K. In particular, we must be careful to choose the arbitrary assignment 'in tandem' for multiple edges between curves with extra automorphisms. This issue arises in larger cycles also. For an explicit example, consider $p = 11$, and $\mathbb{Q}(\sqrt{-23})$, which has a class group of size 3 generated by $\mathfrak{l}$ lying above $\ell = 2$; the rim for the maximal order is of size 3 with j-invariants 1728, 1728, 0. We therefore make a special requirement on the arbitrary assignment which applies to this case.

Definition 3.13 (Safe Arbitrary Assignment) An arbitrary assignment is said to be *safe* if it obeys the following rule: Whenever there is a collection of edges $E_1 \to E_2$ related by precomposition by an automorphism of E_1 besides $[\pm 1]$, we choose their representatives to be of the form $\{\pm\phi\varphi : \varphi \in \mathrm{Aut}(E_1)\}$ for a fixed choice ϕ.

The rule in Definition 3.13 is automatically satisfied at E_1 and E_2 by any arbitrary assignment unless both E_1 and E_2 have extra automorphisms, and there are multiple edges between them. It is also clear that a safe arbitrary assignment is always possible.

With regard to Example 3.12, note that $\frac{1+\pi_p}{2} = -[i]\frac{1-\pi_p}{2}[i]$, which violates Definition 3.13. This is the reason the example 'fails' the bijection.

Proposition 3.14 *Suppose a safe arbitrary assignment has been made for $\mathcal{G}_\ell$. There is a map from rims R of $\mathcal{G}_{K,\ell}$ to closed walks $C(R)$ in $\mathcal{G}_\ell$ with the following properties:*

(1) The curves and isogenies of $C(R)$ are representatives of the isomorphism classes of curves and isogenies of R, in the same order; in particular, there is a map $R \to C(R)$ of directed graphs.
(2) The composition $\pm\theta$ of $C(R)$ at vertex E in $C(R)$ corresponding to (E,ι) in R gives rise to the orientation $\pm\iota$ in the sense that $\theta = \iota(\alpha)$ for some $\alpha \in K$.

Proof **Defining the map.** It is possible to define such a map as follows. Consider a rim R of length r, associated with an ideal $\mathfrak{l}$ above ℓ in the associated ℓ-fundamental order $\mathcal{O}$, with basepoint (E_0,ι_0), written as follows:

$$(E_0, \iota_0) \xrightarrow{\phi_0} (E_1, \iota_1) \xrightarrow{\phi_1} (E_2, \iota_2) \cdots \quad (E_{r-1}, \iota_{r-1}) \xrightarrow{\phi_{r-1}} (E_0, \iota_0) \ .$$

Write $\phi := \phi_{r-1}\phi_{r-2}\cdots\phi_1\phi_0 \in \mathrm{End}(E)$. The result ϕ depends on representative choices of the isomorphism classes of oriented curves (E_i, ι_i) and of the isogenies ϕ_i shown above. We seek to choose such representatives so that

(i) $\phi_*\iota_0 = \pm\iota_0$.
(ii) each ϕ_i is chosen from the safe arbitrary assignment.

Such a choice will be called a *valid set of representatives* during the course of this proof. With such a valid set of representatives, to define a map from rims to isogeny cycles, we then forget orientations to obtain an isogeny cycle in $\mathcal{G}_\ell$. This is now well-defined because of (ii) and will define the map referred to in the statement of the proposition. For any rim R, we will write $C(R)$ for the isogeny cycle which is its image.

Facts About Valid Sets of Representatives. We will need a few facts about valid sets of representatives. First, we will show that if we have a valid set of representatives with respect to a basepoint (E_0, ι_0), then it is also a valid set of representatives for any other (E_i, ι_i) in the cycle. In other words, for any k, we have $(\phi_{r-k}\cdots\phi_{r-k+1})_*\iota_{r-k+1} = \pm\iota_{r-k+1}$ (where indices are taken modulo r). To see this for (E_1, ι_1), use $\iota_1 = (\phi_0)_*\iota_0$ to compute the left-hand side, using the fact that $\phi_*\iota_0 = \pm\iota_0$; repeat for E_2, E_3 etc.

Let $\alpha \in \mathcal{O}$ be such that $\mathfrak{l}^r = (\alpha)$. Without causing confusion, we will write $\phi_\mathfrak{l}\colon (E_i, \iota_i) \to (E_{i+1}, \iota_{i+1})$ for each i (this isogeny is defined up to isomorphism, by the convention that its kernel is $E_i[\iota_i(\mathfrak{l})]$). We will now prove that any valid set of representatives also satisfies

(iii) $\ker\phi_i = \ker\phi_\mathfrak{l} = E_i[\iota_i(\mathfrak{l})]$.
(iv) $\phi = \pm\iota_0(\alpha)$.

To prove this, observe that $\ker\phi_{\mathfrak{l}^r} = E_i[\iota_0(\alpha)] = \ker\iota_0(\alpha)$. In particular, $\iota_0(\alpha)$ is K-isomorphic to $\phi_{\mathfrak{l}^r}$. Furthermore, $(\iota_0(\alpha))_*\iota_0 = \pm\iota_0$ because elements of $\iota_0(K)$ commute. Next, since $\phi_*\iota_0 = \pm\iota_0 = (\iota_0(\alpha))_*\iota_0$ (by (i)), it must be that $\phi = \pm\iota_0(\alpha)$, since this is the only pre-/post-composition ϕ of $\iota_0(\alpha)$ that continues to satisfy $\phi_*\iota_0 = \pm\iota_0$. This implies that $\ker\phi = \ker\phi_{\mathfrak{l}^r}$ and hence $\ker\phi_0 = \ker\phi_\mathfrak{l}$. Since the valid set of representatives is valid at each basepoint, we have $\ker\phi_i = \ker\phi_\mathfrak{l}$ by the same reasoning.

Claim: The map defined on a valid set of representatives satisfies requirements (1) and (2) of Proposition 3.14. By (iv), $\phi = \pm\iota_0(\alpha)$. The composition of the isogeny cycle $C(R)$ is $\pm\phi$, which demonstrates (2) at the vertex (E_0, ι_0). Then (2) holds at the other vertices in R by the fact that a valid set of representatives is valid at each of the vertices of R. The fact that the map satisfies (1) is by definition.

Proof outline. It remains to show that there is a set of representatives satisfying (i)–(ii) (i.e. it is a valid set of representatives) and that any such set leads to the same isogeny cycle (i.e. the map $R \to C(R)$ given above is well defined). To do this, we will divide the possible rims into two cases: exceptional and non-exceptional.

The exceptional case is the following: First, we assume every vertex of the rim has the same j-invariant with extra automorphisms, of which φ is a non-trivial ($\neq [\pm 1]$) element, and then, defining ϕ as above for any basepoint, we assume that no matter which vertex we start with as basepoint, we have $\ker \phi = \ker \phi\varphi$.

The map $R \mapsto C(R)$ is well defined. Let us show that the map $R \mapsto C(R)$ is well defined in the non-exceptional cases. A priori, it may be that there are two choices of representatives satisfying (i)–(ii). Suppose so. Use primes (ι', ϕ_i' etc.) to denote the second valid set of representatives.

First note that if $(E_i, \iota_i') = (E_i, \pm\iota_i)$ at any vertex, then the two choices of representatives must be the same. To see this, note that the vertex representative $(E_i, \pm\iota_i)$ determines the kernel of ϕ_i (by (iii)), which in turn determines ϕ_i up to sign (by (ii)). Therefore $\phi_i' = \pm\phi_i$. But this implies that $(E_{i+1}, \iota_{i+1}') = (E_{i+1}, \pm\iota_{i+1})$. Proceeding in this way, the two valid sets of representatives are the same. Hence the proof is complete if any curve in the rim has no extra automorphisms (in which case there is no choice for ι_i).

Suppose that we have $\iota_0' = \varphi_*\iota_0$ for some non-trivial (not ± 1) automorphism $\varphi \in \mathrm{End}(E_0)$. In particular, E_0 has extra automorphisms. At the same time, $\iota_1' = \eta_*\iota_1$ for some $\eta \in \mathrm{Aut}(E_1)$ (possibly trivial this time). These relationships entail that $\phi_0' = \eta\phi_0\varphi^{-1}$. By (ii), ϕ_0 and ϕ_0' both belong to the safe arbitrary assignment, so $\eta = \pm 1$ and hence $\iota_1' = \pm\iota_1$; thus, we have reduced to the case in the previous paragraph.

Existence of a valid set of representatives in non-exceptional cases. We use the safeness of the safe arbitrary assignment (Definition 3.13). Let ϕ_0 be the map with kernel $E_0[\iota_0(\mathfrak{l})]$ and post-composition determined by the safe arbitrary assignment. Then let $\iota_1 = (\phi_0)_*\iota_0$. Continue in this manner around the entire rim. When we close the rim (reaching (E_0, ι_0) again), we find that $\phi_*\iota_0 = \varphi\iota_0\varphi^{-1}$ for some automorphism φ (since the rim closes, we must reach an oriented curve isomorphic to the initial oriented curve). If E_0 has no extra automorphisms, then the proof of this case is complete. Hence we are done if the cycle contains even one curve without extra automorphisms.

Let us therefore restrict to the case that E_0 has extra automorphisms and so do all other curves in the rim. If item (i) fails, then there is a nontrivial automorphism $\varphi \in \mathrm{Aut}(E_0)$ so that $\phi_*\iota_0 = \varphi_*\iota_0$. Then

$$(\phi\varphi^{-1})_*(\varphi_*\iota_0) = \phi_*\iota_0 = \varphi_*\iota_0.$$

Thus, replacing isomorphism class representatives (E_0, ι_0) with $(E_0, \varphi_*\iota_0)$ and ϕ_0 with $\phi_0\varphi^{-1}$, we recover (i). Provided that ϕ_0 and $\phi_0\varphi^{-1}$ have different kernels, which they do when $E_1 \not\cong E_0$ (Lemma 2.9), then using the conventions of the safe arbitrary assignment (Definition 3.13), we have the freedom to replace ϕ_0 with $\phi_0\varphi^{-1}$ without violating item (ii).

Thus we are reduced to the case that all E_i are isomorphic (assumed equal without loss of generality), and ϕ_0 and $\phi_0\varphi$ have the same kernel (for every choice of basepointing the rim). This is the exceptional case.

Existence of a valid set of representatives in the exceptional case. Suppose we are in the exceptional situation described in the proof outline. That is, all vertices are isomorphic curves with extra automorphisms, and $\ker\phi = \ker\phi\varphi$. Since $\phi \neq \pm\phi\varphi$ (as in the proof of Lemma 2.9), we have $\varphi^k\phi = \pm\phi\varphi$ for some integer k with $\varphi^k \neq \pm 1$. This means either $\phi\varphi = \varphi\phi$ or $\phi\varphi = \varphi\overline{\phi}$. Note that ϕ commutes with φ if and only if $\varphi\phi$ does, so commuting is independent of the safe arbitrary assignment.

The above is true for ϕ_0 as well as ϕ. But $\ell = N(\phi_0) < p$, which by Lemma 2.1 implies that ϕ and φ commute. The same argument applies to each ϕ_i for which $\ker\phi_i = \ker\phi_i\varphi$. But if there is a ϕ_i for which this is not true, we are out of the exceptional case. Hence every ϕ_i commutes with φ.

Therefore we may suppose ϕ commutes with φ. Then $\phi \in \mathbb{Q}(\varphi)$. Regardless of the safe arbitrary assignment, the commutativity implies $\phi_*\iota = \iota$ for the orientation $\iota(\zeta_3) = \varphi$ or $\iota(i) = \varphi$ for $j = 0$ and $j = 1728$, respectively. So suppose ϕ is the arbitrary assignment. Then if R is a rim with $\phi_\mathfrak{l} = \phi$ (up to post-composition by an automorphism), then R is associated with the field $K = \mathbb{Q}(\zeta_3)$ or $\mathbb{Q}(i)$, respectively, and ϕ provides a set of representatives satisfying (i)–(ii). Hence there is a valid set of representatives. □

We observe the following immediate corollary to Theorem 3.2.

Corollary 3.15 *Let $(j_1, \ldots, j_n)$ be a finite sequence of j-invariants of supersingular elliptic curves in $\mathcal{G}_\ell$. Then the number of isogeny cycles in $\mathcal{G}_\ell$ on this ordered sequence of j-invariants is equal to the number of rims of $\bigcup_K \mathcal{G}_{K,\ell}$ on this ordered sequence of j-invariants.*

In fact, an alternate approach to proving Theorem 3.2 would be to observe that it is indeed equivalent to this corollary. One could then break into cases based on the existence of extra automorphisms. In most cases, there are no double edges, and this is a bijection of singleton sets.

3.5 An Undirected Version of the Main Bijection

The arbitrary assignment also allows us to set up a bijection between walks and their 'duals' (walking the same j-invariants backwards), despite the ambiguity at $j = 0$ and $j = 1728$. This means we can discuss the bijection between undirected rims and undirected isogeny cycles.

Lemma 3.16 *Assume we have made a safe arbitrary assignment for $\mathcal{G}_\ell$. Let $(j_0, j_1, \ldots, j_n)$ be a sequence of j-invariants for a closed walk in $\mathcal{G}_\ell$ (where j_i is adjacent to j_{i+1} and j_n to j_0). Then the closed walks through this ordered list of j-invariants can be paired with the closed walks through the list $(j_n, j_{n-1}, \ldots, j_1, j_0)$ (i.e. going backwards) in such a way that the pair of endomorphisms obtained from each pair of walks (by composition) is a pair of dual endomorphisms.*

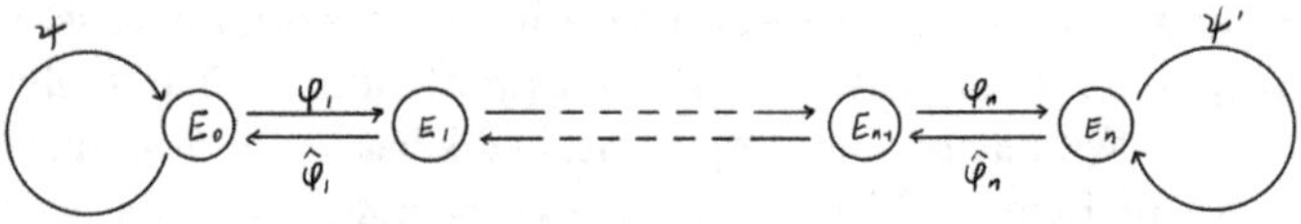

Fig. 4 An example of an isogeny cycle that is a barbell (Definition 3.19). If we walk in the direction of the arrows from E_0, we obtain $\psi\widehat{\varphi_1}\cdots\widehat{\varphi_n}\psi'\varphi_n\cdots\varphi_1$. If we walk the cycle backwards (as defined in Lemma 3.16), we get $\widehat{\varphi_1}\cdots\widehat{\varphi_n}\widehat{\psi'}\varphi_n\cdots\varphi_1\widehat{\psi}$. If we ignore the basepoint and have $\psi = \widehat{\psi}$ and $\psi' = \widehat{\psi'}$, then these are the same cyclic sequence of arrows, in which case this is a barbell

Proof Each walk $\phi_0, \phi_1, \ldots, \phi_n$ with composition θ has a 'dual walk' obtained by replacing the individual isogenies with their duals and reversing the order: the walk $\widehat{\phi_n}, \widehat{\phi_{n-1}}, \ldots, \widehat{\phi_0}$. This results in a walk on the same vertices backwards. If the walk does not pass through $j = 0$ or $j = 1728$, this dual walk gives the dual endomorphism under composition (up to sign). The difficulty arises when choosing a dual for an outgoing edge $\phi_i : E_i \to E_{i+1}$, where E_i has j-invariant 0 or 1728, in which case the dual $\widehat{\phi_i} : E_{i+1} \to E_i$ may not be in the safe arbitrary assignment, and we are forced to use instead $\varphi\widehat{\phi_i} : E_{i+1} \to E_i$ for some $\varphi \in \mathrm{Aut}(E_i)$. In this case the endomorphism obtained by composition is no longer $\pm\widehat{\theta}$ because of the intervening φ between $\widehat{\phi_i}$ and $\widehat{\phi_{i-1}}$. However, this can be adjusted by choosing an appropriate replacement for $\widehat{\phi_{i-1}} : E_i \to E_{i-1}$ from among the available isogenies in the safe arbitrary assignment from E_i to E_{i-1}, which differ by precomposition by an automorphism. In other words, we can obtain the composition $\widehat{\phi_{i-1}}\widehat{\phi_i}$ as $\widehat{\phi_{i-1}}\varphi^{-1}\varphi\widehat{\phi_i}$. In this way, we can obtain a dual path whose composition according to the safe arbitrary assignment is $\pm\widehat{\theta}$. Since this process can be performed in both directions (from clockwise paths to counterclockwise and vice versa), inversely (by Lemma 3.6), the statement is proved. □

The preceding lemma allows us to define, for any closed walk, its *backward walk*.

Lemma 3.17 *If $\overline{R}$ is the conjugate of R, then $\Phi(\overline{R})$ and $\Phi(R)$ are backward isogeny cycles of each other.*

Proof The rim $\overline{R}$ consists of the set of edges and vertices traversed backward, where an isogeny is replaced with its dual. From the proof of Lemma 3.16, this results in the backward isogeny cycle of $\Phi(R)$. □

Corollary 3.18 *There is a bijection between undirected rims of length r of $\mathcal{G}_{K,\ell}$ for all K, and undirected isogeny cycles of length r in $\mathcal{G}_\ell$.*

Definition 3.19 A closed walk in $\mathcal{G}_\ell$ is called a *barbell* if traversing it in the opposite direction as in Lemma 3.16 results in the same walk, up to a choice of basepoint.

An example is shown in Fig. 4, justifying the name of 'barbell'.

Lemma 3.20 *Barbells necessarily involve at least two loops and have an even number r of vertices and edges. The total number of barbells having r edges is at most $\#\mathcal{G}_\ell(\ell+1)\ell^{(r-2)/2}$.*

Proof Any barbell traverses a palindromic sequence of vertices of the following form:

$$j_0, j_1, \ldots, j_{n-1}, j_n, j_n, j_{n-1}, \ldots, j_1, j_0.$$

In particular, the 'turnaround' points must be loops, because otherwise they would involve backtracking. This implies that the number of edges and vertices is even. The count follows by counting the number of walks of length $(r-2)/2$ between two loops and bounding the number of loops by the number of edges of the graph. □

The above bound on the number of barbells is far from tight. There are congruence conditions modulo p depending on ℓ which can rule out loops in $\mathcal{G}_\ell$ and hence rule out the existence of barbells. (These conditions can be worked out by considering the specialization of the modular polynomial $\Phi_\ell(x, x)$ mod p and choosing p such that the roots are not supersingular.)

Theorem 3.21 *Under the bijection of Theorem 3.2, the self-conjugate rims in $\mathcal{G}_{K,\ell}$ are in bijection with the barbells in $\mathcal{G}_\ell$.*

Proof If R is a self-conjugate rim, then conjugation returns the same rim traversed in the opposite direction. Hence $\Phi(R)$ must be a barbell. Conversely, if R is a rim such that $\Phi(R)$ is a barbell, then the rim R' directed in the opposite direction maps to $\Phi(R)$, the cycle $\Phi(R)$ traversed in the opposite direction. Hence $\Phi(R) = \Phi(R')$ and therefore R is self-conjugate. □

4 Supersingular Curves Oriented by $\mathcal{O}$

This section answers the question of when $\mathrm{SS}_{\mathcal{O}}^{\mathrm{pr}} = \rho(\mathcal{E}\ell\ell(\mathcal{O}))$. Its purpose is to prepare for the next section, where we discuss the fibre sizes of the map taking a rim to its corresponding quadratic order. In particular, we wish to understand the action of conjugation on $\mathrm{SS}_{\mathcal{O}}^{\mathrm{pr}}$.

Let p be inert or ramified in K, and let $\mathcal{O}$ be an order of K. In this section, we study the set $\mathrm{SS}_{\mathcal{O}}^{\mathrm{pr}}$ in greater detail, using class field theory. In particular, Theorem 4.4 characterizes when $\rho(\mathcal{E}\ell\ell(\mathcal{O})) = \mathrm{SS}_{\mathcal{O}}^{\mathrm{pr}}$, which holds precisely when p ramifies in K. This also characterizes the scenario when conjugation takes $\rho(\mathcal{E}\ell\ell(\mathcal{O}))$ back to itself.

Let $K = \mathbb{Q}(\beta)$ be an imaginary quadratic field of discriminant Δ_K with a distinguished root β of a fixed minimal polynomial $f_K(x)$ generating $K/\mathbb{Q}$. Let $\mathcal{O}$ be an order of K. Let L/K be the ring class field of $\mathcal{O}$. By the general theory of class fields, the field L has Galois group $\mathrm{Gal}(L/\mathbb{Q}) \cong \mathrm{Cl}(\mathcal{O}) \rtimes \langle\tau\rangle$, where $\mathrm{Cl}(\mathcal{O}) \cong \mathrm{Gal}(L/K)$, and τ is the non-trivial automorphism of $K/\mathbb{Q}$, and where $\sigma\tau = \tau\sigma^{-1}$ for any $\sigma \in \mathrm{Cl}(\mathcal{O})$ [18, Lemma 9.3]. Let $H_{\mathcal{O}}(x)$ be the Hilbert class polynomial (sometimes called the ring class polynomial) of the order $\mathcal{O}$. Define

$$\mathcal{L}_K := \{j \in L : H_{\mathcal{O}}(j) = 0\}, \quad \mathcal{L}_{\mathbb{Q}} := \{(j, \alpha) \in L \times K : H_{\mathcal{O}}(j) = 0 = f_K(\alpha)\}.$$

The Galois groups $\mathrm{Gal}(L/K)$ and $\mathrm{Gal}(L/\mathbb{Q})$ act freely and transitively on $\mathcal{L}_K$ and $\mathcal{L}_{\mathbb{Q}}$, respectively, via $j \mapsto j^\sigma$ and $(j, \alpha) \mapsto (j, \alpha)^\sigma = (j^\sigma, \alpha^\sigma)$. We have $\#\mathcal{L}_K = h_{\mathcal{O}} = \#\mathrm{Gal}(L/K)$ and $\#\mathcal{L}_{\mathbb{Q}} = 2h_{\mathcal{O}} = \#\mathrm{Gal}(L/\mathbb{Q})$.

It is well known that $\mathcal{L}_K$ is in bijection with $\mathcal{E}\ell\ell(\mathcal{O})$, the isomorphism classes of elliptic curves over $\mathbb{C}$ with complex multiplication by $\mathcal{O}$, by the map $j \mapsto E(j)$ taking a j-invariant to its isomorphism class of curves. (We will abuse notation by denoting an element of $\mathcal{E}\ell\ell(\mathcal{O})$ as a curve E without equivalence class brackets $[E]$.) This endows $\mathcal{E}\ell\ell(\mathcal{O})$ with an action of $\mathrm{Gal}(L/K)$.

We now give an analogous result for $\mathcal{L}_{\mathbb{Q}}$. Given $E \in \mathcal{E}\ell\ell(\mathcal{O})$, there are two isomorphisms $\iota\colon \mathcal{O} \to \mathrm{End}_{\mathbb{C}}(E)$ that differ by conjugation. Denote these by ι_E and $\bar{\iota}_E$, where ι_E is normalized so that $\iota_E(\alpha)^*\omega_E = \alpha\omega_E$ for the invariant differential ω_E of E and any $\alpha \in \mathcal{O}$. Let $\mathcal{E}\ell\ell^*(\mathcal{O})$ denote the set of cardinality $2h_{\mathcal{O}}$ of pairs consisting of a curve from $\mathcal{E}\ell\ell(\mathcal{O})$ together with one of the two orientations:

$$\mathcal{E}\ell\ell^*(\mathcal{O}) := \{(E, \iota)\colon E \in \mathcal{E}\ell\ell(\mathcal{O}),\ \iota\colon \mathcal{O} \xrightarrow{\sim} \mathrm{End}(E)\}.$$

Proposition 4.1 *There is a bijection of $\mathcal{L}_{\mathbb{Q}}$ with $\mathcal{E}\ell\ell^*(\mathcal{O})$ via*

$$(j, \delta) \mapsto \left(E = E(j), \begin{cases} \iota_E & \text{if } \delta = \beta \\ \bar{\iota}_E & \text{if } \delta = \bar{\beta} \end{cases}\right).$$

This endows $\mathcal{E}\ell\ell^*(\mathcal{O})$ with an action of $\mathrm{Gal}(L/\mathbb{Q})$.

The bijection is evidently non-canonical; we could make the opposite choice for ι_E.

Choose a finite extension L'/L such that there is a prime $\mathfrak{p}$ of L' above p for which all curves in $\mathcal{E}\ell\ell(\mathcal{O})$ have good reduction at $\mathfrak{p}$. Let $\rho_{\mathfrak{p}}$ denote reduction modulo $\mathfrak{p}$, and put $\tilde{E} = \rho_{\mathfrak{p}}(E)$ for brevity. Reduction provides a map

$$\rho\colon \mathcal{E}\ell\ell^*(\mathcal{O}) \to \mathrm{SS}_{\mathcal{O}}^{\mathrm{pr}} \text{ given by}$$

$$(E, \iota_E) \mapsto (\tilde{E}, \iota_E \bmod \mathfrak{p}), \text{ and } (E, \bar{\iota}_E) \mapsto (\tilde{E}, \bar{\iota}_E \bmod \mathfrak{p}).$$

There is a natural embedding of $\mathcal{E}\ell\ell(\mathcal{O})$ in $\mathcal{E}\ell\ell^*(\mathcal{O})$ given by $E \mapsto (E, \iota_E)$ (i.e. choosing the normalized isomorphism), and to ease notation, we will consider $\mathcal{E}\ell\ell(\mathcal{O})$ a subset of $\mathcal{E}\ell\ell^*(\mathcal{O})$. Consider the following diagram:

$$
\begin{array}{ccc}
\mathrm{Cl}(\mathcal{O}) \cong \mathrm{Gal}(L/K) & \hookrightarrow & \mathrm{Gal}(L/\mathbb{Q}) \cong \mathrm{Cl}(\mathcal{O}) \rtimes \langle \tau \rangle \\
\circlearrowright & & \circlearrowright \\
\mathcal{L}_K & & \mathcal{L}_{\mathbb{Q}} \\
\downarrow\wr & & \downarrow\wr \\
\mathcal{E}\ell\ell(\mathcal{O}) & \hookrightarrow & \mathcal{E}\ell\ell^*(\mathcal{O}) \\
\downarrow \rho_{\mathfrak{p}} & & \downarrow \rho_{\mathfrak{p}} \\
\rho_{\mathfrak{p}}(\mathcal{E}\ell\ell(\mathcal{O})) & \hookrightarrow & \mathrm{SS}_{\mathcal{O}}^{\mathrm{pr}}
\end{array}
$$

The actions of the Galois groups shown are free and transitive. The inclusion map $\mathcal{E}\ell\ell(\mathcal{O}) \hookrightarrow \mathcal{E}\ell\ell^*(\mathcal{O})$ is equivariant with respect to the action of $\mathrm{Gal}(L/K) \subseteq \mathrm{Gal}(L/\mathbb{Q})$.

In the vertical arrows of the lower half of the diagram, we see reduction modulo $\mathfrak{p}$. The square in this portion commutes. The descending arrow on the left is a bijection.

First, we consider the easier unramified case and show that $\rho_{\mathfrak{p}}(\mathcal{E}\ell\ell(\mathcal{O})) \neq \mathrm{SS}_{\mathcal{O}}^{\mathrm{pr}}$. The map $(E, \iota) \mapsto (E, \bar{\iota})$ on $\mathrm{SS}_{\mathcal{O}}^{\mathrm{pr}}$ will be called *conjugation*.

Proposition 4.2 *Let p be inert in K. Then the two isomorphism classes of $\mathcal{O}$-oriented curves represented by (E, ι) and $(E, \bar{\iota})$ cannot both belong to $\rho_{\mathfrak{p}}(\mathcal{E}\ell\ell(\mathcal{O}))$, so $\rho_{\mathfrak{p}}(\mathcal{E}\ell\ell(\mathcal{O})) \neq \mathrm{SS}_{\mathcal{O}}^{pr}$, in which case $2\#\rho_{\mathfrak{p}}(\mathcal{E}\ell\ell(\mathcal{O})) = \#\mathrm{SS}_{\mathcal{O}}^{pr}$.*

Proof Since p is unramified, there exists $\alpha_0 \in \mathcal{O}$ such that $\iota(\alpha_0 - \overline{\alpha_0})$ is separable. Suppose $E_1, E_2 \in \mathcal{E}\ell\ell(\mathcal{O})$. Then there are isomorphisms $[\cdot]_{E_i} : \mathcal{O} \to \mathrm{End}(E_i)$, where we use the normalization of Sect. 2.3, namely we choose the isomorphisms so that $[\alpha]_{E_i}^* \omega_{E_i} = \alpha \omega_{E_i}$ for all $\alpha \in \mathcal{O}$. With notation as in Sect. 2.3, suppose that the reduction $(\tilde{E}_1, \iota_{E_1})$ of $(E_1, [\cdot]_{E_1})$ modulo $\mathfrak{p}$ is in the isomorphism class of (E, ι) and the reduction $(\tilde{E}_2, \iota_{E_2})$ of $(E_2, [\cdot]_{E_2})$ is in the isomorphism class of $(E, \bar{\iota})$.

Let us first assume for simplicity that $(\tilde{E}_1, \iota_{E_1}) = (E, \iota)$ and $(\tilde{E}_2, \iota_{E_2}) = (E, \bar{\iota})$, in which case the proof is more transparent. Then, as discussed in Sect. 2.3, the normalization is preserved by reduction, which is to say,

$$\iota(\alpha)^* \omega_E = \alpha \omega_E = \bar{\iota}(\alpha)^* \omega_E$$

for all $\alpha \in \mathcal{O}$. But then

$$\iota(\alpha)^* \omega_E = \bar{\iota}(\alpha)^* \omega_E = \iota(\overline{\alpha})^* \omega_E,$$

for all $\alpha \in \mathcal{O}$, which implies $\iota(\alpha - \overline{\alpha})^* \omega_E = 0$ for all α, a contradiction for $\alpha = \alpha_0$ since $\iota(\alpha_0 - \overline{\alpha_0})$ is separable.

In general, however, the map ρ is only defined on isomorphism classes, prompting a certain amount of bookkeeping. In particular, assume that $(\tilde{E}_1, \iota_{E_1}) = (\phi_1 E, \phi_1 \iota \phi_1^{-1})$, and $(\tilde{E}_2, \iota_{E_2}) = (\phi_2 E, \phi_2 \bar{\iota} \phi_2^{-1})$, where $\phi_i : E \to \phi_i E, i = 1, 2$, are isomorphisms. Then for all $\alpha \in \mathcal{O}$,

$$\alpha\omega_E = \alpha\phi_2^*\omega_{\phi_2 E} = \phi_2^*\alpha\omega_{\phi_2 E} = \phi_2^*(\phi_2\bar{\iota}(\alpha)\phi_2^{-1})^*\omega_{\phi_2 E} = \bar{\iota}(\alpha)^*\phi_2^*\omega_{\phi_2 E}$$
$$= \bar{\iota}(\alpha)^*\omega_E = \iota(\overline{\alpha})^*\omega_E,$$

and by a similar computation, $\iota(\alpha)^*\omega_E = \alpha\omega_E$. The rest of the proof is as before. □

Next we consider the ramified case, where we show that $\rho_{\mathfrak{p}}(\mathcal{E}\ell\ell(\mathcal{O})) = \mathrm{SS}_{\mathcal{O}}^{\mathrm{pr}}$.

Proposition 4.3 *Suppose p is ramified in K. Then the action of* $\mathrm{Gal}(L/\mathbb{Q})$ *on* $\mathcal{L}_{\mathbb{Q}}$ *induces an action on* $\mathrm{SS}_{\mathcal{O}}^{pr}$. *Under this action, τ acts by conjugation, i.e.* $\tau \cdot (E, \iota) = (E, \bar{\iota})$.

Proof Define an action on $\mathrm{SS}_{\mathcal{O}}^{\mathrm{pr}}$ by $\mathrm{Cl}(\mathcal{O})$ as in Definition 2.22, and by $\langle\tau\rangle$ as $\tau \cdot (E, \iota) = (E, \bar{\iota})$. Let $\mathfrak{P} = \mathfrak{p} \cap L$ be the restriction of $\mathfrak{p}$ to L. Since p is ramified in K, τ is in the inertia group of $\mathfrak{P}$, which implies that $j^\tau \equiv j \pmod{\mathfrak{P}}$ and hence also modulo $\mathfrak{p}$, where $j = j(E)$. Thus

$$\rho_{\mathfrak{p}}(\tau \cdot (E_j, \iota)) = \rho_{\mathfrak{p}}((E_{j^\tau}, \bar{\iota})) = (E_{j^\tau \bmod \mathfrak{p}}, \bar{\iota} \bmod \mathfrak{p}) = (E_{j \bmod \mathfrak{p}}, \bar{\iota} \bmod \mathfrak{p})$$
$$= \tau \cdot (E_{j \bmod \mathfrak{p}}, \iota \bmod \mathfrak{p}) = \tau \cdot \rho_{\mathfrak{p}}((E_j, \iota)).$$

The map ρ is also known to satisfy $\rho_{\mathfrak{p}}(\sigma \cdot (E, \iota)) = \sigma \cdot (\rho_{\mathfrak{p}}((E, \iota)))$ for $\sigma \in \mathrm{Cl}(\mathcal{O})$ [42, Proof of Proposition 3.6]. Therefore, the actions described on $\mathrm{SS}_{\mathcal{O}}^{\mathrm{pr}}$ give a well-defined action of $\mathrm{Gal}(L/\mathbb{Q})$ with respect to which ρ is equivariant. □

Theorem 4.4 *The prime p ramifies in K if and only if $\rho_{\mathfrak{p}}(\mathcal{E}\ell\ell(\mathcal{O})) = \mathrm{SS}_{\mathcal{O}}^{pr}$ if and only if conjugation takes $\rho_{\mathfrak{p}}(\mathcal{E}\ell\ell(\mathcal{O}))$ to itself.*

Proof If $\rho_{\mathfrak{p}}(\mathcal{E}\ell\ell(\mathcal{O})) = \mathrm{SS}_{\mathcal{O}}^{\mathrm{pr}}$, then p ramifies by Proposition 4.2. Conversely, if p ramifies in K, then there is some $\theta \in \iota(\mathcal{O}) \subseteq \mathrm{End}(E)$ such that $\theta^2 = -np$ for some integer n coprime to p. The endomorphism θ is a horizontal isogeny with respect to $\mathcal{O}$ (as an element of $\iota(\mathcal{O})$). Furthermore, θ must factor as

$$E \xrightarrow{\pi_p} E^{(p)} \xrightarrow{\phi} E,$$

where ϕ is of degree n and separable. But since $\theta \in \iota(\mathcal{O})$, $\theta_*\iota = \iota$ and we have the following chain of K-oriented isogenies:

$$(E, \iota) \xrightarrow{\pi_p} (E^{(p)}, \iota^{(p)}) \xrightarrow{\phi} (E, \iota).$$

The maps shown above are horizontal isogenies (π_p is horizontal and so is θ; hence ϕ is also). We wish to show that ϕ represents the action of an element of $\mathrm{Cl}(\mathcal{O})$, up to multiplication by an integer. We can factor n into prime powers and prove it for each one, so without loss of generality, we assume that n is a prime power q^k, for $q \neq p$.

The map ϕ is given by a path in the oriented q-isogeny graph determined by successive kernels, and as it is horizontal, this path must have as many descending as ascending steps. Using the volcano structure of the graph, all the ascending and descending edges can be paired up into consecutive pairs $\phi_2 \circ \phi_1$, where ϕ_1 descends and ϕ_2 ascends. Furthermore, $\phi_2 = \varphi_1 \widehat{\phi_1} \varphi_2$ for some automorphisms φ_1, φ_2. Then $\varphi_1^{-1} \phi_2 \phi_1 = \widehat{\phi_1} \varphi_2 \phi_1$ has trace and norm divisible by q; hence it is divisible by $[q]$. So $\phi_2 \phi_1 = \varphi[q]$ for some automorphism φ. Since we only care about ϕ up to multiplication by an integer, we can replace $\phi_2\phi_1$ with φ, which is horizontal. Repeating this process leaves a purely horizontal path, which must therefore represent an action of an element of $\mathrm{Cl}(\mathcal{O})$ (since $q \neq p$, $\mathrm{Cl}(\mathcal{O})$ acts on $\mathrm{SS}_{\mathcal{O}}^{\mathrm{pr}}$, so any horizontal q-isogeny is the action of an element of the class group). It follows that $(E^{(p)}, \iota^{(p)}) \in \mathrm{Cl}(\mathcal{O}) \cdot (E, \iota)$, which in turn implies that $\rho_{\mathfrak{p}}(\mathcal{E}\ell\ell(\mathcal{O})) = \mathrm{SS}_{\mathcal{O}}^{\mathrm{pr}}$ by [42, Propositions 3.3 and 3.4].

Finally, we show that $\rho_{\mathfrak{p}}(\mathcal{E}\ell\ell(\mathcal{O})) = \mathrm{SS}_{\mathcal{O}}^{\mathrm{pr}}$ if and only if conjugation takes $\rho_{\mathfrak{p}}(\mathcal{E}\ell\ell(\mathcal{O}))$ to itself. Since p is ramified, by Proposition 4.3, the action of $\mathrm{Gal}(L/\mathbb{Q})$ is defined on $\mathrm{SS}_{\mathcal{O}}^{\mathrm{pr}}$ and τ acts by conjugation, fixing the j-invariant $j(E)$ modulo $\mathfrak{p}$. First, note that if (E, ι) and $(E, \bar{\iota})$ are related by an element of the class group, then the entire orbit $\mathrm{Cl}(\mathcal{O}) \cdot (E, \iota)$ consists of elements related to their conjugates by the class group, and therefore, conjugation takes $\rho_{\mathfrak{p}}(\mathcal{E}\ell\ell(\mathcal{O}))$ into itself. Therefore conjugation takes $\rho_{\mathfrak{p}}(\mathcal{E}\ell\ell(\mathcal{O}))$ to itself or else maps it to $\mathrm{SS}_{\mathcal{O}}^{\mathrm{pr}} \setminus \rho_{\mathfrak{p}}(\mathcal{E}\ell\ell(\mathcal{O}))$. We know that $\#\rho_{\mathfrak{p}}(\mathcal{E}\ell\ell(\mathcal{O})) = h_{\mathcal{O}}$, and $\#\mathrm{SS}_{\mathcal{O}}^{\mathrm{pr}} \in \{h_{\mathcal{O}}, 2h_{\mathcal{O}}\}$ (Sect. 2.3). But $\#\mathrm{SS}_{\mathcal{O}}^{\mathrm{pr}} = 2h_{\mathcal{O}}$ if and only if $\rho_{\mathfrak{p}} \colon \mathcal{E}\ell\ell^*(\mathcal{O}) \to \mathrm{SS}_{\mathcal{O}}^{\mathrm{pr}}$ is an isomorphism if and only if $\mathrm{Gal}(L/\mathbb{Q})$ acts freely and transitively on $\mathrm{SS}_{\mathcal{O}}^{\mathrm{pr}}$ (calling upon Proposition 4.3 again). But $\rho_{\mathfrak{p}}(\mathcal{E}\ell\ell(\mathcal{O}))$ forms one $\mathrm{Cl}(\mathcal{O})$-orbit, so the only way the action can be free and transitive is if conjugation takes $\rho_{\mathfrak{p}}(\mathcal{E}\ell\ell(\mathcal{O}))$ outside itself. □

5 Isogeny Cycles and Their Associated Orders

The following theorem describes a way to relate isogeny cycles in $\mathcal{C}_r$ to imaginary quadratic orders. Define

$$\mathcal{I}_r = \left\{ \text{imaginary quadratic orders } \mathcal{O} : \begin{array}{c} p \text{ does not split in the field containing } \mathcal{O} \\ p \text{ does not divide the conductor of } \mathcal{O} \\ \mathcal{O} \text{ is an } \ell\text{-fundamental order,} \\ (\ell) = \mathfrak{l}\bar{\mathfrak{l}} \text{ splits in } \mathcal{O}, \\ \text{and } [\mathfrak{l}] \text{ has order } r \text{ in } \mathrm{Cl}(\mathcal{O}). \end{array} \right\}$$

These are exactly the orders for which $\mathrm{SS}_{\mathcal{O}}^{\mathrm{pr}}$ is non-empty, and the permutation on the vertices arising from the action of a prime ideal in $\mathcal{O}$ above ℓ has a decomposition consisting of cycles of length r only. The following theorem describes how often each order in $\mathcal{I}_r$ is obtained from an isogeny cycle in $\mathcal{G}_\ell$ via the bijection of

Theorem 3.2. If p is inert, one expects to obtain $2h_{\mathcal{O}}/r$ directed isogeny cycles of size r which give rise to an order $\mathcal{O}$, since $\#\,\mathrm{SS}_{\mathcal{O}}^{\mathrm{pr}} = 2h_{\mathcal{O}}$ and the rims are identified in pairs by conjugation, and given a direction. However, in the case that p ramifies in the field K containing $\mathcal{O}$, the exact count may differ from this.

Let $\mathcal{O}$ be a quadratic order with fraction field K. Let L/K be the ring class field of $\mathcal{O}$. Let $g_{\mathcal{O}}$ be the number of genera in $\mathrm{Cl}(\mathcal{O})$. Let $\sigma_\ell := \mathrm{Frob}_\ell \in \mathrm{Gal}(L/K)$ have order r. Define $C_\ell := \langle \sigma_\ell \rangle$ (so $\#C_\ell = r$).

Theorem 5.1 *Let $r > 2$. The map $\mathcal{R}_r \to \mathcal{I}_r$ which takes a rim to the order $\mathcal{O}$ with respect to which all its vertices are primitively $\mathcal{O}$-oriented as in the bijection of Theorem 3.2 has fibres of size $2h_{\mathcal{O}}/r$ or $4h_{\mathcal{O}}/r$, depending on whether or not p ramifies in the field K containing $\mathcal{O}$, respectively. The map factors through $\mathcal{R}_r/\!\sim$, so the induced map $\mathcal{C}_r \to \mathcal{I}_r$ has a fibre of size $\epsilon_{\mathcal{O},\ell} h_{\mathcal{O}}/r$ over $\mathcal{O}$ for some $1 \leq \epsilon_{\mathcal{O},\ell} \leq 2$. Furthermore,*

(1) Suppose p is inert K. Then $\epsilon_{\mathcal{O},\ell} = 2$.
(2) Suppose p ramifies in K. Then there are two possibilities:

(a) There are no self-conjugate rims in $\rho(\mathcal{E\ell\ell}(\mathcal{O}))$. Then $\epsilon_{\mathcal{O},\ell} = 1$.
(b) There are self-conjugate rims in $\rho(\mathcal{E\ell\ell}(\mathcal{O}))$, in which case r is even, and

$$\epsilon_{\mathcal{O},\ell} := 1 + \frac{r g_{\mathcal{O}}}{2h_{\mathcal{O}}}.$$

In this case, $1 < \epsilon_{\mathcal{O},\ell} \leq 2$.

The following commutative diagram illustrates the theorem:

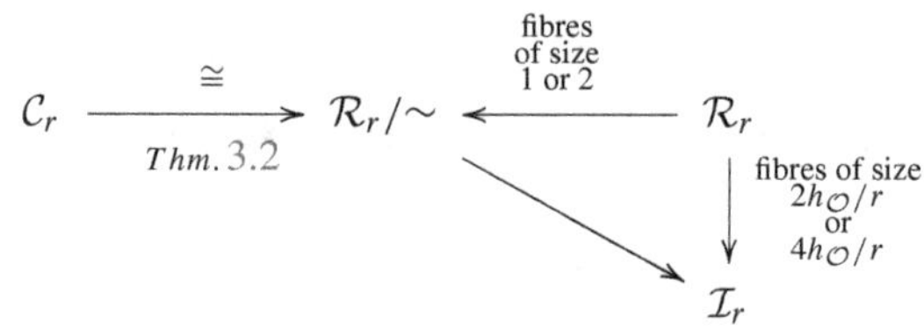

(9)

Proof Define a map $\mathcal{R}_r \to \mathcal{I}_r$ taking a rim to the associated order as described in the statement of the theorem. The size of the fibre above $\mathcal{O}$ is $2\#\,\mathrm{SS}_{\mathcal{O}}^{\mathrm{pr}}/r$ (the factor of 2 arises because rims are directed), which is $2h_{\mathcal{O}}/r$ or $4h_{\mathcal{O}}/r$ in the ramified and inert cases, respectively, by Theorem 4.4 (the conditions on p in the definition of $\mathcal{I}_r$ imply $\mathrm{SS}_{\mathcal{O}}^{\mathrm{pr}}$ is nonempty by Proposition 2.19). Since a conjugate rim is associated with the same order as the original, the map factors through $\mathcal{R}_r/\!\sim$. The fibres of $\mathcal{R}_r \to \mathcal{R}_r/\!\sim$ are of size 1 or 2 (depending upon whether the rim in question was self-conjugate).

In the inert case, $\#\,\mathrm{SS}_{\mathcal{O}}^{\mathrm{pr}} = 2h_{\mathcal{O}}$ by Theorem 4.4. The fibres of $\mathcal{R}_r \to \mathcal{R}_r/\!\sim$ are uniformly of size 2 for rims in $\mathrm{SS}_{\mathcal{O}}^{\mathrm{pr}}$ (since conjugation swaps $\rho(\mathcal{E\ell\ell}(\mathcal{O}))$ and its complement in $\mathrm{SS}_{\mathcal{O}}^{\mathrm{pr}}$ by Theorem 4.4). The map $\mathcal{R}_r \to \mathcal{I}_r$ has a fibre of size $4h_{\mathcal{O}}/r$ over $\mathcal{O}$. Hence the map $\mathcal{C}_r \to \mathcal{I}_r$ has a fibre of size $2h_{\mathcal{O}}/r$ above $\mathcal{O}$.

Now we consider the ramified case. Call upon the conventions and set-up of Sect. 4. Let $G = \mathrm{Gal}(L/\mathbb{Q})$. We consider the stabilizers $\mathrm{Stab}_G((E, \iota))$ of elements of $\mathrm{SS}_{\mathcal{O}}^{\mathrm{pr}}$. All stabilizers are of size 2 since $\#G = 2h_{\mathcal{O}} = 2\#\,\mathrm{SS}_{\mathcal{O}}^{\mathrm{pr}}$ by Theorem 4.4, and no stabilizers are contained in $\mathrm{Cl}(\mathcal{O})$ (whose action is free). Therefore, for some (E, ι), we have $\mathrm{Stab}_G((E, \iota)) = \{id, \sigma\tau\}$, for some $\sigma \in \mathrm{Cl}(\mathcal{O}) \cong \mathrm{Gal}(L/K)$. The set of stabilizers is the orbit of $\mathrm{Stab}_G((E, \iota))$ under conjugation by elements of $G \cong \mathrm{Cl}(\mathcal{O}) \rtimes \langle\tau\rangle$. For $\eta \in \mathrm{Cl}(\mathcal{O})$, we have

$$\eta\, \mathrm{Stab}_G((E, \iota))\eta^{-1} = \{id, \eta\sigma\tau\eta^{-1}\} = \{id, \eta^2\sigma\tau\}.$$

Also $\tau\, \mathrm{Stab}_G((E, \iota))\tau^{-1} = \{id, \tau\sigma\} = \{id, \sigma^{-2}\sigma\tau\}$. Thus, the nontrivial elements of the stabilizers, taken together, form a set $\mathrm{Cl}(\mathcal{O})^2\sigma\tau$. Set the notation $G_\sigma := \mathrm{Cl}(\mathcal{O})^2\sigma$. Then $\sigma \in \mathrm{Cl}(\mathcal{O}) \setminus \mathrm{Cl}(\mathcal{O})^2$, as τ does not stabilize any pairs (E, ι). Hence G_σ is a non-principal genus of $\mathrm{Cl}(\mathcal{O})$.

An element (E, ι) is in the same ℓ-isogeny rim as its conjugate $(E, \bar{\iota})$ if and only if $\sigma_\ell^k\tau \in \mathrm{Stab}_G((E, \iota))$ for some integer k if and only if $C_\ell\tau \cap \mathrm{Stab}_G((E, \iota)) = C_\ell\tau \cap G_\sigma\tau \neq \emptyset$. If $C_\ell \cap G_\sigma = \emptyset$, then there are no stabilizers intersecting $C_\ell\tau$, so there are no self-conjugate rims. Since $\rho(\mathcal{E}\ell\ell(\mathcal{O}))$ is taken to itself by conjugation, in this case all rims pair up in conjugate pairs. Therefore the fibres are of size $h_{\mathcal{O}}/r$.

Suppose there are self-conjugate rims, in which case $\sigma_\ell^k \in G_\sigma$ for some integer k. Then k must be odd, as otherwise $\sigma_\ell^k \in G_\sigma \cap \mathrm{Cl}(\mathcal{O})^2 = \emptyset$. It follows that $\sigma_\ell \in G_\sigma$. A similar argument forces r to be even, as otherwise $\sigma_\ell = \sigma_\ell^{r+1} \in G_\sigma \cap \mathrm{Cl}(\mathcal{O})^2 = \emptyset$.

There are $\#G_\sigma = \#\mathrm{Cl}(\mathcal{O})^2 = h_{\mathcal{O}}/g_{\mathcal{O}}$ distinct stabilizers, of which $\#G_\sigma \cap C_\ell$ belong to elements in the same rim as their conjugates. By the same logic as before, using the fact that r is even, the intersection of G_σ with C_ℓ consists of the odd powers of σ_ℓ and is therefore of size $\#C_\ell/2 = r/2$. Since all rims are the same size, this means the proportion of rims which are self-conjugate is

$$c := \frac{r/2}{\#\mathrm{Cl}(\mathcal{O})^2} = \frac{r g_{\mathcal{O}}}{2h_{\mathcal{O}}}.$$

This represents the proportion of rims (equivalently, of vertices) of $\mathrm{SS}_{\mathcal{O}}^{\mathrm{pr}}$ that are self-conjugate. This proportion is the same when we restrict to $\rho(\mathcal{E}\ell\ell(\mathcal{O}))$.

The proportion $1-c$ of non-self-conjugate rims is double-counted when we wish to count ℓ-isogeny cycles in $\mathcal{G}_\ell$, so we have

$$\epsilon_{\mathcal{O},\ell} = 2c + (1-c) = 1 + c.$$

For the final inequality, note that in the ramified case, the fibres of $\mathcal{R}_r \to \mathcal{I}_r$ are of size $2h_{\mathcal{O}}/r$ and reference diagram (9). □

Example 5.2 The following example illustrates the most complicated type of fibre. Consider $\mathcal{G}_\ell$ with $p = 241$, $\ell = 11$ and the quadratic order $\mathcal{O}$ of discriminant $-4p$, namely $\mathcal{O} = \mathbb{Z}[\sqrt{-p}]$. Since p ramifies, $\mathrm{SS}_{\mathcal{O}}^{\mathrm{pr}} = \rho(\mathcal{E}\ell\ell(\mathcal{O}))$ by Theorem 4.4. Let $r = 4$. The class group of $\mathcal{O}$ is cyclic of order 12. Therefore, there are $12/4 = 3$ rims of size 4 in $\mathrm{SS}_{\mathcal{O}}^{\mathrm{pr}}$ under the action of a prime ideal $\mathfrak{l}$ in $\mathcal{O}$ above ℓ. That means six directed rims. Using an isomorphism $(\mathrm{Cl}(\mathcal{O}), \cdot) \cong (\mathbb{Z}/12\mathbb{Z}, +)$ which takes σ_ℓ of order 4 to the element 3, we have $\mathrm{Cl}(\mathcal{O})^2 = \{0, 2, 4, 6, 8, 10\}$ and $C_\ell = \{0, 3, 6, 9\}$. So $1/3$ of the elements of $\mathrm{Cl}(\mathcal{O})^2$ are in C_ℓ. This implies that $1/3$ of the total directed rims are self-conjugate: two of the six (one of the three undirected rims). Using the formula of Theorem 5.1, with $g_{\mathcal{O}} = 2$, we obtain

$$\epsilon_{\mathcal{O},\ell} = 1 + \frac{4 \cdot 2}{2 \cdot 12} = \frac{4}{3},$$

which gives a fibre size of $\frac{4}{3} \cdot \frac{12}{4} = 4$ for the map $\mathcal{C}_4 \to \mathcal{I}_4$ above $\mathcal{O}$. In other words, there should be four directed isogeny cycles of size 4 (equivalently, two undirected isogeny cycles) in $\mathcal{G}_{11}$ which give rise to $\mathcal{O}$.

The set $\rho(\mathcal{E}\ell\ell(\mathcal{O}))$ has three undirected length 4 isogeny cycles: (64, 93, 216, 240) (twice, conjugate to one another) and (8, 8, 28, 28) (self-conjugate). The subgraph of $\mathcal{G}_{11}$ of $\mathbb{F}_p$-points is shown in Fig. 5, where we observe that there is only one 4-cycle through (964, 93, 216, 240), and that (8, 8, 28, 28) collapses to a 'barbell shaped' isogeny cycle. The commutative diagram (9), if labelled with the (undirected) elements above $\mathbb{Z}[\sqrt{-p}]$, looks something like the following (please compare to Fig. 5):

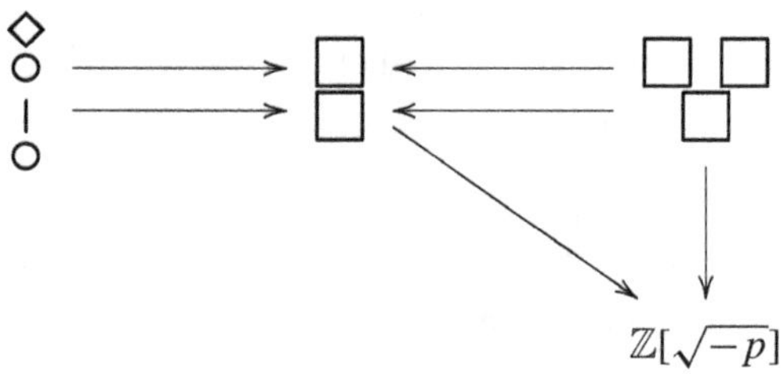

6 An Extended Example

In this section, we give an example that demonstrates our bijection (Theorem 3.2). We choose $p = 179$, $\ell = 2$, and we consider $r = 3$, 4, 5 and 6. Computing the set $\mathcal{R}_r$ involves first finding all ℓ-fundamental imaginary quadratic orders $\mathcal{O}$ such that $\mathcal{O}$ embeds into $B_{p,\infty}$, the ideal $(\ell) = \mathfrak{l}\bar{\mathfrak{l}}$ splits in $\mathcal{O}$ and $[\mathfrak{l}]$ has order r in $\mathrm{Cl}(\mathcal{O})$. A complete list of such orders $\mathcal{O}$ and prime ideals $\mathfrak{l}$, $\bar{\mathfrak{l}}$ can be calculated as follows:

(1) List all representations of ℓ^r by norm forms of sufficiently small discriminant.
(2) Each representation gives rise to an imaginary quadratic element α such that $N(\alpha) = \ell^r$ (up to sign and conjugation).

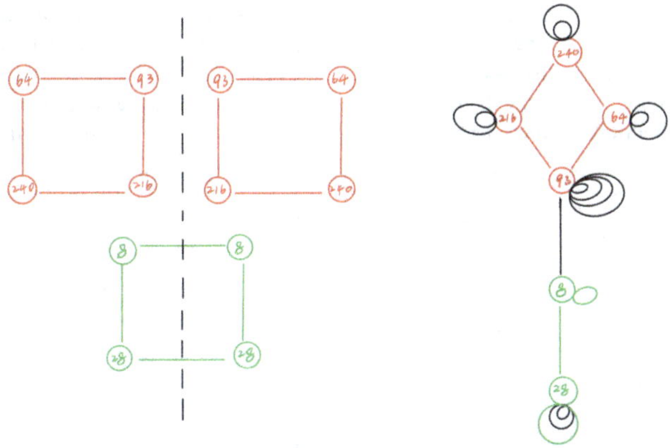

Fig. 5 An illustration of Example 5.2 with $p = 241$, $\ell = 11$ and $\mathcal{O} = \mathbb{Z}[\sqrt{-p}]$, for which $h_{\mathcal{O}} = 12$. On the left, the rims of $\mathcal{G}_{\mathbb{Q}(\sqrt{-p}),11}$ drawn in such a way that conjugation is a mirror symmetry through the dotted line. On the right, the subgraph of $\mathbb{F}_p$-vertices of $\mathcal{G}_{11}$

(3) Check the splitting behaviour of p in $\mathbb{Q}(\alpha)$, discarding α if p splits in $\mathbb{Q}(\alpha)$.
(4) Consider all ℓ-fundamental orders in $\mathbb{Q}(\alpha)$ that contain α and check whether the ideal class $[\mathfrak{l}]$ has order r in $\mathrm{Cl}(\mathcal{O})$, obtaining a list L_r of desirable orders $\mathcal{O}$.

As an example, consider representing $2^3 = 8$. We need only check discriminants Δ with $|\Delta| \leq 4 \cdot 8 = 32$. We find two representations: $1^2 + 31 \cdot 1^2 = 4 \cdot 8$ and $3^2 + 23 \cdot 1^2 = 4 \cdot 8$. These gives rise to two possible α up to conjugation and sign:

$$\frac{\pm 1 \pm \sqrt{-31}}{2}, \frac{\pm 3 \pm \sqrt{-23}}{2}.$$

However, 179 splits in $\mathbb{Q}(\sqrt{-23})$ and not in $\mathbb{Q}(\sqrt{-31})$. Hence we discard the second α. We find that $\mathfrak{l} = \left(2, \frac{1+\sqrt{-31}}{2}\right)$ is not principal and $\mathfrak{l}^3 = (\alpha)$. Hence $L_3 = \left\{\mathbb{Z}\left[\frac{1+\sqrt{-31}}{2}\right]\right\}$.

Each element α corresponds to an orientation ι_α (up to conjugation), which gives a rim (up to conjugation and direction), as in the map Θ of Lemma 3.9. Such a list therefore gives the number of undirected rims: $\#L_r = \frac{1}{2}\#(\mathcal{R}_r/\sim)$.

Theorem 3.2 concerns directed isogeny cycles and rims, but in this example we will ignore the issue of direction, which is allowed by Corollary 3.18. Therefore, we use L_r as a proxy for $\mathcal{R}_r/\sim$ for computational purposes. We compare L_r with $\mathcal{C}_r$, verifying the bijection theorem.

For each $r \in \{3, 4, 5, 6\}$, we compute the set $\mathcal{C}_r$ on the supersingular 2-isogeny graph by reference to Fig. 5 showing $\mathcal{G}_2$. We will actually consider $\mathcal{C}'_r$, the set of isogeny cycles of $\mathcal{C}_r$ but forgetting direction, allowed by Corollary 3.18. Then we determine the associated endomorphisms by composing around the cycles as in the

Table 1 Cycles of lengths three through six, with the associated endomorphisms to which the cycles compose. The two cycles labelled with an asterisk $*$ are not fully specified by giving their j-invariants. However, as discussed in Sect. 3.2, the bijection between these two cycles and the two associated α is not canonical, but if we choose a safe arbitrary assignment, we can make a non-canonical bijection. So there is no reason to specify which cycle is which in the table here

Isogeny cycle	Figure	Length	Endomorphism	Order	Class number
$(j_3, \overline{j_3}, 171)$	Fig. A	3	$\frac{\pm 1 \pm \sqrt{-31}}{2}$	$\mathbb{Z}\left[\frac{1+\sqrt{-31}}{2}\right]$	3
$(61, j_1, 140, \overline{j_1})$	Fig. A	4	$\frac{\pm 5 \pm \sqrt{-39}}{2}$	$\mathbb{Z}\left[\frac{1+\sqrt{-39}}{2}\right]$	4
$(22, \overline{j_2}, \overline{j_3}, j_3, j_2)$	Fig. A	5	$\frac{\pm 9 \pm \sqrt{-47}}{2}$	$\mathbb{Z}\left[\frac{1+\sqrt{-47}}{2}\right]$	5
$(22, \overline{j_2}, \overline{j_1}, 140, j_1, j_2)$	Fig. B	6	$\frac{\pm 13 \pm \sqrt{-87}}{2}$	$\mathbb{Z}\left[\frac{1+\sqrt{-87}}{2}\right]$	6
$(140, j_1, j_2, j_3, 171, 120)$ $(140, \overline{j_1}, \overline{j_2}, \overline{j_3}, 171, 120)$	Fig. C Fig. D	6	$\frac{\pm 5 \pm \sqrt{-231}}{2}$	$\mathbb{Z}\left[\frac{1+\sqrt{-231}}{2}\right]$	12
$(0, 121, 112, 35, 112, 121)^*$	Fig. E	6	$\frac{\pm 3 \pm \sqrt{-247}}{2}$	$\mathbb{Z}\left[\frac{1+\sqrt{-247}}{2}\right]$	6
$(22, j_2, j_3, 171, \bar{j}_3, \bar{j}_2)$ $(0, 121, 112, 35, 112, 121)^*$	Fig. F Fig. E	6	$\frac{\pm 1 \pm \sqrt{-255}}{2}$	$\mathbb{Z}\left[\frac{1+\sqrt{-255}}{2}\right]$	12
$(61, j_1, j_2, 22, \overline{j_2}, \overline{j_1})$	Fig. G	6	$\frac{\pm 11 \pm 3\sqrt{-15}}{2}$	$\mathbb{Z}\left[3\left(\frac{1+\sqrt{-15}}{2}\right)\right]$	6

map Θ; we do this by computing the trace of the composed endomorphism θ using the formulas $\theta + \widehat{\theta}$ or $1 + \deg(\theta) - \deg(1 - \theta)$ in SageMath [53].

For example, there is only one isogeny cycle of length 3 (in the sense of Theorem 3.2, but ignoring direction) in $\mathcal{C}_3'$ for $\mathcal{G}_2$, namely $(j_3, \overline{j_3}, 171)$. Composing around the triangle, one finds an endomorphism of trace ± 1. Since the endomorphism has norm 8, it has minimal polynomial $x^2 \pm x + 8$ and hence is of the form $\frac{\pm 1 \pm \sqrt{-31}}{2}$. This generates an order with class number 3 (the class numbers were computed with SageMath).

In Table 1, we give the lists $\mathcal{C}_r'$, and L_r and put them in correspondence. This represents the first row of the table which puts $\mathcal{C}_3' = \{(j_3, \overline{j_3}, 171)\}$ and $L_3 = \left\{\mathbb{Z}\left[\frac{1+\sqrt{-31}}{2}\right]\right\}$ (as a proxy for $\mathcal{R}_3/\sim$ modulo direction) in bijection.

The table also verifies Corollary 7.3. In our case, we have observed no fields ramified at p, so $\epsilon_{\mathcal{O},\ell} = 2$. We compute

$$2 = 2\#\mathcal{C}_3' = \#\mathcal{C}_3 = \frac{2}{3} h\left(\mathbb{Z}\left[\frac{1+\sqrt{-31}}{2}\right]\right) = \frac{2}{3} \cdot 3 = 2.$$

$$2 = 2\#\mathcal{C}_4' = \#\mathcal{C}_4 = \frac{2}{4} h\left(\mathbb{Z}\left[\frac{1+\sqrt{-39}}{2}\right]\right) = \frac{2}{4} \cdot 4 = 2.$$

$$2 = 2\#\mathcal{C}_5' = \#\mathcal{C}_5 = \frac{2}{5} h\left(\mathbb{Z}\left[\frac{1+\sqrt{-47}}{2}\right]\right) = \frac{2}{5} \cdot 5 = 2.$$

$$
\begin{aligned}
14 = 2\#C_6' = \#C_6 = \frac{2}{6}\Bigg(& h\left(\mathbb{Z}\left[\frac{1+\sqrt{-87}}{2}\right]\right) + h\left(\mathbb{Z}\left[\frac{1+\sqrt{-231}}{2}\right]\right) + \\
& h\left(\mathbb{Z}\left[\frac{1+\sqrt{-247}}{2}\right]\right) + h\left(\mathbb{Z}\left[\frac{1+\sqrt{-255}}{2}\right]\right) + \\
& h\left(\mathbb{Z}\left[3\frac{1+\sqrt{-15}}{2}\right]\right)\Bigg) \\
= \frac{2}{6} \cdot & (6+12+6+12+6) = 14.
\end{aligned}
$$

This verifies Corollary 7.3 in this example.

The Cycle $(0, 121, 112, 35, 112, 121)$ The case of the j-invariant cycle $(0, 121, 112, 35, 112, 121)$ in Fig. 6(E) is particularly interesting. This sequence of j-invariants forms a 6-cycle in 12 possible ways, as follows. There are 4 ways to travel from 112 to 35 and back. There are 3 ways to travel from 121 to 0 and back, producing three distinct endomorphisms of $j = 121$ (but not canonically), each composing to one of ± 2, $\pm\widehat{\varphi} \circ \zeta_3 \circ \varphi$ or $\pm\widehat{\varphi} \circ \zeta_3^2 \circ \varphi$, as described in Fig. 3. Thus, one endomorphism is backtracking and the other two are duals of each other. The isogenies from 112 to 35 each have a dual: Hence two of the 2-cycles from 112 to 35 and back are backtracking (obtained by pairing an isogeny with its dual). There are therefore two non-backtracking 2-cycles to choose from among the four ways to travel from 112 to 35 and back. These two non-backtracking 2-cycles are actually duals of each other and are obtained by simply changing the direction of traversal. If the traces of the paths that include backtracking are computed, we obtain the associated endomorphisms ± 8, $4(\pm 1 \pm \sqrt{-3})$ and $2(\pm 1 \pm \sqrt{-15})$, all of which are divisible by 2. This leaves exactly four non-backtracking 6-cycles on the j-invariants $(0, 121, 112, 35, 112, 121)$, coming in pairs obtained by changing the direction of traversal. These two pairs give the two endomorphisms listed in Table 1.

The Cycle $(61, j_1, j_2, 22, \overline{j_2}, \overline{j_1})$ Another interesting cycle is $(61, j_1, j_2, 22, \overline{j_2}, \overline{j_1})$ in Fig. 6(G). This cycle corresponds to an element generating the non-maximal order $\mathcal{O}_3 := \mathbb{Z}\left[3\left(\frac{1+\sqrt{-15}}{2}\right)\right]$. That element also exists in the order $\mathcal{O}_1 := \mathbb{Z}\left[\frac{1+\sqrt{-15}}{2}\right]$. How do we know which order the volcano rim corresponds to? The easiest answer is that only the class number of $\mathcal{O}_3$ is divisible by 6. However, even without that observation (sometimes several class numbers may be divisible by the correct integer r), we can observe that the orientation ι given by the endomorphism on any of the j-invariants on the cycle is $\mathcal{O}_3$-primitive, not $\mathcal{O}_1$-primitive. This can be seen because if $\mathrm{End}(E)$ contained $\frac{1+\sqrt{-15}}{2}$, which has norm 16, then this element would be realized as a 4-cycle from E, which it is not. (The curves $j_2, \overline{j_2}$ and 22 have no 4-cycle. The curves 61, j_1 and $\overline{j_1}$ have the 4-cycle associated with $\frac{\pm 5 \pm \sqrt{-39}}{2}$.) The moral of this observation is that each cycle can only arise from a single quadratic order, as dictated by Theorem 3.2.

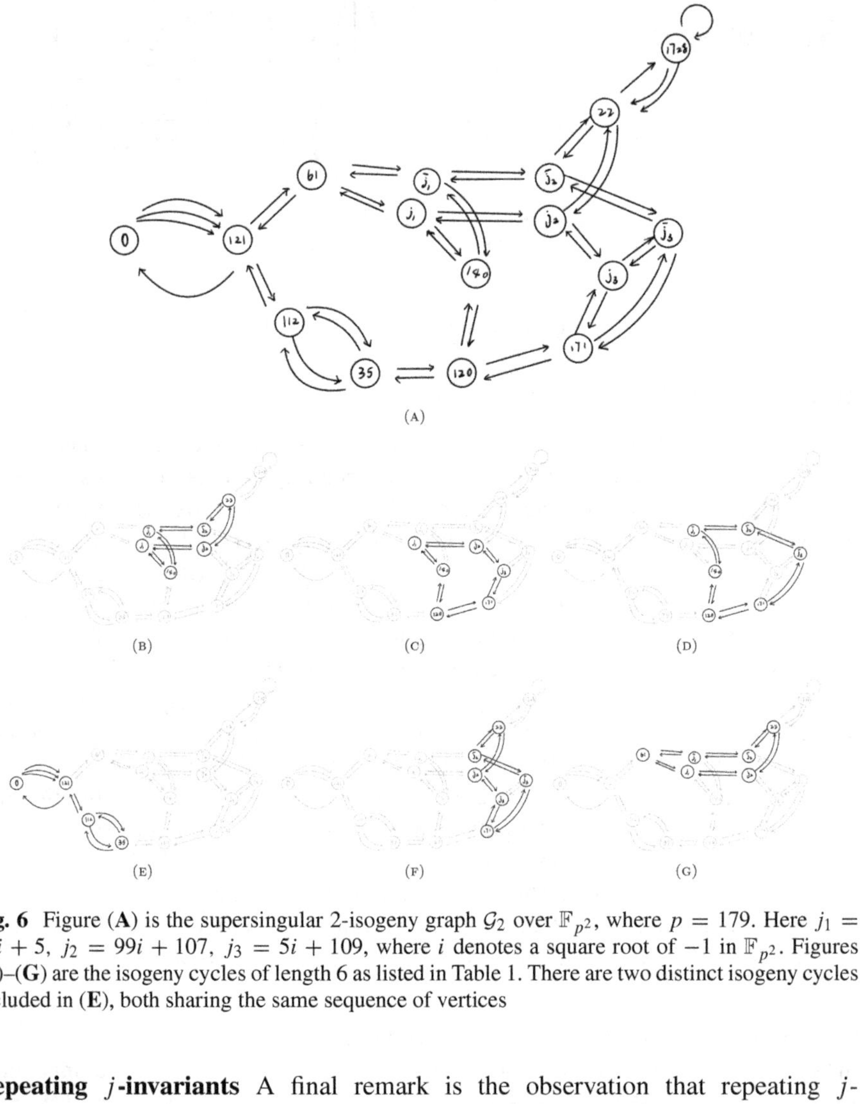

Fig. 6 Figure (**A**) is the supersingular 2-isogeny graph $\mathcal{G}_2$ over $\mathbb{F}_{p^2}$, where $p = 179$. Here $j_1 = 64i + 5$, $j_2 = 99i + 107$, $j_3 = 5i + 109$, where i denotes a square root of -1 in $\mathbb{F}_{p^2}$. Figures (**B**)–(**G**) are the isogeny cycles of length 6 as listed in Table 1. There are two distinct isogeny cycles included in (**E**), both sharing the same sequence of vertices

Repeating j-invariants A final remark is the observation that repeating j-invariants in a cycle can only happen when the discriminant is large with respect to the prime. This is a consequence of a result by Kaneko which gives the necessary condition $|\operatorname{disc}(\mathcal{O})| \geq p$ [30, Theorem 2']. In our case, we do get a cycle with repeating j-invariants, but the associated discriminants are quite large at -247 and -255, both bigger in absolute value than the prime $p = 179$.

Comparison with Gross [28] Working with the same example, Gross counts cycles of length $\log_\ell(m)$ in the supersingular ℓ-isogeny graph $\mathcal{G}_\ell$ via the formula provided in [28, Proposition 1.9]. In the first row of Table 1, we count one isogeny cycle of length 3 in the 2-isogeny graph, namely $(j_3, \overline{j_3}, 171)$ arising from the order of discriminant -31. Gross counts nine via the formula [28, Proposition 1.9]. In particular, six of this count come from the order of discriminant -31. The 3-cycle

$(j_3, \overline{j_3}, 171)$, which we count once, is counted six times by Gross because it is a valid 3-cycle starting at any of j_3, 171, $\overline{j_3}$, in either direction. Two more of Gross's count come the order of discriminant -16. The cycle corresponding to the order of discriminant -16 is necessarily (22, 1728, 1728). This cycle is only without backtracking when we start at 22, so Gross counts it once in each direction. We do not count this cycle, as it contains backtracking as soon as we forget the basepoint. The last one of Gross's count comes from the order of discriminant -4, as 1728 has an endomorphism of degree 8 corresponding to composing the loop at 1728 three times, with appropriate automorphisms in between to avoid backtracking.

7 Counting Isogeny Cycles in $\mathcal{G}_\ell$

Theorem 5.1 makes possible explicit formulas for the number of isogeny cycles of length r in the ℓ-isogeny graph $\mathcal{G}_\ell$. In this section, we do the following:

1. Theorem 7.1 uses spectral graph theory to give an asymptotic for the expected number of length r isogeny cycles, and this estimate is compared to empirical data in Fig. 7. The use of spectral graph theory requires us to restrict to $p \equiv 1 \pmod{12}$, but for any large enough p we still expect the estimates of this theorem to hold.
2. Corollary 7.3 gives an explicit formula for the number of length r isogeny cycles in terms of class numbers of certain quadratic orders, and Theorem 7.4 reformulates this as a sum over these class numbers with simpler conditions on which quadratic orders to include.
3. Theorem 7.4 enables an explicit upper bound on the length r isogeny cycle count, given in Corollary 7.7. Unfortunately a lower bound would seem to be connected to an open problem in number theory (Remark 7.8).

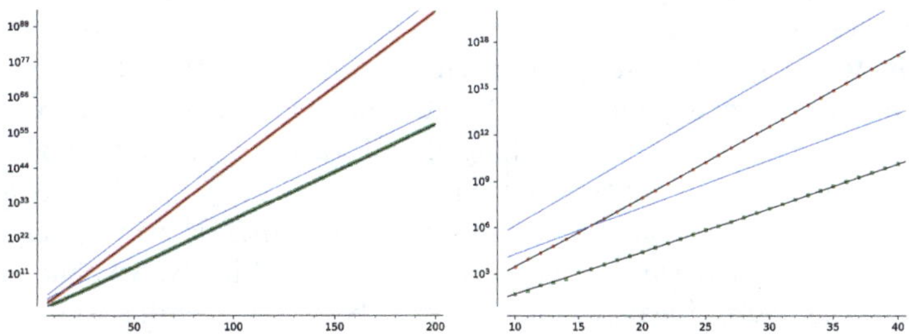

Fig. 7 A semilog plot of cycle counts for $\mathcal{G}_\ell$ for $p = 3361$ and $\ell = 2$ (green squares, lower graph) and for $p = 3229$ and $\ell = 3$ (red circles, upper graph). The black curves upon which the data points appear to sit are $\ell^x/(2x)$. The blue lines above the data in each case indicate the dominant term of Corollary 7.7, indicating our proved upper bound. On the left, the full data. On the right, a close-up of the behaviour for small cycle sizes. The graphs were generated in part using code from [4]

7.1 Asymptotic Number of Isogeny Cycles

It is a standard fact that for $(\ell + 1)$-regular Ramanujan graphs, the number of basepointed non-backtracking closed walks of length r (where the non-backtracking condition does not rule out entering the final vertex along the initial edge of the walk) is

$$(\ell + 1)\ell^{r-1} + \#\mathcal{G}_\ell \, O\left(r(\ell + 1)^{r/2}\right),$$

where the implied constant does not depend on $\#\mathcal{G}_\ell$. This follows, for example, from [19, Theorem 1.4.6]; Some details are provided at [8]. If we disallow entering the final vertex along the initial edge and then forget basepoints, then one would heuristically expect this to become

$$\ell^r/2r + \#\mathcal{G}_\ell \, O\left((\ell + 1)^{r/2}\right). \tag{10}$$

For small r, estimating the number of graph-theoretic r-cycles is similar (recall that a cycle allows no repeated edges or vertices); this is because random d-regular graphs on n vertices have on average $(d-1)^r/2r$ such r-cycles [9, Theorem 2]. In our case, this would give $\ell^r/2r$. The number of r-cycles in some Ramanujan graphs, including LPS graphs, but not supersingular graphs, has been analysed, and these graphs appear to behave as random graphs [54].

In what follows, we give a brief proof of an asymptotic version of (10).

Theorem 7.1 *Let $p \equiv 1 \pmod{12}$. Let $\mathcal{G}_\ell$ be the supersingular ℓ-isogeny graph over $\overline{\mathbb{F}}_p$ and let $\#\mathcal{G}_\ell$ denote its number of vertices. The number of non-backtracking closed walks of length r in $\mathcal{G}_\ell$, taken up to order of traversal and starting point, asymptotically approaches $\ell^r/2r$ as $r \to \infty$. Thus, the number of isogeny cycles in $\mathcal{G}_\ell$ is also asymptotically $\ell^r/2r$ as $r \to \infty$.*

Before giving a formal proof, we give a heuristic description. Choose one of $\#\mathcal{G}_\ell$ initial vertices j_0. Label its $\ell + 1$ neighbours $j_1, j_2, \ldots, j_{\ell+1}$. Choose one of these neighbours for the first step, say j_1 (relabelling without loss of generality). Consider the ℓ^{r-1} non-backtracking length $r - 1$ walks that could follow the edge $j_0 \to j_1$. Such a walk provides a non-backtracking closed walk of length r through $j_0 \to j_1$ if and only if the walk terminates at j_0 via one of $j_2, \ldots, j_{\ell+1}$. The probability of this termination vertex j_0 should be expected to be $1/\#\mathcal{G}_\ell$. However, only ℓ of the $\ell + 1$ incoming edges reach j_0 via one of $j_2, \ldots, j_{\ell+1}$. Finally, we obtain a typical closed non-backtracking walk of length r exactly $2r$ times in this way since there are r choices of initial vertex and two directions. Hence the number of closed non-backtracking walks of length r should be obtained by combining the steps in the outline above:

$$\#\mathcal{G}_\ell(\ell + 1)\ell^{r-1}(1/\#\mathcal{G}_\ell)(\ell/(\ell + 1))/2r = \ell^r/2r.$$

Proof of Theorem 7.1 The natural object for studying non-backtracking walks is a Markov chain on the *non-backtracking matrix* B, which is the adjacency matrix of an auxiliary graph $\mathcal{G}'$ whose vertices are the directed edges of $\mathcal{G}_\ell$ and whose edges indicate adjacency between edges of $\mathcal{G}_\ell$ that respects the directedness and does not backtrack. In other words, the matrix is a scaling of the transition probability matrix of a non-backtracking random walk on the directed edges of $\mathcal{G}_\ell$. In particular, the entries of B^r give the number of non-backtracking walks of length $r + 1$ that start and end on certain edges. This is studied in [31], where it is shown that this Markov chain converges to a uniform stationary distribution (the mixing property). There are $\#\mathcal{G}_\ell(\ell+1)$ directed edges in $\mathcal{G}_\ell$. Choose one of these as the initial directed edge of a walk, originating at a vertex which we may call j_0. From that point, take a random non-backtracking walk of length $r - 1$ (there are ℓ^{r-1} of these). Then we have created a closed non-backtracking walk if and only if we end on one of the ℓ edges that are directed into j_0 without backtracking the initial edge. The mixing property implies that the probability of ending on of these ℓ incoming edges is asymptotically $\ell/(\#\mathcal{G}_\ell(\ell + 1))$. Therefore, asymptotically, we obtain

$$\#\mathcal{G}_\ell(\ell + 1)\ell^{r-1}\left(\frac{\ell}{\#\mathcal{G}_\ell(\ell + 1)}\right) = \ell^r$$

total non-backtracking closed walks with a direction and a basepoint. Most of these can be directed in two ways and a basepoint can be chosen in r ways. The exceptions, namely the barbells (Definition 3.19) are those closed walks that involve the same edges when the direction is reversed. These exceptions are rare enough (Lemma 3.20) that they do not affect the asymptotic. Finally, isogeny cycles are those non-backtracking closed walks which are not powers of smaller non-backtracking closed walks; this requirement clearly also does not affect the asymptotic. □

Remark 7.2 In Theorem 7.1, we restrict to $p \equiv 1 \pmod{12}$ because the graph theory involved requires an $(\ell + 1)$-regular graph. We have remarked at length the difficulties which arise when vertices $j = 0, 1728$ appear in $\mathcal{G}_\ell$. However, these issues are very localized – the $(\ell + 1)$-regularity of the graph may only be affected at vertices adjacent to $j = 0, 1728$. For $p \gg \ell$, this is a very small number of potential deviations from regularity, and we do not expect the estimates provided here to be greatly affected.

By reference to [31] and [2], the mixing rate of the Markov chain could be used to analyse the rate of convergence of the asymptotic and make this theorem more precise.

The proof of Theorem 7.1 lends itself to computation: Using powers of the non-backtracking matrix, we can compute the number of isogeny cycles in $\mathcal{G}_\ell$, at least up to the issue of barbells. Figure 7 shows some data collected in SageMath for the number of non-backtracking closed walks of length r on some small supersingular ℓ-isogeny graphs chosen to avoid loops (so $\mathcal{G}_\ell$ is $(\ell+1)$-regular without any barbells) and its agreement with the asymptotic.

7.2 Counting Isogeny Cycles with Class Numbers

A corollary to Theorems 3.2 and 5.1 counts cycles exactly as a sum of class numbers of certain quadratic orders.

Corollary 7.3 *Let $r > 2$ be fixed and consider primes $\ell \neq p$. Let $\mathcal{I}_r$ be as in Theorem 5.1.*

Let $c_r = |\mathcal{C}_r|$ denote the number of directed isogeny cycles of length r. Then

$$c_r = \frac{1}{r} \sum_{\mathcal{O} \in \mathcal{I}_r} \epsilon_{\mathcal{O},\ell} h_{\mathcal{O}},$$

where $\epsilon_{\mathcal{O},\ell}$ is as defined in Theorem 5.1. In particular, $1 \leq \epsilon_{\mathcal{O},\ell} \leq 2$ and $\epsilon_{\mathcal{O},\ell} = 2$ if p is inert in the quadratic field containing $\mathcal{O}$. If $\ell^r < p$, then $\epsilon_{\mathcal{O},\ell} = 2$ for all $\mathcal{O} \in \mathcal{I}_r$.

Proof This follows from Theorems 3.2 and 5.1. The cycles $\mathcal{C}_r$ of length r in $\mathcal{G}_\ell$ are in bijection with $\mathcal{R}_r/\sim$. Each $[R]$ then maps to some order $\mathcal{O} \in \mathcal{I}_r$. Summing the elements in $\mathcal{I}_r$ weighted by the fibre sizes counts $\#\mathcal{C}_r$. The fibre sizes are given in Theorem 5.1. □

On the graph side, it is clear that this count is finite. On the side of ideal classes, finiteness of $\mathcal{I}_r$ follows from the observation that if $\operatorname{ord}([\mathfrak{l}]) = r$, then ℓ^r is represented by the norm form of the quadratic field. For sufficiently large discriminant, this representation is impossible except as $\ell^r = (\ell^{r/2})^2$ if r is even; but in this case, ℓ is ramified. This observation can be turned into a counting argument based on quadratic residues.

Theorem 7.4 *Assume $p > 2$. For any positive integer N, define*

$$\mathcal{Q}(N) = \left\{ 0 < x < 2\ell^{N/2} : \begin{array}{r} x \not\equiv 0 \pmod{\ell}, \\ x^2 - 4\ell^N \text{ is not a quadratic residue modulo } p, \\ \text{and has valuation } \leq 1 \text{ at } p \end{array} \right\},$$

and

$$Q_N = \sum_{x \in \mathcal{Q}(N)} \sum_{f^2 | (x^2 - 4\ell^N)} \epsilon_{\mathcal{O},\ell}\, h\left(\frac{x^2 - 4\ell^N}{f^2}\right),$$

where $h(D)$ denotes the class number of the order of discriminant D. Let c_N, $N \geq 3$, be as in Corollary 7.3. Then there are values for c_1 and c_2 such that

$$\sum_{r|N} r c_r = Q_N, \quad c_N = \frac{1}{N} \sum_{r|N} \mu(r) Q_{N/r}.$$

Proof Let $r \geq 3$. Recall from Theorems 3.2 and 5.1 that we have maps

$$\mathcal{C}_r \longrightarrow \mathcal{R}_r/\sim \longrightarrow \mathcal{I}_r.$$

The first map is a bijection, and the second map has fibre sizes $\epsilon_{\mathcal{O},\ell} h_{\mathcal{O}}/r$ above $\mathcal{O}$, as described in Theorem 5.1. Using the fibre sizes, we obtain

$$c_N = \frac{1}{N} \sum_{\mathcal{O} \in \mathcal{I}_N} \epsilon_{\mathcal{O},\ell} h_{\mathcal{O}}.$$

Then

$$\sum_{r|N} r c_r = \sum_{\mathcal{O} \in \bigcup_{r|N} \mathcal{I}_r} \epsilon_{\mathcal{O},\ell} h_{\mathcal{O}}.$$

By definition, $\mathcal{O} \in \bigcup_{r|N} \mathcal{I}_r$ if and only if there exists an imaginary quadratic integer $\alpha \in \mathcal{O}$, not divisible by ℓ, such that $N(\alpha) = \ell^r$, p does not split in $\mathbb{Q}(\alpha)$, and ℓ is invertible and splits in $\mathcal{O}$. Since these properties of α are invariant up to conjugation and sign, we can write $\alpha = \frac{x+\sqrt{\Delta}}{2}$, where Δ is the discriminant of the order $\mathbb{Z}[\alpha]$, and $x \geq 0$. The condition that α has norm ℓ^N is equivalent to $x^2 - \Delta = 4\ell^N$. The fact that α is imaginary is equivalent to $0 > \Delta = x^2 - 4\ell^N$, or equivalently, $x < 2\ell^{N/2}$. Next we show that the prime ℓ is invertible and splits in $\mathbb{Z}[\alpha]$ if and only if $x \not\equiv 0 \pmod{\ell}$. To that end, invertibility is equivalent to the condition that ℓ does not divide the conductor of $\mathbb{Z}[\alpha]$, which is equivalent to $x \not\equiv 0 \pmod{\ell}$ (since $x^2 - \Delta = 4\ell^N$), and if these equivalent conditions hold, then $1 = \left(\frac{\Delta}{\ell}\right) = \left(\frac{x^2-4\ell^N}{\ell}\right)$, which holds if and only if ℓ splits in the maximal order containing $\mathcal{O}$ (by Lemma 2.28). The prime p does not split in this maximal order if and only if $x^2 - 4\ell^N$ is not a quadratic residue modulo p. Finally, α is not divisible by ℓ if and only if ℓ does not divide Δ. To conclude then, $\mathcal{O} \in \bigcup_{r|N} \mathcal{I}_r$ if and only if $\Delta_{\mathcal{O}} = x^2 - 4\ell^N$ for $x \in \mathcal{Q}(N)$ of the theorem statement. Note that each $\mathcal{O}$ corresponds to one $\Delta_{\mathcal{O}}$, which in turn corresponds to one value of x. Therefore,

$$\sum_{r|N} r c_r = \sum_{\mathcal{O} \in \bigcup_{r|N} \mathcal{I}_r} \epsilon_{\mathcal{O},\ell} h_{\mathcal{O}} = \sum_{x \in \mathcal{Q}(N)} \sum_{f^2 | x^2 - 4\ell^N} \epsilon_{\mathcal{O},\ell} h\left(\frac{x^2 - 4\ell^N}{f^2}\right) = \mathcal{Q}_N.$$

Calling upon Möbius inversion, we have proved the theorem. □

Remark 7.5 The key observation we make here—that small cycles arise from a combination of splitting behaviour between p and ℓ—is used in [15, Section 5.3.4] to eliminate small cycles by use of congruence conditions on p.

The advantage of Theorem 7.4 is that in contrast to the sum of class numbers in Corollary 7.3, we obtain a sum of class numbers that has no conditions on the class groups. The condition on the order of $[\mathfrak{l}]$ present in Corollary 7.3 is removed when summing over $x \in \mathcal{Q}(N)$, since $\mathcal{Q}_N$ is entirely given in terms of 'elementary' conditions, namely Legendre symbols and valuations at p and ℓ.

Example 7.6 We illustrate the theorem and its proof with the example of Sect. 6, where $\ell = 2$ and $p = 179$. We have, from Table 1, the directed isogeny cycle counts

$$c_3 = 2,\ c_4 = 2,\ c_5 = 2,\ c_6 = 14.$$

Using Sage to compute $\mathcal{Q}(N)$, we obtain, showing the data as tuples $(x, x^2 - 4\ell^n, h_{x^2-4\ell^n})$,

$$\mathcal{Q}(1) = \emptyset,\ \mathcal{Q}(2) = \{(1, -15, 2)\},\ \mathcal{Q}(3) = \{(1, -31, 3)\},$$
$$\mathcal{Q}(4) = \{(5, -39, 4), (7, -15, 2)\},\ \mathcal{Q}(5) = \{(9, -47, 5)\},$$

$$\mathcal{Q}(6) = \{(1, -255, 12), (3, -247, 6), (5, -231, 12), (11, -135, 6), (13, -87, 6), (15, -31, 3)\}.$$

Observe that the last tuple $(15, -31, 3)$ in $\mathcal{Q}(6)$ corresponds to $\alpha = \frac{\pm 15 \pm \sqrt{-31}}{2} = \pm \left(\frac{1 \pm \sqrt{-31}}{2}\right)^2$. The element $\beta = \frac{\pm 1 \pm \sqrt{-31}}{2}$ corresponds to the unique 3-cycle in Table 1. Note that $\mathbb{Z}[\alpha] = \mathbb{Z}[\beta] = \mathcal{O}_{\mathbb{Q}(\sqrt{-31})}$. Accordingly, α corresponds to the 3-cycle repeated twice.

The fourth tuple $(11, -135, 6)$ in $\mathcal{Q}(6)$ corresponds to $\alpha = \frac{\pm 11 \pm 3\sqrt{-15}}{2} = \left(\frac{\pm 1 \pm \sqrt{-15}}{2}\right)^3$. The element $\frac{\pm 1 \pm \sqrt{-15}}{2}$ corresponds to the 2-cycle $(112, 35)$. In this case, $\mathbb{Z}[\beta] = \mathcal{O}_{\mathbb{Q}(\sqrt{-15})}$, but $\mathbb{Z}[\alpha]$ is a suborder of $\mathbb{Z}[\beta]$ of conductor 3. Thus, α corresponds to two different types of 6-cycles: the 2-cycle repeated three times (therefore not an isogeny cycle), and also an isogeny cycle of size 6, as in Table 1.

Therefore,

$$Q_1 = 0,$$

$$Q_2 = 2h\left(\mathbb{Z}\left[\frac{1+\sqrt{-15}}{2}\right]\right) = 4,$$

$$Q_3 = 2h\left(\mathbb{Z}\left[\frac{1+\sqrt{-31}}{2}\right]\right) = 6,$$

$$Q_4 = 2h\left(\mathbb{Z}\left[\frac{1+\sqrt{-15}}{2}\right]\right) + 2h\left(\mathbb{Z}\left[\frac{1+\sqrt{-39}}{2}\right]\right) = 4 + 8 = 12,$$

$$Q_5 = 2h\left(\mathbb{Z}\left[\frac{1+\sqrt{-47}}{2}\right]\right) = 10,$$

$$Q_6 = 2h\left(\mathbb{Z}\left[\frac{1+\sqrt{-255}}{2}\right]\right) + 2h\left(\mathbb{Z}\left[\frac{1+\sqrt{-247}}{2}\right]\right) +$$
$$2h\left(\mathbb{Z}\left[\frac{1+\sqrt{-231}}{2}\right]\right) + 2h\left(\mathbb{Z}\left[\frac{1+\sqrt{-15}}{2}\right]\right) +$$
$$2h\left(\mathbb{Z}\left[3\frac{1+\sqrt{-15}}{2}\right]\right) + 2h\left(\mathbb{Z}\left[\frac{1+\sqrt{-87}}{2}\right]\right) +$$
$$2h\left(\mathbb{Z}\left[\frac{1+\sqrt{-31}}{2}\right]\right) = 24 + 12 + 24 + 4 + 12 + 12 + 6 = 94.$$

Recall that for the Möbius function, $\mu(1) = 1$, $\mu(2) = -1$, $\mu(3) = -1$, $\mu(4) = 0$, $\mu(5) = -1$, $\mu(6) = 1$. So the sum $S(N) := \frac{1}{N}\sum_{r|N} \mu(r)Q_{N/r}$ takes on the following values, which we can verify to match c_N for $3 \leq N \leq 6$:

$$S(1) = Q_1 = 0, \quad S(2) = \frac{1}{2}(Q_2 - Q_1) = 2, \quad S(3) = \frac{1}{3}(Q_3 - Q_1) = 2 = c_3,$$

$$S(4) = \frac{1}{4}(Q_4 - Q_2) = 2 = c_4, \quad S(5) = \frac{1}{5}(Q_5 - Q_1) = 2 = c_5,$$

$$S(6) = \frac{1}{6}(Q_6 - Q_3 - Q_2 + Q_1) = 14 = c_6.$$

We briefly confirm that the conclusion of Theorem 7.4 is in heuristic agreement with Theorem 7.1. Heuristically, we expect $\#\mathcal{Q}(N) \approx \ell^{N/2}$. Estimating the inner sum in Q_N with a Hurwitz class number heuristically of size $\sqrt{4\ell^N - x^2}$ (ignoring log factors), we have

$$Q_N \approx \sum_{x=1}^{\lfloor 2\ell^{N/2} \rfloor} \sqrt{4\ell^N - x^2} \approx \int_1^{2\ell^{N/2}} \sqrt{4\ell^N - x^2}\, dx \approx \ell^N,$$

from which follows $c_N \approx \ell^N/N$ as expected. In fact, a rigorous upper bound of magnitude $\ell^N \log N$ on c_N is given in Corollary 7.7, whose proof is based on this rough idea.

Corollary 7.7 *With the notation of Theorem 7.4,*

$$Q_N < B_N := \frac{2}{3}(e^\gamma \log\log(2\ell^{N/2}) + 7/3)\log(4\ell^N)(\pi\ell^N + 2\ell^{3N/4})$$

and

$$\begin{aligned} c_N &< \frac{B_N}{N} + \left(e^\gamma \log\log N + 7/3 - \frac{1}{N}\right) B_{N/2} \\ &< \frac{2\pi e^\gamma \log(4\ell)}{3}\ell^N(\log N + \log\log(2\sqrt{\ell}) + 7/3) + O(\ell^{3N/4}\log N) \text{ as } N \to \infty, \end{aligned}$$

where $\gamma = 0.577\ldots$ is the Euler–Mascheroni constant.

Proof For brevity, put $X = \lfloor 2\ell^{N/2} \rfloor$. For our bound, we simply ignore all the conditions on x imposed by membership in the set $\mathcal{Q}_N$ of Theorem 7.4 and use $\epsilon_{\mathcal{O},\ell} \le 2$. That is, we upper-bound Q_N by $Q'_N \ge Q_N$, where

$$Q'_N = 2\sum_{x=1}^{X} \sum_{f^2 | x^2 - 4\ell^N} h\left(\frac{x^2 - 4\ell^N}{f^2}\right).$$

Fix an integer x with $0 \le x \le X$ and write $4\ell^N - x^2 = y^2 d$ for positive integers d, y with d square-free. For any imaginary quadratic order $\mathcal{O}$ of discriminant Δ, we have $h_\mathcal{O} < (\sqrt{|\Delta|}\log|\Delta|)/3$ by [39, Equation (8.11)], so

$$\begin{aligned} \sum_{f^2 | (x^2 - 4\ell^N)} h\left(\frac{x^2 - 4\ell^N}{f^2}\right) &< \frac{1}{3}\sum_{f^2 | (x^2 - 4\ell^N)} \frac{\sqrt{4\ell^N - x^2}}{f} \log\left(\frac{4\ell^N - x^2}{f}\right) \\ &< \frac{1}{3}\log(4\ell^N)\sum_{f^2 | y^2 d} \frac{\sqrt{y^2 d}}{f} \\ &= \frac{1}{3}\log(4\ell^N)\sqrt{y^2 d}\sum_{f | y}\frac{1}{f} \\ &= \frac{1}{3}\sqrt{4\ell^N - x^2}\,\log(4\ell^N)\,\frac{\sigma(y)}{y}, \end{aligned}$$

where $\sigma(\cdot)$ denotes the sum of divisors function. It is known (see [3, Exercise 3.9 (a)] for example) that $\sigma(n)\varphi(n) \le n^2$ for all $n \ge 1$, where $\varphi(\cdot)$ is Euler's totient function. By [48, Theorem 15], we have

$$\frac{n}{\varphi(n)} < e^\gamma \log\log n + \frac{2.50637}{\log\log n}$$

for all $n \geq 3$. If $y \geq 19$, then

$$\frac{2.50637}{\log\log y} \leq \frac{2.50637}{\log\log(19)} < \frac{7}{3},$$

so $y = \sqrt{(4\ell^N - x^2)/d} < 2\ell^{N/2}$ yields

$$\frac{\sigma(y)}{y} \leq \frac{y}{\varphi(y)} < e^{\gamma}\log\log y + \frac{2.50637}{\log\log(y)} < e^{\gamma}\log\log(2\ell^{N/2}) + \frac{7}{3}.$$

Simple numerical verification shows that $\sigma(y)/y \leq 7/3$ for $1 \leq y \leq 18$ (the largest value is $\sigma(12)/12 = 7/3$), so this upper bound holds for all $y > 0$. It follows that

$$\sum_{f^2 \mid x^2 - 4\ell^N} h\left(\frac{x^2 - 4\ell^N}{f^2}\right) < \frac{1}{3}(e^{\gamma}\log\log(2\ell^{N/2}) + 7/3)\log(4\ell^N)\sqrt{4\ell^N - x^2}.$$

It remains to bound $\sum_{x=1}^{X} \sqrt{4\ell^N - x^2}$. Since $f(x) = 4\ell^N - x^2$ is monotonically decreasing, we have

$$\sum_{x=1}^{X} \sqrt{4\ell^N - x^2} \leq \int_0^X \sqrt{4\ell^N - x^2}\, dx = \frac{1}{2}X\sqrt{4\ell^N - X^2} + 2\ell^N \arcsin\left(\frac{X}{2\ell^{N/2}}\right).$$

Now $2\arcsin(X/2\ell^{N/2}) \leq 2\arcsin(1) = \pi$. Furthermore, $0 \leq 2\ell^{N/2} - X < 1$ and $0 < 2\ell^{N/2} + X < 4\ell^{N/2}$, so

$$\frac{1}{2}X\sqrt{4\ell^N - X^2} = \frac{1}{2}X\sqrt{(2\ell^{N/2} - X)(2\ell^{N/2} + X)} < \frac{1}{2}2\ell^{N/2}\sqrt{4\ell^{N/2}} = 2\ell^{3N/4}.$$

It follows that $Q'_N < B_N$ where

$$B_N = \frac{2}{3}\left(e^{\gamma}\log\log(2\ell^{N/2}) + 7/3\right)\log(4\ell^N)(\pi\ell^N + 2\ell^{3N/4}).$$

Next, we bound c_N. The quantity B_N is monotonically increasing as a function of N, so $Q_{N/r} < B_{N/r} \leq B_{N/2}$ for all divisors r of N with $r \geq 2$. Thus,

$$c_N = \frac{1}{N}\left(Q_N + \sum_{1<r\mid N} \mu(r)Q_{N/r}\right) \leq \frac{1}{N}\left(Q_N + \sum_{1<r\mid N} Q_{N/r}\right)$$
$$< \frac{1}{N}\left(B_N + (\sigma(N) - 1)B_{N/2}\right).$$

Since $N \geq 3$, we can bound $\sigma(N)$ as before, i.e. $\sigma(N) < N(e^\gamma \log\log N + 7/3)$, to obtain

$$\begin{aligned} c_N &< \frac{1}{N}\left(B_N + (N(e^\gamma \log\log N + 7/3) - 1)B_{N/2}\right) \\ &= \frac{B_N}{N} + \left(e^\gamma \log\log N + 7/3 - \frac{1}{N}\right) B_{N/2}. \end{aligned}$$

The asymptotic bound on c_N is obtained by bounding the quantity $2\ell^{N/2}$ in the double log by $(2\sqrt{\ell})^N$. □

Remark 7.8 The proof of Corollary 7.7 simply ignores the restriction that p should not split and ℓ should split in any of the orders under consideration. If we wish to prove a lower bound, these restrictions must be addressed. There is a widely believed heuristic that a Legendre symbol $\left(\frac{f(x)}{p}\right)$ should take values ± 1 with equal probability, independent of the size of x. However, known bounds on the smallest x for which a polynomial $f(x)$ is a non-residue modulo p are comparatively weak. There is a bound following from work of Weil; for example, Burgess mentioned this in 1967, observing there is a non-residue $x < H$ for $H \gg \sqrt{p}\log p$ [10]. Improvements have only been made in general for polynomials in at least two variables (in contrast to the special case of $f(x) = x$, where this problem is known as the least non-residue problem); for a discussion and literature review, see [45].

8 Path Finding with Oriented Isogeny Volcanoes

In this section, we review the previous work of the present authors [5], which provides inspiration for the subject of this chapter.

As discussed in the introduction, finding a path between two curves in a supersingular ℓ-isogeny graph compromises the security of a variety of cryptographic systems. The bijection of Theorem 3.2 serves as inspiration for producing such a path, by highlighting the many volcano structures present in the graph, hinting that one might navigate the graph by reference to these oriented volcanoes. In fact, we saw that every cycle in the supersingular ℓ-isogeny graph is a rim of a volcano in some oriented isogeny graph. In previous work [5], the authors presented explicit classical and quantum algorithms for finding a path from any elliptic curve E to an initial curve E_{init}. Here, E_{init} is generally taken to be the supersingular elliptic curve of j-invariant 1728, but other choices for E_{init} are also possible. The algorithms use the knowledge of one endomorphism θ on E, or equivalently, an explicit orientation of E by some imaginary quadratic field K, to navigate the oriented ℓ-isogeny graph $\mathcal{G}_{K,\ell}$. Their complexity is subject to certain plausible heuristics, further detailed below, and is in part governed by the time it takes to evaluate θ on points on E.

The reader is encouraged to consult [56] for a different approach to solving the endomorphism ring problem using the data of an orientation.

The runtime of the classical algorithm additionally depends subexponentially on the degree d of θ and linearly on a certain class number that can be significantly smaller than the class number of the quadratic order of discriminant Δ defined by θ. This runtime can be substantially improved in many cases, leading in particular to certain new families of endomorphisms on *every* supersingular elliptic curve, whose exposure gives rise to a classical polynomial time algorithm for finding an ℓ-isogeny path from E to E_{init}. While this fact may on first glance seem alarming, the proof is non-constructive and thus does not provide an efficient strategy for finding such an offending endomorphism. So this result does not pose a threat to the security of isogeny based cryptosystems.

The quantum path finding algorithm relies on two different quantum subroutines: solving an oriented variant of the VECTORIZATION problem, which asks to find a class group element acting to take one oriented curve to another (see also [16, 56]), and a new problem called PRIMITIVEORIENTATION, which asks, for a given endomorphism, to determine the quadratic order for which the orientation it induces is primitive. The resulting algorithm finds a smooth ℓ-isogeny from E to E_{init} such that the smoothness bound and the runtime are subexponential in $\log|\Delta|$.

For completeness, we reproduce the key results of [5] here and give a brief road map of the path finding algorithms. Let $E/\overline{\mathbb{F}}_p$ be a supersingular elliptic curve and θ an endomorphism on E of degree d and discriminant Δ not divisible by p. Denote by $T_\theta(k, p)$ the time it takes to evaluate θ on points on E defined over $\mathbb{F}_{p^k}$. If θ is given by rational maps, then $T_\theta(k, p)$ is a polynomial in d, k and $\log p$. But if θ is given as the product of endomorphisms of smaller degree, then the dependence of $T_\theta(k, p)$ on d is much more favourable. All the algorithms of [5] assume in practice one of these two representations, which are typical in applications involving elliptic curve endomorphisms.

For any discriminant Δ of some quadratic order, the *ℓ-fundamental part* of Δ is the unique discriminant Δ' such that $\Delta = \ell^{2m}\Delta'$, with $v_\ell(\Delta') \leq 1$ when ℓ or Δ are odd and $v_\ell(\Delta'/4) \leq 1$ when $\ell = 2$ and Δ is even. For brevity, write $L_x(1/2) = \exp(O(\sqrt{\log x \log\log x}))$.

Theorem 8.1 ([5, Theorem 11.2]) *Let $E/\overline{\mathbb{F}}_p$ be a supersingular elliptic curve and θ an endomorphism on E of degree d and discriminant Δ coprime to p. Let $\ell \neq p$ be a prime (assumed to be constant in the runtime estimate), Δ' the ℓ-fundamental part of Δ and $h_{\Delta'}$ the class number of the quadratic order of discriminant Δ'. If $L_d(1/2)$ is bounded below by a polynomial in $\log p$ and $|\Delta'| \leq p^2$, then there is a classical algorithm that finds an ℓ-isogeny path of length $O(\log p + h_{\Delta'})$ from E to the curve E_{init} of j-invariant $j = 1728$ in heuristic runtime $T_\theta(L_d(1/2), p) + h_{\Delta'} L_d(1/2)\,\mathrm{poly}(\log p)$.*

Theorem 8.2 ([5, Theorem 11.3]) *Let $E/\overline{\mathbb{F}}_p$ be a supersingular elliptic curve and θ an endomorphism on E of discriminant Δ coprime to p such that $T_\theta(k, p)$ is bounded below by a polynomial in $\log(p^k)$. If $|\Delta| \leq p^2$, then there is a quantum* *algorithm that finds an $L_{|\Delta|}(1/2)$-smooth isogeny of norm $O(\sqrt{|\Delta|})$ from E to E_{init} in heuristic runtime $T_\theta(O(\log^2 d), p)L_{|\Delta|}(1/2)$.*

The basic strategy of the algorithm of Theorem 8.1 [5, Algorithm 8.1] is as follows. Use the orientation induced by θ as an 'orienteering tool' to climb from E to the rim of the oriented ℓ-isogeny volcano containing E. Generate this entire rim via the action of the class group of the associated ℓ-fundamental order $\mathcal{O}$ on the rim curves and store all their j-invariants in a list L. Orient the curve E_{init} (whose endomorphism ring is known) by the same orientation as E and use it to climb its volcano. Hoping to reach the same rim as the path originating at E, i.e. a vertex in L, construct the desired path from E to E_{init} from the two ascending paths and one of the two rim segments connecting them.

Throughout the computation, it is imperative to keep the size of the endomorphisms and the discriminants under consideration reasonable. To that end, endomorphisms are maintained in factored form as a product of endomorphisms of bounded prime power degree, and a technique akin to sieving produces such a representation for which the trace of the endomorphism is also of manageable size. Navigating the oriented ℓ-isogeny graph requires effective subroutines for detecting whether an edge is ascending/descending/horizontal, traversing an edge and carrying along the orientation, dividing an endomorphism by ℓ (to traverse an ascending edge), evaluating an endomorphism on ℓ-torsion points (required for the class group action) and factoring an endomorphism into a product of endomorphisms of prime power degree while controlling the size of its reduced trace. Algorithms for performing all these computational tasks, accompanied by proofs of correctness and a complete complexity analysis, can be found in [5]. Code is available at [6] and was used for a proof-of-concept example to generate an explicit 2-isogeny path of length 5 from $E : y^2 = x^3 + (7i + 86)x + (45i + 174)$ to $E_{\text{init}} : y^2 = x^3 - x$ over $\mathbb{F}_{179^2}$, where $i^2 = -1$.

A crucial ingredient in the path finding technique is orienting E_{init} appropriately. Given an ℓ-fundamental order $\mathcal{O} \subset K$ of discriminant Δ, this requires finding an endomorphism $\theta \in \text{End}(E_{\text{init}})$ that generates an order $\mathcal{O}' \subseteq \mathcal{O}$. The relative index $[\mathcal{O} : \mathcal{O}']$ is a power of ℓ, say ℓ^r for some $r \geq 0$. For reasons of efficiency, r should be small, situating $\mathcal{O}'$ as close to a rim as possible. At the same time, to guarantee the existence of a path from the starting curve E to E_{init} in $\mathcal{G}_{K,\ell}$, the orientation induced by θ must place E_{init} in the same connected component (i.e. volcano) of $\mathcal{G}_{K,\ell}$ as the oriented starting curve. A natural strategy is thus to loop over $r = 0, 1, 2, \ldots$ until the desired endomorphism θ and order $\mathcal{O}'$ are found. This raises the key question of how large r needs to be to ensure that a suitable orientation is encountered, at least with high probability. The answer is governed by how vertices corresponding to a fixed curve are spread around the cordillera. We will return to this issue in a moment.

A number of the algorithms outlined above depend on heuristics. Quoting from the introductory text of [5]:

We rely on a number of heuristic assumptions: (i) The Generalized Riemann Hypothesis. (ii) Powersmoothness in a quadratic sequence or form is as for random integers (a powersmooth analogue of the heuristic assumption underlying the quadratic sieve; see [5, Heuristics 5.10 and 9.3]). (iii) The orientations of a fixed j-invariant are distributed reasonably across all suitable volcanoes [5, Heuristic 3.7]. (iv) This distribution is independent of a certain integer factorization [5, Heuristic 6.7]. (v) The aforementioned integer factorization is prime with the same probability as a random integer ([5, Heuristic 6.4]; this heuristic is similar to those used in [23] and [32]).

Among these, item (i) is standard, and items (ii) and (v) are purely number theoretical questions. This leaves items (iii) and (iv) as the only heuristics which concern oriented isogeny graphs. Unfortunately, item (iv) seems out of reach at the moment, so we will focus on item (iii) which indeed addresses our earlier question. It asks us to consider, for a fixed j-invariant, how its various orientations appear in a cordillera of volcanoes: Does it prefer certain volcanoes over others? At the same time, it is natural to consider the 'opposite' question: For a fixed volcano, how do the various j-invariants lie upon it? Do some j-invariants sit at greater depth than others? This can be rephrased as a question about the covering map from an oriented volcano to the usual supersingular ℓ-isogeny graph $\mathcal{G}_\ell$. In Sect. 9, we address some of these questions.

9 Random Walks and Consequences for Oriented Isogeny Volcanoes

In this section, we use a standard random walk result for Ramanujan graphs to address some questions about the appearance of j-invariants on oriented ℓ-isogeny volcanoes.

9.1 *Random Walks in an ℓ-Isogeny Graph*

In this entire section, as a matter of convenience, **we assume that** $p \equiv 1 \pmod{12}$ **to avoid issues of extra automorphisms**. In particular, strictly speaking, Proposition 9.1 depends upon results about regular undirected Ramanujan graphs. Of course, the result can be expected to hold for all $\mathcal{G}_\ell$, regardless of p, which are at worst very nearly undirected regular graphs, having some perturbations around the two vertices $j = 0$ and $j = 1728$ if these belong to $\mathcal{G}_\ell$.

We refer to the length (number of ℓ-isogenies) of a shortest path between two curves in the ℓ-isogeny graph $\mathcal{G}_\ell$ as the *distance* between the two curves. The diameter of $\mathcal{G}_\ell$ (the maximum distance between pairs of points) is known to be bounded by $2 \log p$ [46, Theorem 1]; see also [38]. One way to obtain such results is using the Ramanujan property and standard methods for expander graphs [29]; our version and proof here closely follow [27, Theorem 1].

Proposition 9.1 *Let* $p \equiv 1 \pmod{12}$. *Let* ℓ *be a prime and* $\mathcal{G}_\ell$ *be the supersingular* ℓ*-isogeny graph over* $\overline{\mathbb{F}}_p$, *with* $\#\mathcal{G}_\ell$ *vertices. Let* $\mathcal{S}$ *represent a subset of* $\#\mathcal{S}$ *initial vertices. Let* j_1 *represent a target vertex. Let* j *represent the vertex reached by a random walk of* t *steps from a random vertex in* $\mathcal{S}$. *We denote by* $\Pr_t$ *the probability of an event over all such random walks of length* t. *Then we have*

$$\left|\Pr_t(j = j_1) - \frac{1}{\#\mathcal{G}_\ell}\right| \leq \frac{1}{\#\mathcal{S}}\left(\frac{2\sqrt{\ell}}{\ell+1}\right)^t.$$

Proof Choose a basis of a vector space of dimension $\#\mathcal{G}_\ell$ to correspond to the vertices of $\mathcal{G}_\ell$. Let A_ℓ represent the $\#\mathcal{G}_\ell \times \#\mathcal{G}_\ell$ adjacency matrix of $\mathcal{G}_\ell$. Let $\mathbf{v}_0$ represent an initial vector whose entry for vertex v is $1/\#\mathcal{S}$ if $v \in \mathcal{S}$ and 0 otherwise. This is the probability vector for the initial vertex of the random walk. Let λ_1 be the largest eigenvalue of A_ℓ and λ_2 the second largest. Recursively define $\mathbf{v}_t = \frac{1}{\lambda_1}A_\ell \mathbf{v}_{t-1}$, so $\mathbf{v}_t$ is the probability vector for the t-th vertex of the random walk.

Let $\mathbf{u}$ have all entries $1/\#\mathcal{G}_\ell$, so $\mathbf{u}$ is the probability vector for the uniform distribution over the vertices of $\mathcal{G}_\ell$. Then $\mathbf{u}$ is an eigenvector for λ_1, which has an eigenspace of dimension 1. Choose an orthogonal basis of eigenvectors $\mathbf{e}_i$ for $\mathbb{R}^{\#\mathcal{G}_\ell}$ such that $\mathbf{e}_1 = \mathbf{u}$. Then since $\mathbf{v}_{t-1}$ is a probability distribution, $\langle \mathbf{v}_{t-1}, \mathbf{1}\rangle = \langle \mathbf{u}, \mathbf{1}\rangle = 1$. Writing $\mathbf{v}_{t-1} = \sum_i \alpha_i \mathbf{e}_i$, this implies that $\alpha_1 = 1$. Hence $\mathbf{v}_{t-1} - \mathbf{u} = \sum_{i=2}^{\#\mathcal{G}_\ell} \alpha_i \mathbf{e}_i$. Therefore,

$$A_\ell(\mathbf{v}_{t-1} - \mathbf{u}) = \sum_{i=2}^{\#\mathcal{G}_\ell} \lambda_i \alpha_i \mathbf{e}_i.$$

Consequently, we have

$$\|\mathbf{v}_t - \mathbf{u}\|_2 = \left\|\frac{1}{\lambda_1}A_\ell(\mathbf{v}_{t-1} - \mathbf{u})\right\|_2 \leq \frac{1}{\lambda_1}\lambda_2\|\mathbf{v}_{t-1} - \mathbf{u}\|_2.$$

We also have

$$\|\mathbf{v}_0 - \mathbf{u}\|_2 = \#\mathcal{S}\left(\frac{1}{\#\mathcal{S}} - \frac{1}{\#\mathcal{G}_\ell}\right)^2 + (\#\mathcal{G}_\ell - \#\mathcal{S})\left(\frac{1}{\#\mathcal{G}_\ell}\right)^2 = \frac{1}{\#\mathcal{S}}\left(1 - \frac{\#\mathcal{S}}{\#\mathcal{G}_\ell}\right) \leq \frac{1}{\#\mathcal{S}}.$$

Iterating, we have

$$\begin{aligned}\left|\Pr(j = j_1) - \frac{1}{\#\mathcal{G}_\ell}\right| &= \|\mathbf{v}_t - \mathbf{u}\|_\infty \leq \|\mathbf{v}_t - \mathbf{u}\|_2 \leq \frac{\lambda_2^t}{\lambda_1^t}||\mathbf{v}_0 - \mathbf{u}||_2 \leq \frac{\lambda_2^t}{\#\mathcal{S}\lambda_1^t} \\ &\leq \frac{1}{\#\mathcal{S}}\left(\frac{2\sqrt{\ell}}{\ell+1}\right)^t.\end{aligned}$$

The final inequality is the bound on the spectral gap (5) for these graphs. □

Corollary 9.2 *Let j_1 be a j-invariant and $\mathcal{S}$ a subset of the vertices of $\mathcal{G}_\ell$. If*

$$t > \frac{\log \#\mathcal{G}_\ell - \log \#\mathcal{S}}{\log((\ell+1)/2\sqrt{\ell})},$$

then there exists a path of length $\leq t$ in $\mathcal{G}_\ell$ starting in $\mathcal{S}$ and ending at j_1. In particular, for

$$t > \log_{\sqrt{\ell}/2} \#\mathcal{G}_\ell - \log_{\sqrt{\ell}/2} \#\mathcal{S},$$

all curves are within distance t of $\mathcal{S}$.

Proof This follows from Proposition 9.1 by multiplying through by $\#\mathcal{G}_\ell$, to obtain

$$|\#\mathcal{G}_\ell \Pr_t(j = j_1) - 1| \leq \frac{\#\mathcal{G}_\ell}{\#\mathcal{S}} \left(\frac{2\sqrt{\ell}}{\ell+1}\right)^t$$

and then letting t be large enough so the right side is less than 1. □

9.2 The Depth of j-Invariants on an Oriented Volcano

Each volcano of the oriented ℓ-isogeny graph includes every j-invariant of the supersingular graph (to see this, fix a j-invariant j_0 of the volcano, and then consider an ℓ^n-isogeny to the desired j-invariant j_1; this generates a path in the oriented graph). It is natural to ask how far down a volcano one must go to see all j-invariants. We may use Corollary 9.2 to answer this.

Corollary 9.3 *Let $\mathcal{V}$ be an oriented volcano, and set $\mathcal{S} := \{j \in \mathcal{G}_\ell : (E_j, \iota)$ is on the rim of $\mathcal{V}$ for some $\iota\}$. Then all j-invariants appear on $\mathcal{V}$ at depth less than or equal to*

$$\lceil \log_{\sqrt{\ell}/2} \#\mathcal{G}_\ell - \log_{\sqrt{\ell}/2} \#\mathcal{S} \rceil.$$

Remark 9.4 The proof of [5, Proposition 6.5], particularly equation (3), provides a sort of computational proof of this behaviour in the case of the Gaussian field.

It is an interesting question to ask about the cardinality of $\mathcal{S}$, which is related to the question of the multiplicity of j-invariants in $\mathrm{SS}_{\mathcal{O}}^{\mathrm{pr}}$. That is, how many $\mathcal{O}$-primitive orientations are there for a fixed curve E? Kaneko showed that a curve cannot have more than one embedding when the discriminant is at most p in absolute value [30, Theorem 2']. On the other hand, [25, Proof of Theorem 1.4] shows that for any fixed curve E and integer t, orders $\mathcal{O}$ of sufficiently large discriminant will always embed into $\mathrm{End}(E)$ at least t times. The number of curves with an $\mathcal{O}$-orientation is bounded below in [34].

9.3 The Volcanoes upon Which the Orientations of a Fixed j-Invariant Lie

Every j-invariant occurs in every volcano (as described at the beginning of Sect. 9.2). In [5], the algorithms depended upon the following heuristic about the relative frequency of a fixed j-invariant across different volcanoes (quoted as item (iii) in Sect. 8 above).

Heuristic 9.5 ([5, Heuristic 3.7]) *Let $\mathcal{O}$ be an ℓ-fundamental order in a quadratic field K. Consider the finite union $\mathcal{S}$ of $\mathcal{O}'$-cordilleras in the oriented supersingular ℓ-isogeny graph $\mathcal{G}_{K,\ell}$ for all $\mathcal{O}' \supseteq \mathcal{O}$. Let $d(v)$ denote the depth of a vertex v on its volcano. Let $j(v)$ denote its j-invariant. Define:*

- *$R_{\mathcal{V}}$, the number of edges descending from the rim of the volcano $\mathcal{V} \in \mathcal{S}$.*
- *$R_{\mathcal{S}}$, the number of edges descending from all rims in $\mathcal{S}$.*

Then for any j-invariant j_0 and any volcano $\mathcal{V} \in \mathcal{S}$, the ratio

$$\frac{\#\{v \in \mathcal{V} : j(v) = j_0, d(v) \leq t\}}{\#\{v \in \mathcal{S} : j(v) = j_0, d(v) \leq t\}}$$

approaches $R_{\mathcal{V}}/R_{\mathcal{S}}$ as $t \to \infty$.

However, naïve methods can only prove a weighted version of this result, where j-invariants at smaller depth are weighted more heavily. Our result also has a restriction on the rim sizes.

Theorem 9.6 *Consider a cordillera $\mathcal{C}$ in the oriented supersingular ℓ-isogeny graph $\mathcal{G}_{K,\ell}$ whose volcanoes are all isomorphic as graphs and $(\ell+1)$-regular (e.g. a single $\mathcal{O}$-cordillera). Let $\mathcal{V}$ be one volcano of the cordillera $\mathcal{C}$ and $\mathcal{S}_{\mathcal{V}}$ the vertex set of its rim. Let $\mathcal{S}_{\mathcal{C}}$ be the set of vertices in the rims of the full cordillera. Let $d(v)$ denote the depth of the vertex v, and let $f_t(d)$ denote the function that counts the number of paths of length t starting at the rim of any volcano and terminating at any fixed vertex at depth d of that volcano (by symmetry, this is independent of the vertex but depends on the depth and rim size). If we define*

$$S_{\mathcal{V},j_1,t} = \frac{\displaystyle\sum_{v\in\mathcal{V},\, j(v)=j_1} f_t(d(v))}{\displaystyle\sum_{v\in\mathcal{C},\, j(v)=j_1} f_t(d(v))}, \tag{11}$$

then

$$\left| S_{\mathcal{V},j_1,t} - \frac{\#\mathcal{S}_{\mathcal{V}}}{\#\mathcal{S}_{\mathcal{C}}} \right| \ll \left(\frac{2\sqrt{\ell}}{\ell+1} \right)^t,$$

where the implied constant depends on $\mathcal{C}$ and ℓ, but not t.

Proof The numerator of (11) is the number of times we encounter $j = j_1$ among all paths of length t originating at the rim of $\mathcal{V}$. The denominator is the corresponding number for all paths of length t originating at the rims of $\mathcal{C}$. The number of paths of length t from the rim set of $\mathcal{V}$ is $(\ell+1)\ell^{t-1}\#\mathcal{S}_{\mathcal{V}}$. So, applying Proposition 9.1 with $\mathcal{S} = \mathcal{S}_{\mathcal{V}}$, we have

$$\frac{1}{(\ell+1)\ell^{t-1}} \sum_{v \in \mathcal{V}, j(v)=j_0} f_t(d(v)) = Pr(j = j_0)\#\mathcal{S}_{\mathcal{V}} = \frac{\#\mathcal{S}_{\mathcal{V}}}{\#\mathcal{G}_\ell} + O\left(\left(\frac{2\sqrt{\ell}}{\ell+1}\right)^t\right).$$

Combining with the analogous fact for the denominator gives the result. □

The restriction to $(\ell+1)$-regularity rules out the rims having oriented curves with automorphisms (see Remark 2.7); it is merely a convenience and the result could be adjusted to accommodate these cases.

This result is insufficient for the purposes of [5] in that it does not rule out the possibility that some volcanoes have few occurrences of $j = j_0$ at small depth, while others have more such occurrences at greater depth. If this is indeed the case, it may be that Algorithm 6.1 of [5], which proceeds by depth, preferentially finds certain volcanoes early on. (Nevertheless, Corollary 9.3 above provides some reassurance that all j-invariants begin to appear after a while.)

Acknowledgments Thank you to David Kohel, Greg Martin, Eli Orvis, Christophe Petit, Lillian B. Pierce, Jan Vonk, and Jonathan Wise for helpful discussions and feedback. We would also like to thank the conference *Women in Numbers 5* for the opportunity to form this research group. We also appreciate the detailed comments of our anonymous referees.

References

1. G. Adj, O. Ahmadi, A. Menezes, On isogeny graphs of supersingular elliptic curves over finite fields. Finite Fields Appl. **55**, 268–283 (2019)
2. N. Alon, I. Benjamini, E. Lubetzky, S. Sodin, Non-backtracking random walks mix faster. Commun. Contemp. Math. **09**(04), 585–603 (2007)
3. T.M. Apostol, *Introduction to Analytic Number Theory*. Undergraduate Texts in Mathematics (Springer, New York, 1976)
4. S. Arpin, C. Camacho-Navarro, K. Lauter, J. Lim, K. Nelson, T. Scholl, J. Sotáková, Adventures in Supersingularland. Exp. Math. **0**(0), 1–28 (2021)
5. S. Arpin, M. Chen, K.E. Lauter, R. Scheidler, K.E. Stange, H.T.N. Tran, Orienteering with one endomorphism (2022). https://arxiv.org/abs/2201.11079
6. S. Arpin, M. Chen, K.E. Lauter, R. Scheidler, K.E. Stange, H.T.N. Tran, Win5 github repository (2022). https://github.com/SarahArpin/WIN5
7. E. Bank, C. Camacho-Navarro, K. Eisenträger, T. Morrison, J. Park, Cycles in the supersingular l-isogeny graph and corresponding endomorphisms, in *Research Directions in Number Theory* (Springer, Berlin, 2019), pp. 41–66

8. BharatRam (https://mathoverflow.net/users/15311/bharatram). What does this connection between Chebyshev, Ramanujan, Ihara and Riemann mean? MathOverflow. https://mathoverflow.net/q/271786 (version: 2018-11-26)
9. B. Bollobás, A probabilistic proof of an asymptotic formula for the number of labelled regular graphs. European J. Combin. **1**(4), 311–316 (1980)
10. D.A. Burgess, On the quadratic character of a polynomial. J. Lond. Math. Soc. **s1–42**(1), 73–80 (1967)
11. W. Castryck, T. Decru, *An Efficient Key Recovery Attack on SIDH* (preliminary version) (2022)
12. W. Castryck, M. Houben, F. Vercauteren, B. Wesolowski, On the decisional Diffie-Hellman problem for class group actions on oriented elliptic curves. Cryptology ePrint Archive, Paper 2022/345 (2022). https://ia.cr/2022/345.
13. W. Castryck, T. Lange, C. Martindale, L. Panny, J. Renes, CSIDH: an efficient post-quantum commutative group action, in *Advances in Cryptology—ASIACRYPT 2018. Part III*. Lecture Notes in Computer Science, vol. 11274 (Springer, Cham, 2018), pp. 395–427
14. W. Castryck, L. Panny, F. Vercauteren, Rational isogenies from irrational endomorphisms, in *Advances in Cryptology—EUROCRYPT 2020. Part II*. Lecture Notes in Computer Science, vol. 12106 (Springer, Cham, 2020), pp. 523–548
15. D.X. Charles, E.Z. Goren, K.E. Lauter, Cryptographic hash functions from expander graphs. J. Cryptology **22**(1), 93–113 (2009). https://eprint.iacr.org/2006/021
16. M. Chenu, B. Smith, Higher-degree supersingular group actions. Math. Cryptology **1**(1), 1–15 (2021)
17. L. Colò, D. Kohel, Orienting supersingular isogeny graphs. J. Math. Cryptol. **14**(1), 414–437 (2020)
18. D.A. Cox, Primes of the form $x^2 + ny^2$, in *Pure and Applied Mathematics (Hoboken)*, 2nd edn. (Wiley, Hoboken, NJ, 2013)
19. G.P. Davidoff, P. Sarnak, A. Valette, *Elementary Number Theory, Group Theory, and Ramanujan Graphs* (Cambridge University, Cambridge, 2003)
20. L. De Feo, C. Delpech de Saint Guilhem, T.B. Fouotsa, P. Kutas, A. Leroux, C. Petit, J. Silva, B. Wesolowski, SÉTA: Supersingular Encryption from Torsion Attacks. *Lecture Notes in Computer Science*, vol. 13090 (Springer International Publishing, Cham, 2021), pp. 249–278
21. L. de Feo, D. Jao, J. Plût, Towards quantum-resistant cryptosystems from supersingular elliptic curve isogenies. J. Math. Cryptol. **8**(3), 209–247 (2014)
22. L. De Feo, D. Kohel, A. Leroux, C. Petit, B. Wesolowski, Sqisign: Compact post-quantum signatures from quaternions and isogenies, in *Advances in Cryptology—ASIACRYPT 2020: 26th International Conference on the Theory and Application of Cryptology and Information Security, Daejeon, South Korea, December 7–11, 2020, Proceedings, Part I* (Springer, Berlin, 2020), pp. 64–93
23. V. de Quehen, P. Kutas, C. Leonardi, C. Martindale, L. Panny, C. Petit, K.E. Stange, Improved Torsion-Point Attacks on SIDH Variants, in *Lecture Notes in Computer Science*, vol. 12827 (Springer International Publishing, Cham, 2021), pp. 432–470
24. K. Eisenträger, S. Hallgren, K. Lauter, T. Morrison, C. Petit, Supersingular isogeny graphs and endomorphism rings: reductions and solutions, in *Advances in Cryptology—EUROCRYPT 2018. Part III*. Lecture Notes in Computer Science, vol. 10822 (Springer, Cham, 2018), pp. 329–368
25. N. Elkies, K. Ono, T. Yang, Reduction of CM elliptic curves and modular function congruences. Int. Math. Res. Not. **2005**(44), 2695–2707 (2005)
26. M. Fouquet, F. Morain, Isogeny volcanoes and the SEA algorithm, in *Algorithmic Number Theory (Sydney, 2002)*. Lecture Notes in Computer Science, vol. 2369 (Springer, Berlin, 2002), pp. 276–291
27. S.D. Galbraith, C. Petit, J. Silva, Identification protocols and signature schemes based on supersingular isogeny problems. J. Cryptology **33**(1), 130–175 (2020)
28. B.H. Gross, Heights and the special values of L-series, in *Number Theory (Montreal, Quebec, 1985)*. CMS Conference Proceedings, vol. 7 (American Mathematical Society, Providence, 1987), pp. 115–187

29. S. Hoory, N. Linial, A. Wigderson, Expander graphs and their applications. Bull. Am. Math. Soc. (N.S.) **43**(4), 439–561 (2006)
30. M. Kaneko, Supersingular j-invariants as singular moduli mod p. Osaka J. Math. **26**(4), 849–855 (1989)
31. M. Kempton, Non-backtracking random walks and a weighted Ihara's theorem. Open J. Discrete Math. **06**, 207–226 (2016)
32. D. Kohel, K. Lauter, C. Petit, J.-P. Tignol, On the quaternion ℓ-isogeny path problem. LMS J. Comput. Math. **17**(A), 418–432 (2014)
33. D.R. Kohel, *Endomorphism Rings of Elliptic Curves Over Finite Fields* (ProQuest LLC, Ann Arbor, 1996). Thesis (Ph.D.)–University of California, Berkeley
34. A. Leroux, An effective lower bound on the number of orientable supersingular elliptic curves. Cryptology ePrint Archive, Paper 2022/357 (2022). https://ia.cr/2022/357
35. S. Li, Y. Ouyang, Z. Xu, Endomorphism rings of supersingular elliptic curves over $\mathbb{F}p$. Finite Fields Appl. **62**, 101619, 24 (2020)
36. S. Li, Y. Ouyang, Z. Xu, Neighborhood of the supersingular elliptic curve isogeny graph at $j = 0$ and 1728. Finite Fields Appl. **61**, 101600, 16 (2020)
37. J. Love, D. Boneh, Supersingular curves with small noninteger endomorphisms, in *ANTS XIV—Proceedings of the Fourteenth Algorithmic Number Theory Symposium*. Open Book Series, vol. 4. (Mathematical Sciences Publishers, Berkeley, CA, 2020), pp. 7–22
38. E. Lubetzky, Y. Peres, Cutoff on all Ramanujan graphs. Geom. Funct. Anal. **26**(4), 1190–1216 (2016)
39. J.M.J. Jacobson, H.C. Williams, *Solving the Pell Equation*. CMS Books in Mathematics/Ouvrages de Mathématiques de la SMC (Springer, New York, 2009)
40. L. Maino, C. Martindale, An attack on sidh with arbitrary starting curve (2022)
41. K. McMurdy, Explicit representation of the endomorphism rings of supersingular elliptic curves (2014). https://phobos.ramapo.edu/~kmcmurdy/research/McMurdy-ssEndoRings.pdf
42. H. Onuki, On oriented supersingular elliptic curves. Finite Fields App. **69**, 101777 (2021)
43. H. Onuki, Y. Aikawa, T. Takagi, The existence of cycles in the supersingular isogeny graphs used in SIKE, in *2020 International Symposium on Information Theory and Its Applications (ISITA)* (2020), pp. 358–362
44. Y. Ouyang, Z. Xu, Loops of isogeny graphs of supersingular elliptic curves at $j = 0$. Finite Fields Appl. **58**, 174–176 (2019)
45. L.B. Pierce, J. Xu, Burgess bounds for short character sums evaluated at forms. Algebra Number Theory **14**(7), 1911–1951 (2020)
46. A.K. Pizer, Ramanujan graphs and Hecke operators. Bull. Am. Math. Soc. (N.S.) **23**(1), 127–137 (1990)
47. D. Robert, Breaking sidh in polynomial time, in *Cryptology ePrint Archive, Paper 2022/1038* (2022). https://eprint.iacr.org/2022/1038
48. J.B. Rosser, L. Schoenfeld, Approximate formulas for some functions of prime numbers. Illinois J. Math. **6**, 64–94 (1962)
49. J.H. Silverman, Advanced topics in the arithmetic of elliptic curves, in *Graduate Texts in Mathematics*, vol. 151 (Springer, New York, 1994)
50. J.H. Silverman, The arithmetic of elliptic curves, in *Graduate Texts in Mathematics*, 2nd edn., vol. 106. (Springer, Dordrecht, 2009)
51. K.E. Stange, Frobenius and the endomorphism ring of $j = 1728$ (2021). http://math.colorado.edu/~kstange/papers/1728.pdf
52. A.V. Sutherland, Isogeny volcanoes, in *ANTS X—Proceedings of the Tenth Algorithmic Number Theory Symposium. Open Book Series*, vol. 1 (Mathematical Sciences Publishers, Berkeley, 2013), pp. 507–530
53. The Sage Developers, in *SageMath, the Sage Mathematics Software System (Version 9.4)* (2022). https://www.sagemath.org
54. J.-P. Tillich, G. Zémor, Optimal cycle codes constructed from Ramanujan graphs. SIAM J. Discrete Math. **10**(3), 447–459 (1997)

55. J. Voight, Quaternion algebras, in *Graduate Texts in Mathematics*, vol. 288 (Springer, Cham, [2021] ©2021)
56. B. Wesolowski, Orientations and the supersingular endomorphism ring problem, in *Advances in Cryptology—EUROCRYPT 2022*. Lecture Notes in Computer Science, vol. 13277 (Springer, Cham, 2022), pp. 345–371
57. B. Wesolowski, The supersingular isogeny path and endomorphism ring problems are equivalent, in *2021 IEEE 62nd Annual Symposium on Foundations of Computer Science—FOCS 2021* (IEEE Computer Society, Los Alamitos, CA, [2022] ©2022), pp. 1100–1111
58. G. Xiao, L. Luo, Y. Deng, Constructing cycles in isogeny graphs of supersingular elliptic curves. J. Math. Cryptol. **15**(1), 454–464 (2021)

Generalized Ramanujan–Sato Series Arising from Modular Forms

Angelica Babei, Lea Beneish, Manami Roy, Holly Swisher, Bella Tobin, and Fang-Ting Tu

2020 Mathematics Subject Classification 11F03, 11F11, 11Y60, 33C05, 33C20

1 Introduction and Statement of Results

Ramanujan [20] gave a list of infinite series identities of the form

$$\sum_{k=0}^{\infty} \frac{(\frac{1}{2})_k(\frac{1}{d})_k(\frac{d-1}{d})_k}{k!^3}(ak+1)(\lambda_d)^k = \frac{\delta}{\pi},$$

The authors thank the Women in Numbers 5 (WIN5) Workshop for the opportunity to initiate this collaboration. The work of the first author was supported by a Simons Collaboration Grant (550029, to John Voight). The fourth author is appreciative of support by the National Science Foundation Grant DMS-2101906.

A. Babei
Department of Mathematics and Statistics, McMaster University, Hamilton, ON, Canada
e-mail: babeia@mcmaster.ca

L. Beneish
Department of Mathematics, University of North Texas, Denton, TX, USA
e-mail: Lea.Beneish@unt.edu

M. Roy
Department of Mathematics, Lafayette College, Easton, PA, USA
e-mail: royma@lafayette.edu

H. Swisher (✉)
Department of Mathematics, Oregon State University, Corvallis, OR, USA
e-mail: swisherh@oregonstate.edu

B. Tobin
Department of Mathematics, Agnes Scott College, Decatur, GA, USA
e-mail: bellatobin@gmail.com

F.-T. Tu
Department of Mathematics, Louisiana State University, Baton Rouge, LA, USA
e-mail: ftu@lsu.edu

A. Bucur et al. (eds.), *Research Directions in Number Theory*, Association for Women in Mathematics Series 33, https://doi.org/10.1007/978-3-031-51677-1_3

for $d = 2, 3, 4, 6$, where λ_d are singular values that correspond to elliptic curves with complex multiplication, a and δ are explicit algebraic numbers, and $(a)_k$ denotes the rising factorial $(a)_k = a(a+1)\cdots(a+k-1)$. In fact, a similar identity was given even earlier by Bauer [2]. Proofs of these formulas were first given by J. Borwein and P. Borwein [3] and D. Chudnovsky and G. Chudnovsky [12], and both approaches rely on the arithmetic of elliptic integrals of the first and second kind, including the Legendre relation at singular values. Since the 1980s series of this type have been at the forefront of algorithms to compute decimal approximations of π. Although Ramanujan indicated that there were general theories behind such series, this remains not fully understood. Deriving new series for $1/\pi^k$ and the unifying theories underlying such series is an active research area (see [7, 30] for example).

In work of Chan, Chan, and Liu [6], motivated by Sato, the authors derive a general series for $1/\pi$ that admits existing series as special cases. They give a table of explicitly computed examples, each relating a Hauptmodul and hypergeometric function connected to an index 2 subgroup pair of triangle groups.

Here, we use the method of Chan, Chan, and Liu [6] to obtain a new general theorem which produces series formulas for $1/\pi$ and as an application construct explicit Ramanujan–Sato type formulas for $1/\pi$ related to other non-compact arithmetic triangle groups as classified by Takeuchi [24, 25]. Some of our examples are new, and some reproduce existing examples.

Throughout the chapter we denote the upper half plane by $\mathfrak{h} = \{\tau = x + iy \in \mathbb{C} : y > 0\}$. Our main theorem is the following.

Theorem 1.1 *Let $Z(\tau)$, $X(\tau)$, $U(\tau)$ be meromorphic on $\mathfrak{h}$ such that $\frac{1}{2\pi i}\frac{d}{d\tau}X(\tau) = U(\tau)X(\tau)Z(\tau)$, and when $\tau \in D$, a domain of $\mathfrak{h}$, $Z(\tau) = X(\tau)^{\varepsilon_0}(1 - X(\tau))^{\varepsilon_1}\sum_{j=0}^{\infty} A_j X(\tau)^j$ for $\varepsilon_0, \varepsilon_1 \in \mathbb{R}$, $A_j \in \mathbb{C}$. Further assume there exist $\alpha, \beta \in \mathbb{C}$, $N, k \in \mathbb{N}$, and elements $\gamma = \left(\begin{smallmatrix} a & b \\ c & -a \end{smallmatrix}\right)$, $\delta = \left(\begin{smallmatrix} 1 & \frac{a}{c}(1-N) \\ 0 & 1 \end{smallmatrix}\right) \in \mathrm{SL}_2(\mathbb{R})$ such that for any $\tau \in \mathfrak{h}$,*

$$\big(Z|_k\gamma\big)(\tau) = \alpha Z(\tau),$$
$$\big(Z|_k\delta\big)(\tau) = \beta Z(\tau).$$

Set $M_N(\tau) := Z(\tau)/Z(N\tau)$, and let $\tau_0 = \frac{a}{c} + \frac{i}{c\sqrt{N}}$. Then it follows that if $\tau_0 \in D$,

$$\frac{ck\sqrt{N}}{2\pi} = U(\tau_0)X(\tau_0)^{\varepsilon_0}(1 - X(\tau_0))^{\varepsilon_1}\sum_{j\geq 0}(2j + a_N)A_jX(\tau_0)^j,$$

where

$$a_N = 2\left(\varepsilon_0 + \varepsilon_1\frac{X(\tau_0)}{X(\tau_0) - 1}\right) - \frac{\alpha i^k}{\beta N^{k/2}}X(\tau_0)\left(\frac{dM_N}{dX}\right)\Big|_{X=X(\tau_0)}.$$

Alternatively if $\gamma\tau_0 \in D$,

$$\frac{ck\sqrt{N}}{2\pi} = X(\gamma\tau_0)^{\varepsilon_0}(1 - X(\gamma\tau_0))^{\varepsilon_1} \sum_{j\geq 0}(b'_N j + a'_N)A_j X(\gamma\tau_0)^j,$$

where

$$b'_N = 2NU(\gamma\tau_0),$$
$$a'_N = 2NU(\gamma\tau_0)\left(\varepsilon_0 + \varepsilon_1 \frac{X(\gamma\tau_0)}{X(\gamma\tau_0)-1}\right) + \frac{1}{\beta}U(\tau_0)X(\tau_0)\left(\frac{dM_N}{dX}\right)|_{X=X(\tau_0)}.$$

The rest of the chapter is organized as follows. In Sect. 2 we provide some relevant background material and give a few lemmas which will be useful for proving Theorem 1.1 and for our applications. In Sect. 3, we prove Theorem 1.1. In Sects. 4 and 5 we give several applications by constructing explicit examples of Ramanujan–Sato series for $1/\pi$ related to certain arithmetic triangle groups. The groups we use to construct examples are among the arithmetic triangle groups commensurable with $\mathrm{PSL}_2(\mathbb{Z}) \cong (2, 3, \infty)$ coming from Takeuchi's class I (those of non-compact type). These have the following subgroup diagram:

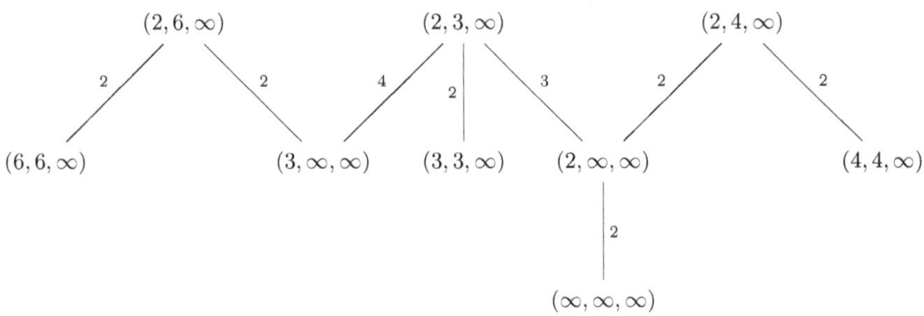

In particular, in Sect. 4 we construct examples of series for $1/\pi$ arising from modular forms for the groups $\Gamma_0(2) \cong (2, \infty, \infty)$, $\Gamma_0(3) \cong (3, \infty, \infty)$, and $\Gamma_0(4) \cong (\infty, \infty, \infty)$. In Sect. 5 we construct examples arising from modular forms for the groups $\mathrm{PSL}_2(\mathbb{Z}) \cong (2, 3, \infty)$, $\Gamma_0(2)^+ \cong (2, 4, \infty)$, and $\Gamma_0(3)^+ \cong (2, 6, \infty)$.

2 Preliminaries

Suppose Γ is a discrete subgroup of $\mathrm{SL}_2(\mathbb{R})$ having genus zero that is commensurable with a subgroup $\Gamma(\mathcal{O})$ of $\mathrm{SL}_2(\mathbb{R})$ arising from a norm 1 group of a quaternion order $\mathcal{O}$. The group Γ acts on the upper half plane $\mathfrak{h}$ and $\mathbb{P}^1(\mathbb{R})$ by linear fractional transformations $\gamma \cdot \tau = \frac{a\tau+b}{c\tau+d}$, where $\gamma = \left(\begin{smallmatrix} a & b \\ c & d \end{smallmatrix}\right) \in \Gamma$. A classical result from the theory of compact Riemann surfaces says that when Γ has genus zero there exist finitely many elliptic or parabolic elements $r_1, \ldots, r_k$ that generate $\Gamma/\{\pm 1\}$ with the relations $r_1 \ldots r_k = 1, r_i^{e_i} = 1$. We call $(0; e_1, \ldots, e_k)$ the *signature* of $\Gamma/\{\pm 1\}$. Work of Yang [28] has shown that the modular forms on Γ can be expressed in terms of a Hauptmodul t of Γ and solutions of the Schwarzian differential equation

$$2Q(t)t'(\tau)^2 + \{t, \tau\} = 0,$$

where

$$\{t, \tau\} = \frac{t'''(\tau)}{t'(\tau)} - \frac{3}{2}\left(\frac{t''(\tau)}{t'(\tau)}\right)^2$$

is the Schwarzian derivative. If Γ has signature $(0; e_1, e_2, e_3)$, then this differential equation is hypergeometric and one can describe the modular forms on Γ using hypergeometric functions.

Takeuchi [24, 25] has classified arithmetic triangle groups, including their quaternion orders and inclusion relations between them.

2.1 Arithmetic Triangle Groups and a Theorem of Yang

Arithmetic triangle groups are certain discrete subgroups of $\mathrm{PSL}_2(\mathbb{R})$. Consider integers $e_1, e_2, e_3 > 1$, possibly ∞, such that $\frac{1}{e_1} + \frac{1}{e_2} + \frac{1}{e_3} < 1$. Then there exists a triangle S in $\mathfrak{h}$ with internal angles $\frac{\pi}{e_1}, \frac{\pi}{e_2}$, and $\frac{\pi}{e_3}$, where $\frac{\pi}{\infty} := 0$. The group of symmetries of the tiling of $\mathfrak{h}$ by triangles congruent to S is called a triangle group and can be presented in terms of generators

$$(e_1, e_2, e_3) = \langle r_1, r_2, r_3 \mid r_1^{e_1} = r_2^{e_2} = r_3^{e_3} = r_1 r_2 r_3 = 1 \rangle,$$

where if one of $e_i = \infty$ for $i = 1, 2, 3$, the relation $r_i^{e_i} = 1$ is trivial. Takeuchi [24, 25] found 85 triples, up to permutation, where the corresponding group $(e_1, e_2, e_3) \le \mathrm{PSL}_2(\mathbb{R})$ is arithmetic. In this context, we define

$$\Gamma_0(N) := \left\{ A = \begin{pmatrix} a & b \\ c & d \end{pmatrix} \in \mathrm{SL}_2(\mathbb{Z}) : A \equiv \begin{pmatrix} * & * \\ 0 & * \end{pmatrix} (\mathrm{mod}\, N) \right\} / \{\pm 1\}.$$

For example, $(2, \infty, \infty) \cong \Gamma_0(2)$ and $(\infty, \infty, \infty) \cong \Gamma_0(4)$. We direct the reader to [24, 25] for more background on arithmetic triangle groups.

Throughout, we call an elliptic point of order ∞ a cusp. The following theorem of Yang allows us to write modular forms on T in terms of ${}_2F_1$ hypergeometric functions evaluated at specific Hauptmoduln $X(\tau)$.

Theorem 2.1 (Yang [28, Thm. 9]) *Assume that Γ has signature $(0; e_1, e_2, e_3)$. Let $X(\tau)$ be the Hauptmodul of Γ with values 0, 1, and ∞ at the elliptic points of orders e_1, e_2, and e_3 (possibly ∞), respectively. Let $k \geq 2$ be an even integer. Then a basis for the space of modular forms of weight k on Γ is given by*

$$X^{\{k(1-1/e_1)/2\}}(1-X)^{\{k(1-1/e_2)/2\}}X^j\left({}_2F_1\left[\begin{matrix} a & b \\ & c \end{matrix}; X\right] + CX_2^{1/e_1}F_1\left[\begin{matrix} a' & b' \\ & c' \end{matrix}; X\right]\right)^k,$$

$j = 0, \ldots, \mathcal{D} - 1$, for some constant C, where for a rational number x, we let $\{x\}$ denote the fractional part of x,

$$a = \frac{1}{2}\left(1 - \frac{1}{e_1} - \frac{1}{e_2} - \frac{1}{e_3}\right), \qquad b = a + \frac{1}{e_3}, \qquad c = 1 - \frac{1}{e_1},$$

$$a' = a + \frac{1}{e_1}, \qquad b' = b + \frac{1}{e_1}, \qquad c' = c + \frac{2}{e_1},$$

and

$$\mathcal{D} = 1 - k + \lfloor k(1 - 1/e_1)/2 \rfloor + \lfloor k(1 - 1/e_2)/2 \rfloor + \lfloor k(1 - 1/e_3)/2 \rfloor$$

is the dimension of the space of weight-k modular forms for Γ.

In Table 1 we catalog the relevant elliptic point data needed to use Theorem 2.1 for the groups Γ we consider in Sects. 4 and 5. In Table 1, the "Stabilizing elements" are the generators of the isotropy subgroups of the group Γ (the stabilizer in the group Γ) that fix the corresponding elliptic points under the group action of Γ on $\mathfrak{h} \cup \mathbb{P}^1(\mathbb{R})$. The "Orders" are the orders of the isotropy subgroups. For example, for the group $\Gamma_0(2)$, there is one in-equivalent elliptic point represented by $\frac{1+i}{2}$ and two in-equivalent cusps, represented by 0 and $i\infty$. The elliptic point $\frac{1+i}{2}$ is fixed by all

Table 1 Elliptic point data

Group Γ	Elliptic points	Stabilizing elements	Orders
$\Gamma_0(2) \cong (2, \infty, \infty)$	$\frac{1+i}{2}, 0, i\infty$	$\left(\begin{smallmatrix} 1 & -1 \\ 2 & -1 \end{smallmatrix}\right), \left(\begin{smallmatrix} -1 & 0 \\ 2 & -1 \end{smallmatrix}\right), \left(\begin{smallmatrix} 1 & 1 \\ 0 & 1 \end{smallmatrix}\right)$	$2, \infty, \infty$
$\Gamma_0(3) \cong (3, \infty, \infty)$	$\frac{3+i\sqrt{3}}{6}, 0, i\infty$	$\left(\begin{smallmatrix} -1 & 1 \\ -3 & 2 \end{smallmatrix}\right), \left(\begin{smallmatrix} -1 & 0 \\ 3 & -1 \end{smallmatrix}\right), \left(\begin{smallmatrix} 1 & 1 \\ 0 & 1 \end{smallmatrix}\right)$	$3, \infty, \infty$
$\Gamma_0(4) \cong (\infty, \infty, \infty)$	$0, \frac{1}{2}, i\infty$	$\left(\begin{smallmatrix} 1 & 0 \\ 4 & 1 \end{smallmatrix}\right), \left(\begin{smallmatrix} 1 & -1 \\ 4 & -3 \end{smallmatrix}\right), \left(\begin{smallmatrix} 1 & 1 \\ 0 & 1 \end{smallmatrix}\right)$	∞, ∞, ∞
$PSL_2(\mathbb{Z}) \cong (2, 3, \infty)$	$i, \frac{-1+i\sqrt{3}}{2}, i\infty$	$\left(\begin{smallmatrix} 0 & -1 \\ 1 & 0 \end{smallmatrix}\right), \left(\begin{smallmatrix} 0 & -1 \\ 1 & 1 \end{smallmatrix}\right), \left(\begin{smallmatrix} 1 & 1 \\ 0 & 1 \end{smallmatrix}\right)$	$2, 3, \infty$
$\Gamma_0(2)^+ \cong (2, 4, \infty)$	$\frac{i}{\sqrt{2}}, \frac{1+i}{2}, i\infty$	$\frac{1}{\sqrt{2}}\left(\begin{smallmatrix} 0 & -1 \\ 2 & 0 \end{smallmatrix}\right), \sqrt{2}\left(\begin{smallmatrix} 0 & -\frac{1}{2} \\ 1 & -1 \end{smallmatrix}\right), \left(\begin{smallmatrix} 1 & 1 \\ 0 & 1 \end{smallmatrix}\right)$	$2, 4, \infty$
$\Gamma_0(3)^+ \cong (2, 6, \infty)$	$\frac{i}{\sqrt{3}}, \frac{3+i\sqrt{3}}{6}, i\infty$	$\frac{1}{\sqrt{3}}\left(\begin{smallmatrix} 0 & -1 \\ 3 & 0 \end{smallmatrix}\right), \sqrt{3}\left(\begin{smallmatrix} 0 & -\frac{1}{3} \\ 1 & -1 \end{smallmatrix}\right), \left(\begin{smallmatrix} 1 & 1 \\ 0 & 1 \end{smallmatrix}\right)$	$2, 6, \infty$

Table 2 Choice of Hauptmoduln

Group Γ	Hauptmodul	e_1, e_2, e_3
$\Gamma_0(2) \cong (2, \infty, \infty)$	$t_2(\tau) = -64\eta(2\tau)^{24}/\eta(\tau)^{24}$	$\infty, 2, \infty$
$\Gamma_0(3) \cong (3, \infty, \infty)$	$t_3(\tau) = -27\eta(3\tau)^{12}/\eta(\tau)^{12}$	$\infty, 3, \infty$
$\Gamma_0(4) \cong (\infty, \infty, \infty)$	$t_\infty(\tau) = 16\eta(\tau)^8\eta(4\tau)^{16}/\eta(2\tau)^{24}$	∞, ∞, ∞
$\mathrm{PSL}_2(\mathbb{Z}) \cong (2, 3, \infty)$	$t_{2,3}(\tau) = 1728/j(\tau)$	$\infty, 2, 3$
$\Gamma_0(2)^+ \cong (2, 4, \infty)$	$t_{2,4}(\tau) = 256\eta(\tau)^{24}\eta(2\tau)^{24}/(\eta(\tau)^{24} + 64\eta(2\tau)^{24})^2$	$\infty, 2, 4$
$\Gamma_0(3)^+ \cong (2, 6, \infty)$	$t_{2,6}(\tau) = 108\eta(\tau)^{12}\eta(3\tau)^{12}/(\eta(\tau)^{12} + 27\eta(3\tau)^{12})^2$	$\infty, 2, 6$

elements of the subgroup $\langle \left(\begin{smallmatrix} 1 & -1 \\ 2 & -1 \end{smallmatrix}\right) \rangle$, the cusp 0 is fixed by all elements of $\langle \left(\begin{smallmatrix} -1 & 0 \\ 2 & -1 \end{smallmatrix}\right) \rangle$, and $i\infty$ is fixed by all elements of $\langle \left(\begin{smallmatrix} 1 & 1 \\ 0 & 1 \end{smallmatrix}\right) \rangle$.

Each of the groups Γ listed in Table 1 yields a modular curve of genus 0, and thus the function field of this modular curve is generated by a single modular function for Γ, a Hauptmodul. By the theory of Riemann surfaces [19, Prop. 1.21], any nonconstant modular function for Γ that has a single simple pole is a Hauptmodul for Γ.

In order to apply Theorem 2.1 to our groups Γ, we need to choose a Hauptmodul which takes the values 0, 1, and ∞ at the elliptic points of Γ. Then assign e_1, e_2, and e_3 to be the orders of the elliptic points yielding the values 0, 1, and ∞, respectively. We list this choice of Hauptmodul $X(\tau)$ for each group Γ, along with the values of e_1, e_2, and e_3 in Table 2, where $\eta(\tau)$ is the Dedekind eta function $\eta(\tau) := e^{\pi i\tau/12}\prod_{n=1}^{\infty}(1 - e^{2\pi i n\tau})$. That these are Hauptmoduln can be checked directly or observed in [18] and [13].

We note the following relationship between the j-function and t_2, as defined in Table 2, which was observed by Maier [18][1]

$$j = \frac{64(4t_2 - 1)^3}{t_2}. \tag{1}$$

We now use Theorem 2.1 and the Hauptmoduln in Table 2 to compute generators of spaces of modular forms for our groups Γ. For each group, we choose the least weight k that yields nontrivial modular forms. The generators obtained are listed in Table 3. To save space, we use the notation

$$F[a, b; c; x] := {}_2F_1\left[\begin{matrix} a & b \\ & c \end{matrix} ; x\right].$$

[1] The t_2 in [18] is equal to a constant multiple of -2^6 from our t_2.

Table 3 Hypergeometric Modular Form Generators

Group Γ	Weight k	Generator(s) of weight k modular forms for Γ
$\Gamma_0(2) \cong (2, \infty, \infty)$	2	$(1-t_2)^{\frac{1}{2}} F[\frac{1}{4}, \frac{1}{4}; 1; t_2]^2$
$\Gamma_0(3) \cong (3, \infty, \infty)$	2	$(1-t_3)^{\frac{2}{3}} F[\frac{1}{3}, \frac{1}{3}; 1; t_3]^2$
$\Gamma_0(4) \cong (\infty, \infty, \infty)$	2	$F[\frac{1}{2}, \frac{1}{2}; 1; t_\infty]^2, t_\infty F[\frac{1}{2}, \frac{1}{2}; 1; t_\infty]^2$
$\mathrm{PSL}_2(\mathbb{Z}) \cong (2, 3, \infty)$	4	$F[\frac{1}{12}, \frac{5}{12}; 1; t_{2,3}]^4$
$\Gamma_0(2)^+ \cong (2, 4, \infty)$	4	$F[\frac{1}{8}, \frac{3}{8}; 1; t_{2,4}]^4$
$\Gamma_0(3)^+ \cong (2, 6, \infty)$	4	$F[\frac{1}{6}, \frac{1}{3}; 1; t_{2,6}]^4$

2.2 *Connection to Known Modular Forms and Their Properties*

In our examples, we will use some known modular forms. For even $k \geq 4$, the Eisenstein series of weight k is a modular form (of weight k) for $\mathrm{PSL}_2(\mathbb{Z})$ defined [15] as

$$E_k(\tau) := \frac{1}{2\zeta(k)} \sum_{(m,n)\in\mathbb{Z}^2-(0,0)} \frac{1}{(m\tau+n)^k} = 1 - \frac{2k}{B_k} \sum_{n\geq 1} \sigma_{k-1}(n) e^{2\pi i n\tau}, \tag{2}$$

where $\zeta(s)$ is the Riemann zeta function, B_k is the kth Bernoulli number, and $\sigma_{k-1}(n)$ is the arithmetic function

$$\sigma_{k-1}(n) = \sum_{d|n} d^{k-1}.$$

The latter equality in (2) follows since $E_k(\tau)$ converges absolutely for $k \geq 4$.

When $k = 2$, we define

$$E_2(\tau) = 1 - 24 \sum_{n\geq 1} \sigma_1(n) e^{2\pi i n\tau}.$$

For any positive number N,

$$E_{2,N} := E_2(\tau) - N E_2(N\tau) \tag{3}$$

is a modular form of weight 2 for $\Gamma_0(N)$ [15, Exercise 1.2.8].

The Jacobi θ-functions are defined as

$$\theta_2(\tau) := \sum_{n\in\mathbb{Z}} e^{\pi i (n+1/2)^2\tau}, \quad \theta_3(\tau) := \sum_{n\in\mathbb{Z}} e^{\pi i n^2\tau}, \quad \theta_4(\tau) := \sum_{n\in\mathbb{Z}} (-1)^n e^{\pi i n^2\tau}.$$

They can be expressed as eta quotients via [5, p.29]

$$\theta_2(\tau) = 2\frac{\eta(2\tau)^2}{\eta(\tau)}, \quad \theta_3(\tau) = \frac{\eta(\tau)^5}{\eta(\tau/2)^2\eta(2\tau)^2}, \quad \theta_4(\tau) = \frac{\eta(\tau/2)^2}{\eta(\tau)}, \tag{4}$$

and satisfy the Jacobi identity [5, p.28]

$$\theta_2^4 + \theta_4^4 = \theta_3^4. \tag{5}$$

Moreover, the modular λ-function is given by

$$\lambda = \left(\frac{\theta_2}{\theta_3}\right)^4, \quad 1 - \lambda = \left(\frac{\theta_4}{\theta_3}\right)^4 \tag{6}$$

and satisfies the transformation formula [9, p.109]

$$\lambda\left(\frac{-1}{\tau}\right) = 1 - \lambda(\tau). \tag{7}$$

The function $\lambda(\tau)$ is a Hauptmodul for the group $\Gamma(2)$, which has cusps 0, 1, and $i\infty$ at which the values of λ are 1, ∞, and 0, respectively [9, Chapter VII, §7-8]. Moreover, the function $\lambda(2\tau)$ is a Hauptmodul for $\Gamma_0(4)$, and one can check if $\lambda(2\tau)$ equals the choice of Hauptmodul t_∞ for $\Gamma_0(4)$ given in Table 2. Furthermore, each $\theta_i(2\tau)$ is a modular form of weight $1/2$ for $\Gamma_0(4)$ [5, p.28].

We also note the following classical results:

$$\theta_3(2\tau)^4 = \frac{4E_2(4\tau) - E_2(\tau)}{3}, \; \theta_3(2\tau)^4 + \theta_2(2\tau)^4 = \frac{1}{2}\left(\theta_3(\tau)^4 + \theta_4(\tau)^4\right) = -\, E_{2,2}(\tau), \tag{8}$$

$$(\theta_3(2\tau)\theta_3(6\tau) + \theta_2(2\tau)\theta_2(6\tau))^2 = \frac{1}{2}E_{2,3}(\tau) = -\frac{(3\eta(3\tau)^3 + \eta(\tau/3)^3)^2}{\eta(\tau)^2}, \tag{9}$$

$${}_2F_1\left[\begin{matrix}\frac{1}{2} & \frac{1}{2} \\ & 1\end{matrix} ; \lambda(2\tau)\right]^2 = \theta_3(2\tau)^4, \quad (1-\lambda(\tau))\,{}_2F_1\left[\begin{matrix}\frac{1}{2} & \frac{1}{2} \\ & 1\end{matrix} ; \lambda(2\tau)\right]^2 = \theta_4(2\tau)^4. \tag{10}$$

Identities such as those above can be checked using Sturm-type bound arguments, which we employ throughout the chapter. A theorem by Choi and Kim [11] gives a Sturm-type bound for computationally determining when modular forms on genus 0 groups $\Gamma_0(N)^+$ are equal. Since their bound is less than Sturm's bound [22] for $\Gamma_0(N)$, we give special cases of these results as follows.

Given a formal sum $f = \sum_{n \gg -\infty} c(n)q^n$, define

$$\mathrm{ord}_{i\infty} f := \inf\{n \in \mathbb{Z} \mid c(n) \neq 0\}.$$

Theorem 2.2 (Sturm [22] and Choi, Kim [11]) *Let $N \in \mathbb{N}$ such that $\Gamma_0(N)^+$ has genus 0. Suppose f is a meromorphic modular form of weight k for $\Gamma \in \{\Gamma_0(N), \Gamma_0(N)^+\}$ having poles only at the cusp $i\infty$. If f has integer Fourier coefficients and*

$$\begin{cases} ord_{i\infty} f > \frac{k}{12}[\mathrm{SL}_2(\mathbb{Z}) : \Gamma_0(N)] & when\ \Gamma = \Gamma_0(N), \\ ord_{i\infty} f > \frac{k}{24}[\mathrm{SL}_2(\mathbb{Z}) : \Gamma_0(N)] & when\ \Gamma = \Gamma_0(N)^+, \end{cases}$$

then $f = 0$.

2.3 Some Useful Results

The following formula due to Clausen [1] states that

$$ {}_2F_1\left[\begin{matrix} a & b & \\ & a+b+\frac{1}{2} & ; z \end{matrix}\right]^2 = {}_3F_2\left[\begin{matrix} 2a & 2b & a+b & \\ & 2a+2b & a+b+\frac{1}{2} & ; z \end{matrix}\right]. \tag{11}$$

We next give two useful lemmas. The first we use in the proof of Theorem 1.1.

Lemma 2.3 *Suppose $f(\tau)$ is a meromorphic function on $\mathfrak{h}$. If there exist $A = \left(\begin{smallmatrix} a & b \\ c & d \end{smallmatrix}\right) \in \mathrm{SL}_2(\mathbb{R})$, $k \in \mathbb{N}$, and $\beta \in \mathbb{C}$ such that for all $\tau \in \mathfrak{h}$,*

$$(f|_k A)(\tau) = \beta f(\tau),$$

then for any $\tau \in \mathfrak{h}$ that is not a pole of f,

$$\frac{1}{\beta(c\tau+d)^2}\frac{df}{d\tau}(A\tau) = (c\tau+d)^k\frac{df}{d\tau}(\tau) + kc(c\tau+d)^{k-1}f(\tau).$$

Proof By hypothesis,

$$f(A\tau) = \beta(c\tau+d)^k f(\tau).$$

The proof follows immediately from taking $\frac{d}{d\tau}$ of both sides of this equation. □

The next lemma is used in our construction of explicit examples in Sect. 5.

Lemma 2.4 *Let $Z(\tau)$ be a modular form for Γ of even weight k. If $\gamma_s = \frac{1}{\sqrt{s}}\begin{pmatrix} 0 & -1 \\ s & 0 \end{pmatrix} \in \Gamma$ for some real $s > 0$, then for any real $r > 0$, we have*

$$Z\left(\frac{i}{\sqrt{rs}}\right) = (-r)^{k/2} Z\left(i\sqrt{\frac{r}{s}}\right).$$

□

Proof By the modularity of $Z(\tau)$, for any $\tau \in \mathfrak{h}$,

$$Z(-1/s\tau) = (\sqrt{s}\tau)^k Z(\tau).$$

Setting $\tau = i\sqrt{\frac{r}{s}}$ gives the desired result. □

Remark 2.5 We note that $\mathrm{PSL}_2(\mathbb{Z}) \cong (2, 3, \infty)$ contains γ_1, $\Gamma_0(2)^+ \cong (2, 4, \infty)$ contains γ_2, and $\Gamma_0(3)^+ \cong (2, 6, \infty)$ contains γ_3, with γ_s as described in Lemma 2.4.

3 Proof of Theorem 1.1

For convenience we use the notation $f'(\tau) := \frac{1}{2\pi i}\frac{df}{d\tau}(\tau)$ throughout this section.

Proof of Theorem 1.1 By hypothesis, we have for $\tau \in \mathfrak{h}$ that

$$Z(\gamma\tau) = \alpha(c\tau - a)^k Z(\tau). \tag{12}$$

Thus by Lemma 2.3,

$$Z'(\gamma\tau) = \alpha(c\tau - a)^{k+2} Z'(\tau) + \frac{\alpha ck}{2\pi i}(c\tau - a)^{k+1} Z(\tau).$$

Plugging in $\tau_0 = \frac{a}{c} + \frac{i}{c\sqrt{N}}$ yields

$$\frac{ck\sqrt{N}}{2\pi} \cdot Z(\tau_0) = Z'(\tau_0) + \alpha^{-1} i^{-k} N^{\frac{k+2}{2}} Z'(\gamma\tau_0).$$

Dividing by $Z(\tau_0)$ and using (12), we obtain

$$\frac{ck\sqrt{N}}{2\pi} = \frac{Z'(\tau_0)}{Z(\tau_0)} + N\frac{Z'(\gamma\tau_0)}{Z(\gamma\tau_0)}. \tag{13}$$

On the other hand, changing the variable $\tau \mapsto N\tau$, we also have by hypothesis that

$$Z(\delta \cdot N\tau) = \beta Z(N\tau),$$

so applying Lemma 2.3 with $c = 0$ yields that

$$Z'(\delta \cdot N\tau) = \beta Z'(N\tau).$$

Therefore, given $M_N(\tau) = \frac{Z(\tau)}{Z(N\tau)}$, we obtain

$$\frac{M_N'(\tau)}{M_N(\tau)} = \frac{Z'(\tau)}{Z(\tau)} - N\frac{Z'(N\tau)}{Z(N\tau)} = \frac{Z'(\tau)}{Z(\tau)} - N\frac{Z'(\delta \cdot N\tau)}{Z(\delta \cdot N\tau)}. \tag{14}$$

Note that $\gamma \cdot \tau_0 = \delta \cdot N\tau_0$, so from (12) we obtain

$$\beta Z(N\tau_0) = Z(\delta \cdot N\tau_0) = Z(\gamma \cdot \tau_0) = \alpha \frac{i^k}{N^{k/2}} Z(\tau_0),$$

and therefore $M_N(\tau_0) = \frac{\beta N^{k/2}}{\alpha i^k}$. So (14) at τ_0 becomes

$$\frac{\alpha i^k}{\beta N^{k/2}} M_N'(\tau_0) = \frac{Z'(\tau_0)}{Z(\tau_0)} - N\frac{Z'(\gamma\tau_0)}{Z(\gamma\tau_0)}. \tag{15}$$

Using (15) we can rewrite (13) as both

$$\frac{ck\sqrt{N}}{2\pi} = 2\frac{Z'(\tau_0)}{Z(\tau_0)} - \frac{\alpha i^k}{\beta N^{k/2}} M_N'(\tau_0), \tag{16}$$

$$\frac{ck\sqrt{N}}{2\pi} = 2N\frac{Z'(\gamma\tau_0)}{Z(\gamma\tau_0)} + \frac{\alpha i^k}{\beta N^{k/2}} M_N'(\tau_0). \tag{17}$$

From our hypotheses, we have that $X'(\tau) = U(\tau)X(\tau)Z(\tau)$ for $\tau \in \mathfrak{h}$ and $Z(\tau) = X(\tau)^{\varepsilon_0}(1 - X(\tau))^{\varepsilon_1} \sum_{j\geq 0} A_j X(\tau)^j$ for $\tau \in D$. Using the fact that

$$\frac{X(\tau)}{X'(\tau)} \cdot \frac{1}{2\pi i}\frac{d}{d\tau} X(\tau)^{\varepsilon_0}(1 - X(\tau))^{\varepsilon_1} = X(\tau)^{\varepsilon_0}(1 - X(\tau))^{\varepsilon_1}\left(\varepsilon_0 + \varepsilon_1 \frac{X(\tau)}{X(\tau) - 1}\right),$$

it follows that for $\tau \in D$,

$$\frac{Z'(\tau)}{Z(\tau)} = U(\tau)X(\tau)^{\varepsilon_0}(1 - X(\tau))^{\varepsilon_1} \sum_{j\geq 0}\left(j + \varepsilon_0 + \varepsilon_1 \frac{X(\tau)}{X(\tau) - 1}\right) A_j X(\tau)^j. \tag{18}$$

On the other hand for $\tau \in D$, $M_N(\tau)$ can be expressed as a function of $X(\tau)$ and $X(N\tau)$, so

$$\frac{1}{2\pi i}\frac{dM_N}{d\tau}(\tau) = \frac{dM_N}{dX}(\tau)\cdot U(\tau)X(\tau)Z(\tau). \tag{19}$$

Thus for $\tau \in D$,

$$\frac{1}{2\pi i}\frac{dM_N}{d\tau}(\tau) = U(\tau)X(\tau)^{\varepsilon_0+1}(1-X(\tau))^{\varepsilon_1}\frac{dM_N}{dX}(\tau)\sum_{j\geq 0}A_jX(\tau)^j. \tag{20}$$

Hence when $\tau_0 \in D$, we can rewrite (16) using (18) and (20) to obtain

$$\begin{aligned}\frac{ck\sqrt{N}}{2\pi} &= 2U(\tau_0)X(\tau_0)^{\varepsilon_0}(1-X(\tau_0))^{\varepsilon_1}\sum_{j\geq 0}\left(j+\varepsilon_0+\varepsilon_1\frac{X(\tau_0)}{X(\tau_0)-1}\right)A_jX(\tau_0)^j\\ &-\frac{\alpha i^k}{\beta N^{k/2}}U(\tau_0)X(\tau_0)^{\varepsilon_0+1}(1-X(\tau_0))^{\varepsilon_1}\left(\frac{dM_N}{dX}\right)|_{X=X(\tau_0)}\sum_{j\geq 0}A_jX(\tau_0)^j,\end{aligned}$$

which yields our first identity. Similarly when $\gamma\tau_0 \in D$ we rewrite (17) using (18) and (19) to obtain

$$\begin{aligned}\frac{ck\sqrt{N}}{2\pi} &= 2NU(\gamma\tau_0)X(\gamma\tau_0)^{\varepsilon_0}(1-X(\gamma\tau_0))^{\varepsilon_1}\\ &\times\sum_{j\geq 0}\left(j+\varepsilon_0+\varepsilon_1\frac{X(\gamma\tau_0)}{X(\gamma\tau_0)-1}\right)A_jX(\gamma\tau_0)^j\\ &+\frac{\alpha i^k}{\beta N^{k/2}}\left(\frac{dM_N}{dX}\right)|_{X=X(\tau_0)}\cdot U(\tau_0)X(\tau_0)Z(\tau_0).\end{aligned}$$

Since by (12) we have $Z(\tau_0) = \frac{N^{k/2}}{\alpha i^k}Z(\gamma\tau_0)$, this becomes

$$\begin{aligned}\frac{ck\sqrt{N}}{2\pi} &= 2NU(\gamma\tau_0)X(\gamma\tau_0)^{\varepsilon_0}(1-X(\gamma\tau_0))^{\varepsilon_1}\\ &\times\sum_{j\geq 0}\left(j+\varepsilon_0+\varepsilon_1\frac{X(\gamma\tau_0)}{X(\gamma\tau_0)-1}\right)A_jX(\gamma\tau_0)^j\\ &+\frac{1}{\beta}U(\tau_0)X(\tau_0)\left(\frac{dM_N}{dX}\right)|_{X=X(\tau_0)}X(\gamma\tau_0)^{\varepsilon_0}(1-X(\gamma\tau_0))^{\varepsilon_1}\\ &\times\sum_{j\geq 0}A_jX(\gamma\tau_0)^j,\end{aligned}$$

which yields our second identity. □

We conclude this section with a lemma that will be useful when we compute examples in the following sections.

Lemma 3.1 *Suppose $X(\tau), Z(\tau), U(\tau)$ satisfy the conditions in Theorem 1.1. Then, writing $X := X(\tau)$ and $Y := X(N\tau)$ gives*

$$M_N = N\frac{dX}{dY}\cdot\frac{Y}{X}\cdot\frac{U(N\tau)}{U(\tau)}.$$

Proof Since M_N is a function of X and Y,

$$\frac{dY}{dX} = \frac{dY}{d\tau}\cdot\frac{d\tau}{dX} = N\frac{dX}{d\tau}(N\tau)\cdot\frac{d\tau}{dX} = N\frac{\frac{dX}{d\tau}(N\tau)}{\frac{dX}{d\tau}} = N\frac{YU(N\tau)Z(N\tau)}{XU(\tau)Z(\tau)}$$
$$= \frac{N}{M_N}\frac{Y}{X}\frac{U(N\tau)}{U(\tau)},$$

which gives the result. □

4 Examples of Theorem 1.1 of First Type

In this section we obtain six different Ramanujan–Sato series for $1/\pi$ as examples of Theorem 1.1. One is already available in the literature and the others are new according to our knowledge.

To construct the examples in this section we use the following corollary for modular forms on certain groups where the cusp $i\infty$ has width 1, which follows immediately from Theorem 1.1 with $h = 1$ and $\beta = 1$.

Corollary 4.1 *Let Γ be a discrete subgroup of $\mathrm{SL}_2(\mathbb{R})$ commensurable with $\mathrm{SL}_2(\mathbb{Z})$ such that $\left(\begin{smallmatrix}1&1\\0&1\end{smallmatrix}\right) \in \Gamma$ (i.e., the cusp $i\infty$ of Γ has width 1), and let $X(\tau)$ be a Hauptmodul of Γ. Let $Z(\tau)$ be a weight-k modular form for Γ such that $\frac{1}{2\pi i}\frac{dX}{d\tau} = U(\tau)X(\tau)Z(\tau)$, and when $\tau \in D$, a domain of $\mathfrak{h}$, $Z(\tau) = X(\tau)^{\varepsilon_0}(1 - X(\tau))^{\varepsilon_1}\sum_{j=0}^{\infty} A_jX(\tau)^j$ for $\varepsilon_0, \varepsilon_1 \in \mathbb{R}$, $A_j \in \mathbb{C}$. Further assume there exist $\alpha \in \mathbb{C}$ and $\gamma = \left(\begin{smallmatrix}a&b\\c&-a\end{smallmatrix}\right) \in \mathrm{SL}_2(\mathbb{R})$ such that*

$$\big(Z|_k\gamma\big)(\tau) = \alpha Z(\tau).$$

Set $M_N(\tau) := Z(\tau)/Z(N\tau)$ for $N \in \mathbb{N}$ satisfying $\frac{a}{c}(1-N) \in \mathbb{Z}$, and let $\tau_0 = \frac{a}{c} + \frac{i}{c\sqrt{N}}$. Then if $\gamma\tau_0 = \frac{a}{c} + \frac{i\sqrt{N}}{c} \in D$, we have

$$\frac{ck\sqrt{N}}{2\pi} = X(\gamma\tau_0)^{\varepsilon_0}(1 - X(\gamma\tau_0))^{\varepsilon_1} \sum_{j\geq 0}(b_N j + a_N)A_j X(\gamma\tau_0)^j,$$

where

$$b_N = 2NU(\gamma\tau_0),$$
$$a_N = 2NU(\gamma\tau_0)\left(\varepsilon_0 + \varepsilon_1 \frac{X(\gamma\tau_0)}{X(\gamma\tau_0) - 1}\right) + U(\tau_0)X(\tau_0)\left(\frac{dM_N}{dX}\right)\Big|_{X=X(\tau_0)}.$$

Note that the arithmetic triangle groups $(m, \infty, \infty) \cong \Gamma_0(m)$ for $m \in \{2, 3\}$ and $(\infty, \infty, \infty) \cong \Gamma_0(4)$ contain the element $\left(\begin{smallmatrix}1 & 1\\ 0 & 1\end{smallmatrix}\right)$. In this section we construct examples of Corollary 4.1 for these groups. We choose the Hauptmodul t_m for $\Gamma_0(m)$ with $m \in \{2, 3\}$ and t_∞ for $\Gamma_0(4)$ given in Table 2. Furthermore, from Table 3 we have that the space of modular forms of weight 2 for $\Gamma_0(m)$ for $m = 2, 3, 4$ is generated by Z_2, Z_3, and Z_∞, respectively; these are given by

$$Z_m(\tau) = (1 - t_m)^{1-\frac{1}{m}}\, {}_2F_1\left[\begin{matrix}\frac{1}{2} - \frac{1}{2m} & \frac{1}{2} - \frac{1}{2m} \\ & 1\end{matrix} ; t_m(\tau)\right]^2, \tag{21}$$

and

$$Z_\infty(\tau) = {}_2F_1\left[\begin{matrix}\frac{1}{2} & \frac{1}{2} \\ & 1\end{matrix} ; t_\infty(\tau)\right]^2. \tag{22}$$

Write $f' := \frac{1}{2\pi i}\frac{df}{d\tau}$. We next compute each U_m for t_m with $m \in \{2, 3\}$ so that

$$t_m' = t_m Z_m U_m. \tag{23}$$

From Table 2, the Hauptmodul t_m for $m \in \{2, 3\}$ can be written as

$$t_m(\tau) = -\alpha \frac{\eta\,(m\tau)^{k_m}}{\eta(\tau)^{k_m}}, \tag{24}$$

where $\alpha = m^{\frac{k_m}{4}}$ and $k_m = \frac{24}{m-1}$. Using the property $\eta'/\eta = \frac{1}{24}E_2$, we get that

$$\frac{t_m'}{t_m}(\tau) = k_m\left(m\frac{\eta'(m\tau)}{\eta(m\tau)} - \frac{\eta'(\tau)}{\eta(\tau)}\right) = \left(\frac{-1}{m-1}\right) E_{2,m}.$$

Moreover, since $E_{2,m}$ is a modular form of weight 2 for $\Gamma_0(m)$ by (3) and Z_m generates the space of weight 2 modular forms for $\Gamma_0(m)$ as seen in Table 3, Theorem 2.2 gives $Z_m(\tau) = \left(\frac{1}{1-m}\right) E_{2,m}$. Therefore for $m = 2, 3$,

$$t'_m = t_m Z_m, \text{ and } U_m = 1. \tag{25}$$

In order to apply Corollary 4.1 in these cases, we need $\frac{dM_N}{dX}$ where $X(\tau) = t_m(\tau)$ and $Y(\tau) = t_m(N\tau)$. Observe from Lemma 3.1 that when $U = 1$, we have

$$M_N = N\frac{dX}{dY} \cdot \frac{Y}{X}.$$

Thus taking the derivative with respect to X yields

$$\frac{dM_N}{dX} = N\left(\frac{d\left(\frac{dX}{dY}\right)}{dX} \cdot \frac{Y}{X} + \frac{1}{X} - \frac{dX}{dY} \cdot \frac{Y}{X^2}\right), \tag{26}$$

which will be useful in subsequent subsections.

The computation of U and $\frac{dM_N}{dX}$ for the group $\Gamma_0(4)$ is given in Sect. 4.3 where we obtain an example for this group.

4.1 $\Gamma_0(2) \cong (2, \infty, \infty)$

As in Table 2 we choose the Hauptmodul $X(\tau) = t_2(\tau) = -64\frac{\eta(2\tau)^{24}}{\eta(\tau)^{24}}$, and for the corresponding domain D we choose the intersection of $\{\tau \in \mathfrak{h} : |X(\tau)| < 1\}$ and the following fundamental domain FD of $\Gamma_0(2)$:

$$FD = \{\tau \in \mathfrak{h} : |\mathrm{Re}(\tau)| \le 1/2,\ |\tau - 1/2| > 1/2,\ |\tau + 1/2| > 1/2\}.$$

Recall that $Z = Z_2$ from (21). It follows from (11) that

$$Z(\tau) = (1-X(\tau))^{\frac{1}{2}}\,{}_2F_1\left[\begin{matrix}\frac{1}{4} & \frac{1}{4} \\ & 1\end{matrix} ; X(\tau)\right]^2 = (1-X(\tau))^{\frac{1}{2}}\,{}_3F_2\left[\begin{matrix}\frac{1}{2} & \frac{1}{2} & \frac{1}{2} \\ & 1 & 1\end{matrix} ; X(\tau)\right]. \tag{27}$$

Thus,

$$Z(\tau) = (1 - X(\tau))^{\frac{1}{2}} \sum_{j=0}^{\infty} \frac{\left(\frac{1}{2}\right)_j \left(\frac{1}{2}\right)_j \left(\frac{1}{2}\right)_j}{(j!)^3} X^j(\tau). \tag{28}$$

For Examples 4.2 and 4.3 below, we use $\gamma = w_2 = \frac{1}{\sqrt{2}}\left(\begin{smallmatrix} 0 & -1 \\ 2 & 0 \end{smallmatrix}\right)$. To find the transformation property of Z with respect to w_2, we first note that Theorem 2.2 implies

$$Z(\tau) = \frac{1}{2}(\theta_3(\tau)^4 + \theta_4(\tau)^4).$$

Also, using the transformation formula $\eta(-1/\tau) = \sqrt{-i\tau}\eta(\tau)$ [15, p.20] and (4), we have

$$\theta_3\left(\frac{-1}{2\tau}\right)^4 = -4\tau^2\theta_3(2\tau)^4,$$

$$\theta_4\left(\frac{-1}{2\tau}\right)^4 = -4\tau^2 \cdot 16\frac{\eta(4\tau)^8}{\eta(2\tau)^4} = -4\tau^2\theta_2(2\tau)^4.$$

Using the last two equalities in (8), we thus obtain the following transformation property:

$$(Z|w_2)(\tau) = -Z(\tau), \text{ i.e., } Z(-1/2\tau) = -2\tau^2 Z(\tau). \tag{29}$$

We see that Z and X above satisfy the conditions in Corollary 4.1 with $U = 1$, $\gamma = w_2$, $k = 2$, $\varepsilon_0 = 0$, $\varepsilon_1 = \frac{1}{2}$, $\alpha = -1$, and any $N \in \mathbb{N}$.

For Examples 4.4 and 4.5 below, we consider $\gamma = \left(\begin{smallmatrix} 1 & -1 \\ 2 & -1 \end{smallmatrix}\right)$. Since $Z = Z_2$ is a weight 2 modular form on $\Gamma_0(2)$, we have the following transformation property

$$(Z|\gamma)(\tau) = Z(\tau), \text{ i.e., } Z\left(\frac{\tau - 1}{2\tau - 1}\right) = (2\tau - 1)^2 Z(\tau). \tag{30}$$

We see that Z and X above satisfy conditions in Corollary 4.1 with $U = 1$, $\gamma = \left(\begin{smallmatrix} 1 & -1 \\ 2 & -1 \end{smallmatrix}\right)$, $k = 2$, $\varepsilon_0 = 0$, $\varepsilon_1 = \frac{1}{2}$, and $\alpha = 1$. Note that, for this choice of γ, we need to choose N such that $\frac{1-N}{2} \in \mathbb{Z}$.

We now proceed with examples for the group $\Gamma_0(2) \cong (2, \infty, \infty)$.

Example 4.2 Let $N = 3$, $\gamma = w_2$ above, and $\tau_0 = i/\sqrt{6}$. Using the values of $\eta(i/\sqrt{6})$, $\eta(i\sqrt{6})$, $\eta(i\sqrt{2/3})$, and $\eta(i\sqrt{3/2})$ from Table 5, we get

$$X(i/\sqrt{6}) = -17 - 12\sqrt{2},$$

$$Y(i/\sqrt{6}) = X(i\sqrt{3/2}) = -17 + 12\sqrt{2}.$$

Using (26) and the polynomial relationship $\Phi_3(X, Y) = 0$ between $X(\tau) = t_2(\tau)$ and $Y(\tau) = t_2(\gamma\tau)$ given in Lemma 6.2, we obtain

$$\left(\frac{dM_3}{dX}\right)\Big|_{X=X(i/\sqrt{6})} = 12 - \frac{17}{\sqrt{2}}.$$

From Corollary 4.1 and (28), we then have

$$\frac{\sqrt{6}}{\pi} = (1 - X(i\sqrt{3/2}))^{1/2}\sum_{j\geq 0}(6j + a_3)\frac{\left(\frac{1}{2}\right)_j\left(\frac{1}{2}\right)_j\left(\frac{1}{2}\right)_j}{(j!)^3}X(i\sqrt{3/2})^j,$$

where

$$a_3 = 3\left(\frac{X\left(i\sqrt{\frac{3}{2}}\right)}{X\left(i\sqrt{\frac{3}{2}}\right)-1}\right) + X(i/\sqrt{6})\left(\frac{dM_3}{dX}\right)\Big|_{X=X(i/\sqrt{6})} = \frac{3}{2} - \frac{1}{\sqrt{2}}.$$

Finally we have

$$\frac{2}{\pi} = (\sqrt{2}-1)\sum_{j\geq 0}\left(12j + 3 - \sqrt{2}\right)\frac{\left(\frac{1}{2}\right)_j\left(\frac{1}{2}\right)_j\left(\frac{1}{2}\right)_j}{(j!)^3}\left(12\sqrt{2}-17\right)^j.$$

Example 4.3 Let $N = 5$, $\gamma = w_2$, and $\tau_0 = i/\sqrt{10}$. Using the values of $j(i/\sqrt{10})$ and $j\left(i\sqrt{\frac{5}{2}}\right)$ from Table 4 and the relationship $j = \frac{64(4X-1)^3}{X}$ from (1), we get the values

$$X(i/\sqrt{10}) = -161 - 72\sqrt{5},$$
$$X\left(i\sqrt{\frac{5}{2}}\right) = -161 + 72\sqrt{5}.$$

Using the polynomial relationship $\Phi_5(X,Y) = 0$ between $X(\tau) = t_2(\tau)$ and $Y(\tau) = t_2(\gamma\tau)$ from Lemma 6.2, we obtain

$$\frac{dM_5}{dX}\Big|_{X=X(i/\sqrt{10})} = \frac{1440 - 644\sqrt{5}}{9}.$$

From Corollary 4.1 and (28), we then have

$$\frac{\sqrt{10}}{\pi} = \left(1 - X\left(i\sqrt{\frac{5}{2}}\right)\right)^{1/2}\sum_{j\geq 0}(10j + a_5)\frac{\left(\frac{1}{2}\right)_j\left(\frac{1}{2}\right)_j\left(\frac{1}{2}\right)_j}{(j!)^3}X\left(i\sqrt{\frac{5}{2}}\right)^j,$$

where

$$a_5 = 5\left(\frac{X\left(i\sqrt{\frac{5}{2}}\right)}{X\left(i\sqrt{\frac{5}{2}}\right)-1}\right) + X(i/\sqrt{10})\left(\frac{dM_5}{dX}\right)\Big|_{X=X(i/\sqrt{10})} = \frac{5}{2} - \frac{2\sqrt{5}}{3}.$$

So, finally we have

$$\frac{2\sqrt{5}}{\pi} = (\sqrt{5}-2)\sum_{j\geq 0}\left(60j+15-4\sqrt{5}\right)\frac{\left(\frac{1}{2}\right)_j\left(\frac{1}{2}\right)_j\left(\frac{1}{2}\right)_j}{(j!)^3}\left(72\sqrt{5}-161\right)^j.$$

Example 4.4 Let $N = 3$, $\gamma = \left(\begin{smallmatrix}1 & -1\\ 2 & -1\end{smallmatrix}\right)$, and $\tau_0 = \frac{1}{2} + \frac{i}{2\sqrt{3}}$. Using the values $\eta\left(\frac{-1+i\sqrt{3}}{2}\right)$, $\eta(i\sqrt{3})$ from Table 5, and the transformation formula $\eta(\gamma\tau)^{24} = (c\tau+d)^{12}\eta(\tau)^{24}$ [15, p. 20], we have

$$X(\tau_0) = X(\gamma\tau_0) = \frac{1}{4}.$$

Using the polynomial relationship $\Phi_3(X, Y) = 0$ between $X(\tau) = t_2(\tau)$ and $Y(\tau) = t_2(\gamma\tau)$ from Lemma 6.2 and (26), we obtain

$$\left(\frac{dM_3}{dX}\right)\Big|_{X=X(\tau_0)} = 8. \tag{31}$$

Corollary 4.1 and (28) yield the series

$$\frac{2\sqrt{3}}{\pi} = (1-X(\gamma\tau_0))^{\frac{1}{2}}\sum_{j\geq 0}(6j+a_3)\frac{\left(\frac{1}{2}\right)_j\left(\frac{1}{2}\right)_j\left(\frac{1}{2}\right)_j}{(j!)^3}X(\gamma\tau_0)^j,$$

where

$$a_3 = 3\left(\frac{X(\gamma\tau_0)}{X(\gamma\tau_0)-1}\right) + X(\tau_0)\left(\frac{dM_3}{dX}\right)\Big|_{X=X(\tau_0)} = 1.$$

Hence, we obtain

$$\frac{4}{\pi} = \sum_{j=0}^{\infty}(1+6j)\frac{\left(\frac{1}{2}\right)_j\left(\frac{1}{2}\right)_j\left(\frac{1}{2}\right)_j}{(j!)^3}(1/4)^j.$$

This series is one of the well-known Ramanujan series in [20] for $1/\pi$, which also arises from Chan, Chan, and Liu [6, (1.1)]).

Example 4.5 Let $N = 5$, $\gamma = \begin{pmatrix} 1 & -1 \\ 2 & -1 \end{pmatrix}$, and $\tau_0 = \frac{1}{2} + \frac{i}{2\sqrt{5}}$. Then, by using the values of $j(\tau_0)$ and $j(\gamma\tau_0)$ from Table 4 and the relationship $j = \frac{64(4X-1)^3}{X}$ from (1), we get

$$X(\tau_0) = X(\gamma\tau_0) = 9 - 4\sqrt{5}.$$

Next, using the polynomial relationship $\Phi_5(X, Y) = 0$ between $X(\tau) = t_2(\tau)$ and $Y(\tau) = t_2(\gamma\tau)$ from Lemma 6.2, we obtain

$$\left(\frac{dM_5}{dX}\right)|_{X=X(\tau_0)} = 15 + \frac{27\sqrt{5}}{4}.$$

Then, from Corollary 4.1 and (28), the series is of the form

$$\frac{2\sqrt{5}}{\pi} = (1 - X(\gamma\tau_0))^{\frac{1}{2}} \sum_{j \geq 0} (10j + a_5) \frac{\left(\frac{1}{2}\right)_j \left(\frac{1}{2}\right)_j \left(\frac{1}{2}\right)_j}{(j!)^3} X(\gamma\tau_0)^j,$$

where

$$a_5 = 5\left(\frac{X(\gamma\tau_0)}{X(\gamma\tau_0) - 1}\right) + X(\tau_0)\left(\frac{dM_5}{dX}\right)|_{X=X(\tau_0)} = \frac{5 - \sqrt{5}}{2}.$$

So, we finally obtain

$$\frac{2\sqrt{5}}{\pi} = (\sqrt{5} - 2)^{1/2} \sum_{j \geq 0} \left(20j + 5 - \sqrt{5}\right) \frac{\left(\frac{1}{2}\right)_j \left(\frac{1}{2}\right)_j \left(\frac{1}{2}\right)_j}{(j!)^3} (9 - 4\sqrt{5})^j.$$

4.2 $\Gamma_0(3) \cong (3, \infty, \infty)$

As in Table 2 we choose the Hauptmodul

$$X(\tau) = t_3(\tau) = -27\frac{\eta(3\tau)^{12}}{\eta(\tau)^{12}},$$

and for the corresponding domain D we choose the intersection of $\{\tau \in \mathfrak{h} : |X(\tau)| < 1\}$ and the following fundamental domain FD of $\Gamma_0(3)$:

$$FD = \{\tau \in \mathfrak{h} : |\mathrm{Re}(\tau)| \leq 1/2,\ |\tau - 1/3| > 1/3,\ |\tau + 1/3| > 1/3\}.$$

Recall $Z = Z_3$ as given in (21). From the hypergeometric product formula [16, Theorem 2.3], we get that

$$Z(\tau) = (1 - X(\tau))^{\frac{2}{3}} \sum_{j=0}^{\infty} A_j X(\tau)^j,$$

where

$$A_j = \frac{\left(\frac{1}{3}\right)_j^2}{j!^2} \sum_{k=0}^{j} \frac{(-j)_k^2 \left(\frac{1}{3}\right)_k^2}{k!^2 \left(\frac{2}{3} - j\right)_k^2}. \tag{32}$$

Recalling (25), we see that Z and X above satisfy the conditions in Corollary 4.1 with $U = 1$, $k = 2$, $\varepsilon_0 = 0$, $\varepsilon_1 = \frac{2}{3}$, and $\alpha = 1$.

Example 4.6 Consider $N = 2$, $\gamma = w_3 = \frac{1}{\sqrt{3}} \left(\begin{smallmatrix} 0 & -1 \\ 3 & 0 \end{smallmatrix}\right)$, and $\tau_0 = \frac{i}{\sqrt{6}}$. Using the values $\eta\left(i\sqrt{3/2}\right)$, $\eta\left(i/\sqrt{6}\right)$, $\eta\left(i\sqrt{6}\right)$, and $\eta\left(i\sqrt{2/3}\right)$ from Table 5, we have

$$t_3\left(\frac{i}{\sqrt{6}}\right) = -3 - 2\sqrt{2},$$

$$t_3\left(i\sqrt{\frac{2}{3}}\right) = -3 + 2\sqrt{2}.$$

Using the polynomial relationship $\Phi_2(X, Y) = 0$ between $X(\tau) = t_3(\tau)$ and $Y(\tau) = t_3(\gamma\tau)$ from Lemma 6.3, we get

$$\left(\frac{dM_2}{dX}\right)\Big|_{X=X(\tau_0)} = \frac{4 - 3\sqrt{2}}{3}.$$

Then, from Corollary 4.1 and (28), we get the following series:

$$\frac{\sqrt{6}}{\pi} = (1 - X(\gamma\tau_0))^{2/3} \sum_{j\geq 0} (4j + a_2) A_j X(\gamma\tau_0)^j,$$

where A_j is given in (32) and

$$a_2 = 4\left(\frac{2}{3} \frac{X(\gamma\tau_0)}{X(\gamma\tau_0) - 1}\right) + X(\tau_0)\left(\frac{dM_2}{dX}\right)\Big|_{X=X(\tau_0)} = \frac{4 - \sqrt{2}}{3}.$$

Then, finally we obtain

$$\frac{3\sqrt{6}}{\pi} = (4-2\sqrt{2})^{2/3}\sum_{j\geq 0}(12j+4-\sqrt{2})A_j(2\sqrt{2}-3)^j,$$

where A_j is given in (32).

4.3 $\Gamma_0(4) \cong (\infty, \infty, \infty)$

By Table 2, we choose the Hauptmodul

$$X(\tau) = t_\infty(\tau) = 16\eta(\tau)^8\eta(4\tau)^{16}/\eta(2\tau)^{24}, \tag{33}$$

which can also be written $\lambda(2\tau)$, where $\lambda(\tau)$ is the modular lambda function introduced in (6). For the corresponding fundamental domain, we use

$$\{\tau \in \mathfrak{h} : |\mathrm{Re}(\tau)| \leq 1, |\tau - 1/4| > 1/4, |\tau - 3/4| > 1/4\}.$$

Then using (22) and (10), Z is a weight 2 modular form for $\Gamma_0(4)$ given by

$$Z(\tau) = {}_2F_1\left[\begin{matrix}\frac{1}{2} & \frac{1}{2} \\ & 1\end{matrix}\; ;\; X(\tau)\right]^2 = \theta_3(2\tau)^4.$$

Using the hypergeometric product formula [16, Theorem 2.3],

$$Z(\tau) = \sum_{j=0}^{\infty} A_j X(\tau)^j,$$

where

$$A_j = \frac{\left(\frac{1}{2}\right)_j^2}{j!^2}\sum_{k=0}^{j}\frac{(j)_k^2\left(\frac{1}{2}\right)_k^2}{k!^2\left(\frac{1}{2}-j\right)_k^2}. \tag{34}$$

As before, we write $f' := \frac{1}{2\pi i}\frac{df}{d\tau}$. We next compute U so that

$$X' = XZU. \tag{35}$$

Differentiating $X(\tau) = 16\eta(\tau)^8\eta(4\tau)^{16}/\eta(2\tau)^{24}$ and using the classical fact that $\eta' = \frac{1}{24}\eta E_2$, we compute that

$$X' = (1 - X)XZ.$$

Here we needed to use the identity $E_{2,2}(\tau) = -(\theta_2(2\tau)^4 + \theta_3(2\tau)^4)$ from (8). Thus

$$U = 1 - X.$$

From Lemma 3.1, we obtain the following by differentiating M_N with respect to X:

$$\frac{dM_N}{dX} = N\frac{Y}{X}\frac{1-Y}{1-X}$$
$$\times\left(\frac{d\left(\frac{dX}{dY}\right)}{dX} + \frac{dX}{dY}\frac{X\left(\frac{dY}{dX}\right) - Y}{XY} + \frac{dX}{dY}\frac{(1-X)\left(-\frac{dY}{dX}\right) + (1-Y)}{(1-Y)(1-X)}\right).$$

We choose $\gamma = \begin{pmatrix} 0 & -1/2 \\ 2 & 0 \end{pmatrix}$. Using transformation properties of theta functions, we can show for γ that

$$Z(-1/4\tau) = \theta_3(-1/2\tau)^4 = -4\tau^2 \cdot \theta_3(2\tau)^4 = -4\tau^2 Z(\tau).$$

Thus Z and X above satisfy the conditions in Corollary 4.1 with $U = 1 - X$, $\gamma = \begin{pmatrix} 0 & -1/2 \\ 2 & 0 \end{pmatrix}$, $k = 2$, $\varepsilon_0 = 0$, $\varepsilon_1 = 0$, and $\alpha = -1$.

Example 4.7 Let $N = 2$, $\gamma = \begin{pmatrix} 0 & -1/2 \\ 2 & 0 \end{pmatrix}$, and $\tau_0 = i/2\sqrt{2}$. Using the value $\lambda(\sqrt{2}i) = (\sqrt{2} - 1)^2$ [3, (4.6.10)], we obtain

$$X(i/\sqrt{2}) = 3 - 2\sqrt{2},$$
$$U(i/\sqrt{2}) = 2(\sqrt{2} - 1).$$

Moreover, the transformation property of λ (7) yields that

$$X(i/2\sqrt{2}) = \lambda(i/\sqrt{2}) = 1 - (\sqrt{2} - 1)^2 = 2(\sqrt{2} - 1),$$
$$U(i/2\sqrt{2}) = 3 - 2\sqrt{2}.$$

Next, using the polynomial relationship $\Phi_2(X, Y) = 0$ satisfied by $X(\tau) = t_\infty(\tau)$ and $Y(\tau) = t_\infty(\gamma\tau)$ from Lemma 6.4, we get

$$\frac{dM_2}{dX}\Big|_{X=X(i/2\sqrt{2})} = \sqrt{2} + 2.$$

Thus applying Corollary 4.1 yields that

$$\frac{\sqrt{2}}{\pi}=\sum_{j\geq 0}(4(\sqrt{2}-1)j-4+3\sqrt{2})A_j(3-2\sqrt{2})^j,$$

where A_j is given in (34).

5 Examples of Theorem 1.1 of Second Type

In this section we obtain five additional Ramanujan–Sato series for $1/\pi$ as examples of Theorem 1.1. To construct the examples we use the following corollary for modular forms on certain groups where the cusp $i\infty$ has general width h, which follows immediately from Theorem 1.1 with $\gamma = \gamma_s$, $\delta = I$, $\alpha = 1$, $\beta = 1$, and $N \in \mathbb{N}$ is arbitrary.

Corollary 5.1 *Let* Γ *be a discrete subgroup of* $\mathrm{SL}_2(\mathbb{R})$ *commensurable with* $\mathrm{SL}_2(\mathbb{Z})$ *such that* $\gamma_s = \frac{1}{\sqrt{s}}\left(\begin{smallmatrix}0 & -1\\ s & 0\end{smallmatrix}\right) \in \Gamma$ *for some real* $s > 0$*, and let* h *be the width of the cusp* $i\infty$ *of* Γ*. Let* $X(\tau)$ *be a Hauptmodul of* Γ *and* $Z(\tau)$ *a weight-*k *modular form for* Γ *such that* $\frac{h}{2\pi i}\frac{dX}{d\tau} = U(\tau)X(\tau)Z(\tau)$*, and when* $\tau \in D$*, a domain of* $\mathfrak{h}$*,* $Z(\tau) = X(\tau)^{\varepsilon_0}(1-X(\tau))^{\varepsilon_1}\sum_{j=0}^{\infty} A_j X(\tau)^j$ *for* $\varepsilon_0, \varepsilon_1 \in \mathbb{R}$*,* $A_j \in \mathbb{C}$*. Set* $M_N(\tau) := Z(\tau)/Z(N\tau)$ *for* $N \in \mathbb{N}$ *and let* $\tau_0 = \frac{i}{\sqrt{sN}}$*. Then if* $\tau_0 \in D$*,*

$$\frac{kh\sqrt{sN}}{2\pi}=U(i/\sqrt{sN})X(i/\sqrt{sN})^{\varepsilon_0}\left(1-X(i/\sqrt{sN})\right)^{\varepsilon_1}$$
$$\times\sum_{j\geq 0}(2j+a_N)A_jX\left(i/\sqrt{sN}\right)^j,$$

where

$$a_N=2\left(\varepsilon_0+\varepsilon_1\frac{X(i/\sqrt{sN})}{X(i/\sqrt{sN})-1}\right)-\frac{i^k}{N^{k/2}}X(i/\sqrt{sN})\left(\frac{dM_N}{dX}\right)\Big|_{X=X(i/\sqrt{sN})}.$$

Alternatively if $\gamma\tau_0 \in D$*,*

$$\frac{kh\sqrt{sN}}{2\pi}=X(i\sqrt{N}/\sqrt{s})^{\varepsilon_0}(1-X(i\sqrt{N}/\sqrt{s}))^{\varepsilon_1}\sum_{j\geq 0}(b_N'j+a_N')A_jX(i\sqrt{N}/\sqrt{s})^j,$$

where

$$b'_N = 2NU(i\sqrt{N}/\sqrt{s}),$$
$$a'_N = 2NU(i\sqrt{N}/\sqrt{s})\left(\varepsilon_0 + \varepsilon_1\frac{X(i\sqrt{N}/\sqrt{s})}{X(i\sqrt{N}/\sqrt{s})-1}\right)$$
$$+\,U(i/\sqrt{sN})X(i/\sqrt{sN})\left(\frac{dM_N}{dX}\right)\Big|_{X=X(i/\sqrt{sN})}.$$

From Remark 2.5, we know that the arithmetic triangle groups $(2, m, \infty)$ for $m \in \{3, 4, 6\}$ contain the element γ_s for $s = \lfloor \frac{m}{2} \rfloor$, respectively. In this section we construct examples of Corollary 5.1 for these groups.

Fix $m \in \{3, 4, 6\}$ and let $\Gamma_m = (2, m, \infty)$. Then the width of the cusp $i\infty$ of Γ_m is $h = 1$, and we have seen in Tables 2 and 3 that $t_{2,m}$ is a Hauptmodul for Γ_m and that the space of modular forms of weight 4 for Γ_m is generated by $Z_m(\tau)$. Using Clausen's formula (11),

$$Z_m(\tau) := {}_2F_1\left[\begin{matrix}\frac{1}{4}-\frac{1}{2m} & \frac{1}{4}+\frac{1}{2m} \\ & 1\end{matrix}\ ;\ t_{2,m}(\tau)\right]^4$$
$$= {}_3F_2\left[\begin{matrix}\frac{1}{2} & \frac{1}{2}-\frac{1}{m} & \frac{1}{2}+\frac{1}{m} \\ & 1 & 1\end{matrix}\ ;\ t_{2,m}(\tau)\right]^2.$$

Thus we see that

$$Z_m(\tau) = \sum_{j=0}^{\infty} A_{m,j}\, t_{2,m}(\tau)^j, \tag{36}$$

where by the hypergeometric product formula [16, Theorem 2.3], we have[2]

$$A_{m,j} = \frac{\left(\frac{1}{2}\right)_j\left(\frac{1}{2}-\frac{1}{m}\right)_j\left(\frac{1}{2}+\frac{1}{m}\right)_j}{j!^3}\sum_{n=0}^{j}\frac{(-j)_n^3(\frac{1}{2})_n(\frac{1}{2}-\frac{1}{m})_n(\frac{1}{2}+\frac{1}{m})_n}{(\frac{1}{2}-j)_n(\frac{1}{2}+\frac{1}{m}-j)_n(\frac{1}{2}-\frac{1}{m}-j)_n n!^3}. \tag{37}$$

Thus for each $m \in \{3, 4, 6\}$, Z_m meets the conditions for Corollary 5.1 with $\varepsilon_0 = \varepsilon_1 = 0$.

We use Theorem 2.2 to recognize Z_m in terms of common modular forms. In each case the weight is $k = 4$, so it suffices to check the Fourier coefficients up to the q^1 term. We obtain that

[2] Note that the ${}_6F_5$ series arising from the formula in [16] naturally truncates at j.

$$Z_3(\tau) = E_4(\tau), \tag{38}$$

$$Z_4(\tau) = E_{2,2}(\tau)^2, \tag{39}$$

$$Z_6(\tau) = \frac{1}{4} E_{2,3}(\tau)^2. \tag{40}$$

Write $f' := q\frac{df}{dq} = \frac{1}{2\pi i}\frac{df}{d\tau}$. We next compute each U_m so that

$$t'_{2,m} = t_{2,m} Z_m U_m. \tag{41}$$

When $m = 3$, we have $t_{2,3} = 1728/j$. Thus,

$$t'_{2,3} = \left(\frac{-j'}{j}\right) t_{2,3},$$

and using Theorem 2.2 with $k = 6$ it is easy to check that

$$-j' E_4 = j E_6,$$

so from (38) we obtain that

$$U_3(\tau) = \frac{E_6(\tau)}{E_4(\tau)^2}. \tag{42}$$

For $m = 4, 6$ we first observe that

$$t_{2,m} = \frac{4\alpha f g}{(f + \alpha g)^2},$$

where $f(\tau) = \eta(\tau)^{k_m}$, $g(\tau) = \eta(m\tau/2)^{k_m}$, $k_m = 48/(m-2)$, and $\alpha = (m/2)^{12/(m-2)}$. Using the fact that $\eta' = \frac{1}{24}\eta E_2$, differentiating and simplifying yields

$$\begin{aligned} t'_{2,m} &= \frac{4\alpha[-fg(f' + \alpha g' + (f^2 g' + \alpha f' g^2)]}{(f + \alpha g)^3} \\ &= t_{2,m} \frac{(f - \alpha g)}{(f + \alpha g)} \left[\frac{g'}{g} - \frac{f'}{f}\right] \\ &= t_{2,m} \frac{(f - \alpha g)}{(f + \alpha g)} \left(\frac{-k_m}{24}\right) E_{2,\frac{m}{2}} \\ &= t_{2,m} Z_m \frac{(2 - m)}{2E_{2,\frac{m}{2}}} \frac{(f - \alpha g)}{(f + \alpha g)}. \end{aligned}$$

Thus we have that

$$U_4(\tau) = \frac{-1}{E_{2,2}(\tau)} \cdot \frac{(\Delta(\tau) - 64\Delta(2\tau))}{(\Delta(\tau) + 64\Delta(2\tau))}, \tag{43}$$

$$U_6(\tau) = \frac{-2}{E_{2,3}(\tau)} \cdot \frac{(\eta(\tau)^{12} - 27\eta(3\tau)^{12})}{(\eta(\tau)^{12} + 27\eta(3\tau)^{12})}. \tag{44}$$

Note that in each case we can see by Theorem 2.2 that

$$U_m(\tau)^2 = \frac{1 - t_{2,m}(\tau)}{Z_m(\tau)}. \tag{45}$$

Moreover, we have that $1/U(\tau)$ is a meromorphic modular form of weight 2 on $\mathrm{SL}_2(\mathbb{Z})$ so

$$U_m(-1/s\tau) = (\sqrt{s}\tau)^{-2} U_m(\tau). \tag{46}$$

For each $m \in \{3, 4, 6\}$, let $X = t_{2,m}$, $Y = X(N\tau)$, and $U = U_m$. In order to compute examples via Corollary 5.1, we need to know the derivative $\frac{dM_N}{dX}$. By Lemma 3.1 we have that

$$M_N(\tau) = N \frac{dX}{dY} \frac{Y}{X}(\tau) \frac{U(N\tau)}{U(\tau)}, \tag{47}$$

so

$$M_N(\tau)^2 = N^2 \left(\frac{dX}{dY} \frac{Y}{X}(\tau) \right)^2 \frac{U(N\tau)^2}{U(\tau)^2}. \tag{48}$$

From (45) we have

$$\frac{U(N\tau)^2}{U(\tau)^2} = \left(\frac{1 - Y(\tau)}{Z_m(N\tau)} \cdot \frac{Z_m(\tau)}{1 - X(\tau)} \right) = \frac{1 - Y}{1 - X} M_N(\tau). \tag{49}$$

Therefore

$$M_N(\tau)^2 = N^2 \left(\frac{dX}{dY} \frac{Y}{X} \right)^2 \frac{1 - Y}{1 - X} M_N(\tau),$$

and so

$$M_N = N^2 \left(\frac{dX}{dY} \frac{Y}{X} \right)^2 \frac{1 - Y}{1 - X}. \tag{50}$$

Differentiating with respect to X, we get

$$\frac{dM_N}{dX} = 2N^2\left(\frac{dX}{dY}\frac{Y}{X}\right)\left(\frac{1-Y}{1-X}\right)\left(\frac{d\left(\frac{dX}{dY}\right)}{dX}\frac{Y}{X} + \frac{dX}{dY}\frac{X\left(\frac{dY}{dX}\right)-Y}{X^2}\right)+$$
$$N^2\left(\frac{dX}{dY}\frac{Y}{X}\right)^2\frac{(1-X)\left(-\frac{dY}{dX}\right)+(1-Y)}{(1-X)^2}. \tag{51}$$

For each $m \in \{3, 4, 6\}$ we construct specific examples of Corollary 5.1 for some choices of N. For each example we need to compute the special values of X, Y, and U_m and find an explicit polynomial relationship between $X(\tau) := t_{2,m}(\tau)$ and $Y(\tau) := X(N\tau)$ in order to compute the special value of $\frac{dM_N}{dX}$.

5.1 $\mathrm{PSL}_2(\mathbb{Z}) \cong (2, 3, \infty)$

For Examples 5.2 and 5.3 we consider $m = 3$, $s = 1$, and $\gamma = \left(\begin{smallmatrix}0 & -1\\ 1 & 0\end{smallmatrix}\right)$.

As in Tables 2 and 3, we define

$$X(\tau) := t_{2,3}(\tau) = \frac{1728}{j(\tau)},$$

and for the corresponding domain D we choose the standard fundamental domain

$$\{\tau \in \mathfrak{h} : |\mathrm{Re}(\tau)| \le 1/2,\ |\tau| > 1\}.$$

Let

$$Z(\tau) := Z_3(\tau) = \sum_{j=0}^{\infty} A_j X^j(\tau) = E_4(\tau),$$

where by (37),

$$A_j = \frac{(\frac{1}{6})_j(\frac{5}{6})_j(\frac{1}{2})_j}{j!^3}\cdot\sum_{n=0}^{j}\frac{(-j)_n(\frac{1}{6})_n(\frac{5}{6})_n(\frac{1}{2})_n}{(\frac{5}{6}-j)_n(\frac{1}{6}-j)_n(\frac{1}{2}-j)_n n!^3}. \tag{52}$$

Furthermore, define $Y(\tau) := X(N\tau)$ and $U(\tau) := U_3(\tau)$.

Example 5.2 Let $N = 2$ and $\tau_0 = \frac{i}{\sqrt{2}}$. We need to determine $X(i/\sqrt{2})$ and $Y(i/\sqrt{2}) = X(i\sqrt{2})$. From the values of $j(i/\sqrt{2})$ and $j(i\sqrt{2})$ in Table 4, we have

$$X(i/\sqrt{2}) = Y(i/\sqrt{2}) = \frac{27}{125}.$$

Using the polynomial relationship $\Phi_2(X, Y) = 0$ from Lemma 6.1 and (51), we find $\frac{dM_2}{dX}\Big|_{X(\tau_0)} = -\frac{500}{63}$.

Letting $\tau = \sqrt{2}i$ in (46) gives that

$$U(i\sqrt{2}) = \frac{1}{2}U(i/\sqrt{2}). \tag{53}$$

Using (45), we can determine $U(i\sqrt{2})$ using the value of $X(i\sqrt{2})$ from above and $Z_3(i\sqrt{2}) = E_4(i\sqrt{2})$ from Table 6 to determine that

$$U(\sqrt{2}i) = \pi^3 \frac{2^5 \cdot 7}{5^2} \Gamma\left(\frac{1}{8}\right)^{-2} \Gamma\left(\frac{3}{8}\right)^{-2},$$

and thus from (46)

$$U(i/\sqrt{2}) = -\pi^3 \frac{2^4 \cdot 7}{5^2} \Gamma\left(\frac{1}{8}\right)^{-2} \Gamma\left(\frac{3}{8}\right)^{-2}.$$

Since $\tau_0 = \frac{i}{\sqrt{2}} \notin D$ but $\gamma\tau_0 \in D$, Corollary 5.1 implies that

$$\frac{2\sqrt{2}}{\pi} = \sum_{j=0}^{\infty} (a_2 + b_2 j)\, A_j X^j(\sqrt{2}i),$$

where

$$a_N = U(i/\sqrt{2})X(i/\sqrt{2}) \left.\frac{dM_N}{dX}\right|_{X=X(i/\sqrt{2})} = \frac{\pi^3 \cdot 3 \cdot 2^6}{5^2} \Gamma\left(\frac{1}{8}\right)^{-2} \Gamma\left(\frac{3}{8}\right)^{-2},$$

$$b_N = 4U(\sqrt{2}i) = \frac{\pi^3 \cdot 2^7 \cdot 7}{5^2} \Gamma\left(\frac{1}{8}\right)^{-2} \Gamma\left(\frac{3}{8}\right)^{-2}.$$

This can be written as

$$\frac{5^2}{\pi^4} = 2^{9/2} \Gamma\left(\frac{1}{8}\right)^{-2} \Gamma\left(\frac{3}{8}\right)^{-2} \sum_{j=0}^{\infty} (3 + 14j)\, A_j \left(\frac{27}{125}\right)^j,$$

where A_j is as in (52).

Example 5.3 Consider $N = 3$ so $\tau_0 = \frac{i}{\sqrt{3}}$. Using the values $j(i/\sqrt{3})$ and $j(i\sqrt{3})$ from Table 4, we get

$$X(i/\sqrt{3}) = Y(i/\sqrt{3}) = \frac{4}{125}.$$

Using the polynomial relationship $\Phi_3(X, Y) = 0$ from Lemma 6.1 and (51), we find $\frac{dM_3}{dX}\Big|_{X=X(i/\sqrt{3})} = -\frac{1125}{11}$. Furthermore, using (45), the value $X(i\sqrt{3})$ above, and the value of $E_4(i\sqrt{3})$ from Table 6, we have

$$U(\sqrt{3}i) = \sqrt{\frac{1-\frac{4}{125}}{E_4(\sqrt{3}i)}} = \frac{2^{14/3}\cdot 11\cdot \pi^4}{3\cdot 5^2\cdot \Gamma\left(\frac{1}{3}\right)^6}.$$

Thus from (46) we obtain

$$U(i/\sqrt{3}) = -\frac{1}{3}U(\sqrt{3}i) = -\frac{2^{14/3}\cdot 11\cdot \pi^4}{3^2\cdot 5^2\cdot \Gamma\left(\frac{1}{3}\right)^6}.$$

Since $\tau = \frac{i}{\sqrt{3}} \notin D$ and $\gamma\tau = \sqrt{3}i \in D$, Corollary 5.1 implies that

$$\frac{2\sqrt{3}}{\pi} = \sum_{j=0}^{\infty}(a_3 + b_3 j)A_j X(e^{-(2/\sqrt{3})\pi})^j,$$

where

$$a_3 = \frac{2^{20/3}\cdot \pi^4}{5^2}\Gamma\left(\frac{1}{3}\right)^{-6},$$

$$b_3 = \frac{2^{17/3}\cdot 11\cdot \pi^4}{5^2}\Gamma\left(\frac{1}{3}\right)^{-6}.$$

We can write this as

$$\frac{\sqrt{3}\cdot 5^2}{\pi^5} = 2^{14/3}\Gamma\left(\frac{1}{3}\right)^{-6}\sum_{j=0}^{\infty}(11j+2)A_j\left(\frac{4}{125}\right)^j, \tag{54}$$

where A_j is as in (52).

5.2 $\Gamma_0^+(2) \cong (2, 4, \infty)$

For example 5.4 we consider the case where $m = 4$. Here we have that $s = 2$ so $\gamma = \frac{1}{\sqrt{2}}\begin{pmatrix} 0 & -1 \\ 2 & 0 \end{pmatrix}$.

As in Tables 2 and 3, we define

$$X(\tau) := t_{2,4} = \frac{256\eta(\tau)^{24}\eta(2\tau)^{24}}{(\eta(\tau)^{24} + 64\eta(2\tau)^{12})^2},$$

and for the corresponding domain D we choose the intersection of $\{\tau \in \mathfrak{h} : |X(\tau)| < 1\}$ and the following fundamental domain FD of $\Gamma_0^+(2)$:

$$FD = \{\tau \in \mathfrak{h} : |\mathrm{Re}(\tau)| \le 1/2,\ |\tau| > 1/\sqrt{2}\}.$$

Let

$$Z(\tau) := Z_4(\tau) = \sum_{j=0}^{\infty} A_j X(\tau)^j = E_{2,2}(\tau)^2,$$

where by (37),

$$A_j = \frac{\left(\frac{1}{2}\right)_j \left(\frac{1}{2} - \frac{1}{4}\right)_j \left(\frac{1}{2} + \frac{1}{4}\right)_j}{j!^3} \sum_{n=0}^{j} \frac{(-j)_n^3 (\frac{1}{2})_n (\frac{1}{4})_n (\frac{3}{4})_n}{(\frac{1}{2} - j)_n (\frac{3}{4} - j)_n (\frac{1}{4} - j)_n n!^3}. \tag{55}$$

Furthermore, define $Y(\tau) := X(N\tau)$ and $U(\tau) := U_4(\tau)$.

Example 5.4 Let $N = 3$ and $\tau_0 = i/\sqrt{6}$. Using the values of $\eta(i/\sqrt{6})$, $\eta(i\sqrt{2/3})$, $\eta(i\sqrt{3/2})$, and $\eta(i\sqrt{6})$ from Table 5, we get that

$$X(i/\sqrt{6}) = Y(i/\sqrt{6}) = \frac{1}{9}.$$

Using the polynomial relationship $\Phi_3(X, Y) = 0$ from Lemma 6.5 and (51), we find $\frac{dM_3}{dX}\Big|_{X=X(i/\sqrt{6})} = -\frac{81}{2}$. Recall from (45) that

$$U(\tau) = \frac{\sqrt{1 - X(\tau)}}{E_{2,2}(\tau)}.$$

Using the value of $E_{2,2}(i/\sqrt{6})$ from Table 7, we determine that

$$U\left(\frac{i}{\sqrt{6}}\right) = -\frac{32\pi^3}{3\sqrt{3}} \left(\Gamma\left(\frac{1}{24}\right)\Gamma\left(\frac{5}{24}\right)\Gamma\left(\frac{7}{24}\right)\Gamma\left(\frac{11}{24}\right)\right)^{-1}, \tag{56}$$

and furthermore from (46),

$$U\left(i\sqrt{\frac{3}{2}}\right) = \frac{32\pi^3}{\sqrt{3}} \left(\Gamma\left(\frac{1}{24}\right)\Gamma\left(\frac{5}{24}\right)\Gamma\left(\frac{7}{24}\right)\Gamma\left(\frac{11}{24}\right)\right)^{-1}. \tag{57}$$

Thus Corollary 5.1 yields that

$$\frac{2\sqrt{6}}{\pi} = \sum_{j=0}^{\infty}(b_3 j + a_3)A_j\left(\frac{1}{9}\right)^j, \tag{58}$$

where

$$a_3 = 2^4\sqrt{3}\pi^3\left(\Gamma\left(\frac{1}{24}\right)\Gamma\left(\frac{5}{24}\right)\Gamma\left(\frac{7}{24}\right)\Gamma\left(\frac{11}{24}\right)\right)^{-1},$$

$$b_3 = 2^6\sqrt{3}\pi^3\left(\Gamma\left(\frac{1}{24}\right)\Gamma\left(\frac{5}{24}\right)\Gamma\left(\frac{7}{24}\right)\Gamma\left(\frac{11}{24}\right)\right)^{-1},$$

and A_j is as in (55).

We can write this as

$$\frac{1}{\pi^4} = 2^{5/2}\left(\Gamma\left(\frac{1}{24}\right)\Gamma\left(\frac{5}{24}\right)\Gamma\left(\frac{7}{24}\right)\Gamma\left(\frac{11}{24}\right)\right)^{-1}\sum_{j=0}^{\infty}(4j+1)A_j\left(\frac{1}{9}\right)^j,$$

where A_j is as in (55).

5.3 $\Gamma_0^+(3) \cong (2, 6, \infty)$

For Examples 5.5 and 5.6 we consider $m = 6$, $s = 3$, and $\gamma = \frac{1}{\sqrt{3}}\left(\begin{smallmatrix} 0 & -1 \\ 3 & 0 \end{smallmatrix}\right)$. As in Tables 2 and 3, we define

$$X(\tau) := t_{2,6}(\tau) = \frac{108\eta(\tau)^{12}\eta(3\tau)^{12}}{(\eta(\tau)^{12} + 27\eta(3\tau)^{12})^2},$$

and for the corresponding domain D we choose the intersection of $\{\tau \in \mathfrak{h} : |X(\tau)| < 1\}$ and the following fundamental domain FD of $\Gamma_0^+(3)$:

$$\{\tau \in \mathfrak{h} : |\mathrm{Re}(\tau)| \leq 1/2,\ |\tau| > 1/\sqrt{3}\}.$$

Let

$$Z(\tau) := Z_6(\tau) = \sum_{j=0}^{\infty} A_j X(\tau)^j = \frac{1}{4}E_{2,3}(\tau)^2,$$

where

$$A_j = \frac{\left(\frac{1}{2}\right)_j \left(\frac{1}{3}\right)_j \left(\frac{2}{3}\right)_j}{j!^3} \sum_{n=0}^{j} \frac{(-j)_n \left(\frac{1}{2}\right)_n \left(\frac{1}{3}\right)_n \left(\frac{2}{3}\right)_n}{\left(\frac{1}{2}-j\right)_n \left(\frac{2}{3}-j\right)_n \left(\frac{1}{3}-j\right)_n n!^3}. \tag{59}$$

Furthermore, define $Y(\tau) := X(N\tau)$ and $U(\tau) := U_6(\tau)$. Recall from (44) that

$$U(\tau) = \frac{-2}{E_{2,3}(\tau)} \cdot \frac{(\eta(\tau)^{12} - 27\eta(3\tau)^{12})}{(\eta(\tau)^{12} + 27\eta(3\tau)^{12})}.$$

Using (9) we obtain

$$U(\tau) = \frac{(\eta(\tau)^{12} - 27\eta(3\tau)^{12})\eta(\tau)^2}{(\eta(\tau)^{12} + 27\eta(3\tau)^{12})(3\eta(3\tau)^3 + \eta(\tau/3)^3)^2}. \tag{60}$$

Example 5.5 Let $N = 2$ and $\tau = \frac{i}{\sqrt{6}}$. We use the values of $\eta\left(i/\sqrt{6}\right)$, $\eta\left(i\sqrt{6}\right)$, $\eta\left(i\sqrt{2/3}\right)$, $\eta\left(i\sqrt{3/2}\right)$, and $\eta\left(i/3\sqrt{6}\right)$ from Table 5 to compute $X(i/\sqrt{6}) = Y(i/\sqrt{6}) = \frac{1}{2}$. Using the polynomial relationship $\Phi_2(X, Y) = 0$ from Lemma 6.6 and (51), we get that $\frac{dM_2}{dX}\Big|_{X=X(i/\sqrt{6})} = -\frac{16}{3}$.

Using (60) and the necessary η−values from Table 5, we obtain

$$U\left(\frac{i}{\sqrt{6}}\right) = -\frac{8\sqrt{2}\pi^3}{\sqrt{3}},$$

and from the modularity of U (46),

$$U\left(i\sqrt{\frac{2}{3}}\right) = \frac{16 \cdot \sqrt{2}\pi^3}{\sqrt{3}}.$$

Since $2\tau = i\sqrt{\frac{2}{3}} \in D$, Corollary 5.1 implies that

$$\frac{2\sqrt{6}}{\pi} = \sum_{j=0}^{\infty} (b_2 j + a_2) A_j \left(\frac{1}{2}\right)^j, \tag{61}$$

where

$$a_2 = -\frac{8}{3} U\left(\frac{i}{\sqrt{6}}\right),$$

$$b_2 = 4U\left(i\sqrt{\frac{2}{3}}\right) = -8U\left(\frac{i}{\sqrt{6}}\right).$$

We can write this as

$$\frac{9}{\pi^4} = 2^5 \cdot \left(\Gamma\left(\frac{1}{24}\right)\Gamma\left(\frac{5}{24}\right)\Gamma\left(\frac{7}{24}\right)\Gamma\left(\frac{11}{24}\right)\right)^{-1} \sum_{j=0}^{\infty}(3j+1)A_j\left(\frac{1}{2}\right)^j, \tag{62}$$

where A_j is as in (59).

Example 5.6 Let $N = 5$ and $\tau_0 = i/\sqrt{15}$. Using the values of $\eta(i/\sqrt{15})$, $\eta(i\sqrt{3/5})$, $\eta(i\sqrt{5/3})$ and $\eta(i\sqrt{10})$ from Table 5, we find that

$$X(i/\sqrt{15}) = Y(i/\sqrt{15}) = \frac{4}{125}.$$

Using the polynomial relationship $\Phi_5(X, Y) = 0$ from Lemma 6.6 and (51), we get $\frac{dM_5}{dX}\Big|_{X=X(i/\sqrt{15})} = \frac{-12500}{33}$. Furthermore, from (60) and (46), we use the values of $\eta(i/\sqrt{15})$, $\eta(i\sqrt{3/5})$, and $\eta(i/3\sqrt{15})$ from Table 5 to obtain

$$U(i/\sqrt{15}) = -\frac{352\pi^3}{25\sqrt{15}}\left(\Gamma\left(\frac{1}{15}\right)\Gamma\left(\frac{2}{15}\right)\Gamma\left(\frac{4}{15}\right)\Gamma\left(\frac{8}{15}\right)\right)^{-1},$$

and

$$U(i\sqrt{5/3}) = -5U(i/\sqrt{10}) = \frac{352\pi^3}{5\sqrt{15}}\left(\Gamma\left(\frac{1}{15}\right)\Gamma\left(\frac{2}{15}\right)\Gamma\left(\frac{4}{15}\right)\Gamma\left(\frac{8}{15}\right)\right)^{-1}.$$

By Corollary 5.1 we have that

$$\frac{2\sqrt{15}}{\pi} = \sum_{j=0}^{\infty}(b_5 j + a_5)A_j\left(\frac{4}{125}\right)^j,$$

where

$$b_5 = 10U(i\sqrt{5/3}) = \frac{10 \cdot 352\pi^3}{5\sqrt{15}}\left(\Gamma\left(\frac{1}{15}\right)\Gamma\left(\frac{2}{15}\right)\Gamma\left(\frac{4}{15}\right)\Gamma\left(\frac{8}{15}\right)\right)^{-1},$$

$$a_5 = \frac{400}{33} \cdot \frac{10 \cdot 352\pi^3}{5\sqrt{15}}\left(\Gamma\left(\frac{1}{15}\right)\Gamma\left(\frac{2}{15}\right)\Gamma\left(\frac{4}{15}\right)\Gamma\left(\frac{8}{15}\right)\right)^{-1}.$$

We can write this as

$$\frac{3^2 \cdot 5}{\pi^4} = 2^5\left(\Gamma\left(\frac{1}{15}\right)\Gamma\left(\frac{2}{15}\right)\Gamma\left(\frac{4}{15}\right)\Gamma\left(\frac{8}{15}\right)\right)^{-1} \cdot \sum_{j=0}^{\infty}(33j+8)\,A_j\left(\frac{4}{125}\right)^j,$$

where A_j is as in (59).

6 Appendix: Modular Polynomials

Throughout, let Γ be a discrete subgroup of $\mathrm{SL}_2(\mathbb{R})$ commensurable with $\mathrm{SL}_2(\mathbb{Z})$. Further assume Γ is of genus zero and contains a principal congruence subgroup $\Gamma(N)$. Let t be a Hauptmodul of Γ and N be the smallest positive integer such that Γ contains $\Gamma(N)$. For a positive integer m coprime to N, the modular polynomial of level m is defined to be the polynomial $\Phi_m(x, y)$ of minimal degree (up to scalar) such that for $\alpha \in \mathrm{GL}_2(\mathbb{Q})$ with $\det \alpha = m$,

$$\Phi_m(x, t(\tau)) = \prod_{\gamma \in \Gamma \backslash \Gamma \alpha \Gamma} (x - t(\gamma\tau)).$$

Below we state as lemmas each of the modular polynomials that we use in this chapter. In addition to using the Fourier expansion of t to find modular polynomials computationally, we prove some of the lemmas to illustrate how we can obtain modular polynomials using known modular polynomials computed using the method of Bröker, Lauter, and Sutherland [4] and the covering maps between modular curves.

6.1 Modular Polynomials for $t_{2,3}$, t_2, t_3, and t_∞

We first recall results for groups which are subgroups of $\mathrm{PSL}_2(\mathbb{Z})$.

Lemma 6.1 *For $t_{2,3}(\tau) = 1728/j(\tau)$, the level-2 and level-3 modular polynomials are, respectively,*

$$\begin{aligned}\Phi_2(X, Y) &= 1728(X^3+Y^3) - 162000(X^3Y+XY^3)+2571264(X^2Y+XY^2) - 2985984XY\\ &\quad + 5062500(X^3Y^2 + X^2Y^3) + 40773375X^2Y^2 - 52734375X^3Y^3,\\ \Phi_3(X, Y) &= (1728)^2(X^4 + Y^4) - (1728)^4XY + (1728)^3(2232)(X^2Y + XY^2)\\ &\quad - (1728)^2(1069956)(X^3Y + XY^3) + (1728)(36864000)(X^4Y + XY^4)\\ &\quad + (1728)^2(2587918086)X^2Y^2 + (1728)(8900222976000)(X^2Y^3 + X^3Y^2)\\ &\quad + 452984832000000(X^2Y^4 + X^4Y^2) - 770845966336000000X^3Y^3\\ &\quad + 1073741824000000000(X^4Y^3 + X^3Y^4).\end{aligned}$$

Proof Let $x = j(\tau)$ and $y = j(2\tau)$. From Sutherland [23] we have that x and y satisfy

$$0=(x^3+y^3)-162000(x^2+y^2)+1488(x^2y+xy^2)-x^2y^2 \\ +8748000000(x+y)+40773375xy-157464000000000. \qquad (63)$$

Since $X(\tau) = 1728/x$, multiplying through by $(1728)^4/x^3y^3$ gives us the first result.

Similarly, for $x = j(\tau)$ and $y = j(3\tau)$, from Sutherland [23] we have that x and y satisfy

$$\begin{aligned}0=(x^4+y^4)-x^3y^3+2232(x^3y^2+x^2y^3)-1069956(x^3y+xy^3)\\ +36864000(x^3+y^3)+2587918086x^2y^2\\ +8900222976000(x^2y+xy^2)+452984832000000(x^2+y^2)\\ -770845966336000000xy+1855425871872000000000(x+y).\end{aligned}$$

Multiplying through by $(1728)^6/x^4y^4$ and simplifying give us the second result. □

Lemma 6.2 *For* $t_2(\tau) = -64\frac{\eta(2\tau)^{24}}{\eta(\tau)^{24}}$, *the level-3 and level-5 modular polynomials are, respectively,*

$$\begin{aligned}\Phi_3(X,Y)&=X^4+Y^4-4096X^3Y^3-900\left(X^3Y+XY^3\right)+28422X^2Y^2\\ &\quad+4608\left(X^3Y^2+X^2Y^3+X^2Y+XY^2\right)-4096XY,\\ \Phi_5(X,Y)&=X^6+Y^6-16777216(XY+X^5Y^5)\\ &\quad+31457280(X^2Y+XY^2+X^4Y^5+X^5Y^4)\\ &\quad-17940480(X^3Y+XY^3+X^3Y^5+X^5Y^3)\\ &\quad+3143680(X^4Y+X^2Y^5+X^5Y^2+XY^4)\\ &\quad-90630(X^5Y+XY^5)+3709829120(X^2Y^2+X^4Y^4)\\ &\quad+746465295(X^4Y^2+X^2Y^4)\\ &\quad+6259476480(X^3Y^2+X^2Y^3+X^4Y^3+X^3Y^4)-33983400980X^3Y^3.\end{aligned}$$

Proof Recall from Sutherland [23] that the level-3 modular polynomial for the elliptic j-function is

$$\begin{aligned}\Phi_3(x,y)&=x^4+y^4-x^3y^3+2232(x^3y^2+x^2y^3)-1069956(x^3y+xy^3)\\ &\quad+2587918086x^2y^2\\ &\quad+36864000(x^3+y^3)+8900222976000(x^2y+xy^2)\end{aligned}$$

$$+ 452984832000000(x^2 + y^2)$$
$$- 770845966336000000xy + 1855425871872000000000(x + y).$$

Moreover the relation between j and t_2 is $j = \frac{64(4t_2-1)^3}{t_2}$. Hence, the functions $s := t_2(\tau)$ and $t := t_2(3\tau)$ satisfy the equation

$$\Phi_3\left(\frac{64(4s-1)^3}{s}, \frac{64(4t-1)^3}{t}\right) = 0.$$

Together with the Fourier expansions of s and t, we obtain

$$- 4096s^3t^3 + 4608(s^3t^2 + s^2t^3) + s^4 + t^4 - 900(s^3t + st^3) + 28422s^2t^2$$
$$+ 4608(s^2t + st^2) - 4096st = 0,$$

which gives the modular polynomial of level-3 for t_2. The proof for level-5 follows similarly. □

Lemma 6.3 *For* $t_3(\tau) := -27\frac{\eta(3\tau)^{12}}{\eta(\tau)^{12}}$, *the level-2 modular polynomial is*

$$\Phi_2(X, Y) = X^3 + Y^3 + 27X^2Y^2 - 24(X^2Y + XY^2) + 27XY.$$

Proof The proof follows similarly to that of Lemmas 6.1 and 6.2, using the relation

$$j = -27\frac{(t_3 - 1)(9t_3 - 1)^3}{t_3}.$$

□

Lemma 6.4 *For* $t_\infty(\tau) := -16\eta(\tau)^8\eta(4\tau)^{16}/\eta(2\tau)^{24}$, *the functions* $t_\infty(\tau)$ *and* $t_2(2\tau)$ *satisfy the equation*

$$0 = X^2Y^2 - 2X^2Y + X^2 + 16XY - 16Y.$$

6.2 Modular Polynomials for $t_{2,4}$ and $t_{2,6}$

Lemma 6.5 *For* $t_{2,4}(\tau) = \frac{256\eta(\tau)^{24}\eta(2\tau)^{24}}{(\eta(\tau)^{24}+64\eta(2\tau)^{24})^2}$, *the level-3 and level-5 modular polynomials are, respectively,*

$$\Phi_3(X, Y)$$
$$= X^4+Y^4+5308416X^4Y^4+442368(X^4Y^3+X^3Y^4)+13824(X^4Y^2 + X^2Y^4)$$

$$+192(X^4Y+XY^4)-14015488X^3Y^3+2058048(X^3Y^2+X^2Y^3)$$
$$-19332(X^3Y+XY^3)+3622662X^2Y^2+79872(X^2Y+XY^2)-65536XY,$$

$$\begin{aligned}\Phi_5(X,Y)&\\ &=X^6+Y^6+451377585192960000(X^6Y^4+X^4Y^6)+7612031596694077744X^5Y^5\\ &\quad+69657034752000(X^6Y^3+X^3Y^6)-609930927695462400(X^5Y^4+X^4Y^5)\\ &\quad+4031078400(X^6Y^2+X^2Y^6)-20244489582182400(X^5Y^3+X^3Y^5)\\ &\quad+15444168822005760 0X^4Y^4+103680(X^6Y+XY^6)\\ &\quad+4666060857600(X^5Y^2+X^2Y^5)\\ &\quad+36839200367577600(X^4Y^3+X^3Y^4)-65094150(X^5Y+XY^5)\\ &\quad+98471158056975(X^4Y^2+X^2Y^4)-13453926179834900X^3Y^3\\ &\quad+1256857600(X^4Y+XY^4)+173582058905600(X^3Y^2+X^2Y^3)\\ &\quad-5655756800(X^3Y+XY^3)+24370885427200X^2Y^2\\ &\quad+8724152320(X^2Y+XY^2)-4294967296XY.\end{aligned}$$

Proof In this case the group we are considering is $\Gamma_0(2)^{+2} := \langle\Gamma_0(2),\omega_2\rangle$. However, we first consider $\Gamma_0(6)^{+2} := \langle\Gamma_0(6),\omega_2\rangle$, which is an index-4 subgroup of $\Gamma_0(6)^{+2}$. It is known that $u = \left(\frac{\eta(6\tau)\eta(3\tau)}{\eta(\tau)\eta(2\tau)}\right)^4$ is a Hauptmodul on $\Gamma_0(6)^{+2}$ (see [8] for example), so X can be written as a rational function of degree 4 in u. In particular, one can check that

$$X=\frac{256u}{(1+27u)^4}.$$

Next, we observe that since ω_3 normalizes $\Gamma_0(6)^{+2}$, $u(\omega_3\tau)$ is also a Hauptmodul and $u(\omega_3\tau) = \frac{au+b}{cu+d}(\tau)$ for some $\left(\begin{smallmatrix} a & b\\ c & d\end{smallmatrix}\right) \in \mathrm{GL}_2(\mathbb{C})$. In particular, for $\omega_3 = \frac{1}{\sqrt{3}}\left(\begin{smallmatrix} 3 & -2\\ 6 & -3\end{smallmatrix}\right)$, we have $\omega_3\tau = \frac{3\tau-2}{6\tau-3}$. Thus using the transformation law for the η-function, we can relate $u(\omega_3\tau)$ and $u(\tau)$, namely,

$$u(\omega_3\tau)=\frac{\eta^4\left(6\frac{3\tau-2}{6\tau-3}\right)\eta^4\left(3\frac{3\tau-2}{6\tau-3}\right)}{\eta^4\left(\frac{3\tau-2}{6\tau-3}\right)\eta^4\left(2\frac{3\tau-2}{6\tau-3}\right)}=\frac{\eta^4\left(\left(\begin{smallmatrix} 3 & -2\\ 2 & -1\end{smallmatrix}\right)\tau\right)\eta^4\left(\left(\begin{smallmatrix} 3 & -4\\ 1 & -1\end{smallmatrix}\right)2\tau\right)}{\eta^4\left(\left(\begin{smallmatrix} 1 & -2\\ 2 & -3\end{smallmatrix}\right)3\tau\right)\eta^4\left(\left(\begin{smallmatrix} 1 & -4\\ 1 & -3\end{smallmatrix}\right)6\tau\right)}=\frac{1}{81u(\tau)}.$$

Hence,

$$Y(\tau):=X(\omega_3\tau)=\frac{256u^3}{(1+3u)^4},$$

and the rational function determined by the relations $X = \frac{256u}{(1+27u)^4}$ and $Y = \frac{256u^3}{(1+3u)^4}$ gives rise to the desired level-3 polynomial for X.

Next we obtain the polynomial for level-5. Consider the following Atkin–Lehner involutions for $\Gamma_0(10)$:

$$\omega_2 = \frac{1}{\sqrt{2}}\begin{pmatrix} 2 & -1 \\ 10 & -4 \end{pmatrix}, \quad \omega_5 = \frac{1}{\sqrt{5}}\begin{pmatrix} 5 & 2 \\ 10 & 5 \end{pmatrix}.$$

Similar to the previous case, we start with the relation between X and the given Hauptmodul $u = \left(\frac{\eta(10\tau)\eta(5\tau)}{\eta(\tau)\eta(2\tau)}\right)^2$ for $\Gamma_0(10)^{+2} := \langle \Gamma_0(10), \omega_2 \rangle$. One can check that

$$X = \frac{256u}{(1+25u)^4(25u^2+6u+1)},$$

and

$$u(\omega_5\tau) = \frac{1}{25}\left(\frac{\eta(\tau)\eta(2\tau)}{\eta(10\tau)\eta(5\tau)}\right)^2 = \frac{1}{25}\frac{1}{u(\tau)}.$$

Hence,

$$Y(\tau) := X(\omega_5\tau) = \frac{256u^5}{(u+1)^4(25u^2+6u+1)},$$

from which we obtain the desired level-5 polynomial. □

Lemma 6.6 *For* $t_{2,6}(\tau) = \frac{108\eta(\tau)^{12}\eta(3\tau)^{12}}{(\eta(\tau)^{12}+27\eta(3\tau)^{12})^2}$*, the level-2 and level-5 modular polynomials are, respectively,*

$$\begin{aligned}
\Phi_2(X, Y) = {} & 4X^3Y^3 - 12(X^3Y^2 + X^2Y^3) + 12(X^3Y + XY^3) - 381X^2Y^2 \\
& - 4(X^3 + Y^3) - 336(X^2Y + XY^2) + 432XY, \\
\Phi_5(X, Y) = {} & 262144000000X^6Y^6 + 19660800000X^6Y^5 \\
& + 19660800000X^5Y^6 + 614400000X^6Y^4 \\
& - 2550877126656X^5Y^5 + 614400000X^4Y^6 + 10240000X^6Y^3 \\
& + 2094980505600X^5Y^4 \\
& + 2094980505600X^4Y^5 + 10240000X^3Y^6 + 96000X^6Y^2 \\
& - 128213414400X^5Y^3 \\
& - 4716435974400X^4Y^4 - 128213414400X^3Y^5 + 96000X^2Y^6
\end{aligned}$$

$$+480X^6Y+1141065600X^5Y^2+3568236045600X^4Y^3$$
$$+3568236045600X^3Y^4+1141065600X^2Y^5$$
$$+480XY^6+X^6-1221150X^5Y+75265374975X^4Y^2$$
$$-4489016056900X^3Y^3$$
$$+75265374975X^2Y^4-1221150XY^5+Y^6+31422600X^4Y$$
$$+309367560600X^3Y^2$$
$$+309367560600X^2Y^3+31422600XY^4-160088400X^3Y$$
$$+101058937200X^2Y^2$$
$$-160088400XY^3+264539520X^2Y+264539520XY^2$$
$$-136048896XY.$$

7 Appendix: Special Values

In this appendix, we list the special values used in this chapter in Tables 4, 6, 7, and 5. Most of the values are obtained by either applying the Chowla–Selberg formula (see for example [10, Eq. (1)]) or using known or obtainable values together with relations between modular functions. We first give an example of each of these methods below.

Example 7.1 Let $\Delta(\tau) := \eta(\tau)^{24}$, the normalized weight-12 Hecke eigenform on $\mathrm{PSL}_2(\mathbb{Z})$. Then

$$\Delta(\sqrt{2}i) = \frac{1}{2^{33}\pi^{18}}\Gamma\left(\frac{1}{8}\right)^{12}\Gamma\left(\frac{3}{8}\right)^{12}.$$

Proof Using the Chowla–Selberg formula [10, Eq. (1)], we compute $\Delta(\sqrt{2}i)$ in terms of Gamma functions. In particular, $\mathbb{Q}(\sqrt{-8}) = \mathbb{Q}(\sqrt{-2})$ has class number 1, and the unique reduced binary quadratic form with discriminant -8 is x^2+2y^2. Moreover, ± 1 are the only roots of unity in $\mathbb{Q}(\sqrt{-2})$. Thus the Chowla–Selberg formula gives that

$$\Delta(\sqrt{2}i) = (16\pi)^{-6}\prod_{m=1}^{8}\Gamma\left(\frac{m}{8}\right)^{6\left(\frac{-8}{m}\right)},$$

where in the powers of the Gamma values are Kronecker symbols. Evaluating the Kronecker symbols, we have that $\left(\frac{a}{m}\right) = 0$ when a and m are both even, and $\left(\frac{-8}{1}\right) = \left(\frac{-8}{3}\right) = 1$, while $\left(\frac{-8}{5}\right) = \left(\frac{-8}{7}\right) = -1$. Thus we obtain

$$\Delta(\sqrt{2}i) = \frac{1}{2^{24}\pi^6} \frac{\Gamma\left(\frac{1}{8}\right)^6 \Gamma\left(\frac{3}{8}\right)^6}{\Gamma\left(\frac{5}{8}\right)^6 \Gamma\left(\frac{7}{8}\right)^6}.$$

By the Γ-reflection formula, $\Gamma(1/8)\Gamma(7/8) = \frac{\pi}{\sin(\pi/8)}$ and $\Gamma(3/8)\Gamma(5/8) = \frac{\pi}{\sin(3\pi/8)}$. Thus we have

$$\Delta(\sqrt{2}i) = \frac{(\sin(\pi/8)\sin(3\pi/8))^6}{2^{24}\pi^{18}} \Gamma\left(\frac{1}{8}\right)^{12} \Gamma\left(\frac{3}{8}\right)^{12}.$$

Using half-angle trigonometric formulas, we calculate that $\sin(\pi/8) = (1/2)\sqrt{2-\sqrt{2}}$ and $\sin(3\pi/8) = (1/2)\sqrt{2+\sqrt{2}}$ so that $(\sin(\pi/8)\sin(3\pi/8))^6 = 2^{-9}$, which gives the desired value. □

Example 7.2 Let $X(\tau)$ be the Hauptmodul $-64\frac{\eta(2\tau)^{24}}{\eta(\tau)^{24}}$ for $\Gamma_0(2)$. We have

$$X(i/\sqrt{6}) = -17 - 12\sqrt{2},$$

$$X(i\sqrt{6}) = 6\sqrt{2} + \frac{71}{8} - \frac{21}{4}\sqrt{3} - \frac{27}{8}\sqrt{6},$$

$$X(i\sqrt{3/2}) = -17 + 12\sqrt{2}.$$

Proof We first establish the j-value

$$j(i\sqrt{6}) = j(i/\sqrt{6}) = 1728(1399 + 988\sqrt{2}). \tag{64}$$

Since $i\sqrt{6}$ is a CM point and $\mathbb{Q}(i\sqrt{6})$ a CM field of discriminant -24 and class number 2, Class Field Theory gives that $\mathbb{Q}(j(i\sqrt{6})) = \mathbb{Q}(\sqrt{2})$ and $j(i\sqrt{6}) \in \mathbb{Z}[\sqrt{2}]$. Moreover, since the lattices $\mathbb{Z} + i\sqrt{6}\mathbb{Z}$ and $2\mathbb{Z} + i\sqrt{6}\mathbb{Z}$ are inequivalent as ideals of $\mathbb{Z}[i\sqrt{6}]$, the j-values $j(i\sqrt{6})$ and $j(i\sqrt{6}/2)$ are Galois conjugates, so there exist $a, b \in \mathbb{Z}$ such that $j(i\sqrt{6}) = a + b\sqrt{2}$ and $j(i\sqrt{6}/2) = a - b\sqrt{2}$. These are the j-invariants of elliptic curves with CM discriminant -24 over a quadratic field, and the L-functions and Modular Forms Database [26] show that the only j-values of such curves are $1728(1399+988\sqrt{2})$ and $1728(1399-988\sqrt{2})$. As the j-values are finite at these points, the Fourier expansion of j allows the use of numerical approximation to determine which value is which. (See [14, 21] for example.)

The relation between j and X is

$$j = \frac{64(4X-1)^3}{X}, \tag{65}$$

and solving the equation $\frac{64(4X-1)^3}{X} = 1728(1399 + 988\sqrt{2})$ yields the three solutions

$$6\sqrt{2}+\frac{71}{8}-\frac{21}{4}\sqrt{3}-\frac{27}{8}\sqrt{6},\quad 6\sqrt{2}+\frac{71}{8}+\frac{21}{4}\sqrt{3}+\frac{27}{8}\sqrt{6},\quad -17-12\sqrt{2}.$$

To nail down which of the three values above are $X(i/\sqrt{6})$ and $X(i\sqrt{6})$, one can plug $i/\sqrt{6}$ and $i\sqrt{6}$, respectively, into the Fourier expansion of X to approximate and recognize the desired values.

To find the third value $X(i\sqrt{3/2}) = X(i\sqrt{6}/2)$, we can use either of the following two methods. First, recall from (63) that level-2 polynomial satisfied by $x = j(\tau)$ and $y = j(2\tau)$ is

$$\begin{aligned}\Phi_2(x, y) = x^3 + y^3 - x^2y^2 + 1488(xy^2 + x^2y) - 162000(x^2 + y^2)\\ + 40773375xy + 8748000000(x + y) - 157464000000000.\end{aligned}$$

Thus using (64) to solve the equation

$$\Phi_2(x, j(i\sqrt{6})) = 0$$

yields the possible values of $j(i\sqrt{6}/2)$, so we can use the Fourier expansion of $j(\tau)$ as above to approximate $j(i\sqrt{6}/2)$ and determine which is the correct value. Then using (65) we can similarly deduce the desired value of $X(i\sqrt{3/2})$.

Alternatively, using the transformation law for the η-function, we note that

$$X\left(\frac{-1}{2\tau}\right) = -64\frac{\eta\left(\frac{-1}{\tau}\right)^{24}}{\eta\left(\frac{-1}{2\tau}\right)^{24}} = \frac{1}{X(\tau)}.$$

Therefore, letting $\tau = i/\sqrt{6}$ gives that

$$X(i\sqrt{3/2}) = 1/(-17 - 12\sqrt{2}) = -17 + 12\sqrt{2}. \tag{66}$$

□

7.1 Tables of Special Values

In Tables 4 and 5 we indicate references for known or determined values. We also indicate when a value is obtained directly or from previous values in the table using one or more of the following methods labeled A-K below

Table 4 j-values

τ	$j(\tau)$	Method
$i\sqrt{2}$	8000	[29]
$i/\sqrt{2}$	8000	F
$(1+i\sqrt{3})/2$	0	[29]
$i\sqrt{3}$	54000	E, G
$i/\sqrt{3}$	54000	F
$2i\sqrt{3}$	$40500(35010+20213\sqrt{3})$	E
$i\sqrt{3}/2$	$40500(35010-20213\sqrt{3})$	E
$i\sqrt{6}$	$1728(1399+988\sqrt{2})$	K
$i/\sqrt{6}$	$1728(1399+988\sqrt{2})$	F
$i\sqrt{6}/2$	$1728(1399-988\sqrt{2})$	K
$i\sqrt{2/3}$	$1728(1399-988\sqrt{2})$	E
$(1+i\sqrt{6})/2$	$216(27014055899 + 19101822064\sqrt{2} - 15596572446\sqrt{3} - 11028442113\sqrt{6})$	E, G
$i\sqrt{10}$	$8640(24635+11016\sqrt{5})$	K
$i/\sqrt{10}$	$8640(24635+11016\sqrt{5})$	F
$i\sqrt{5/2}$	$8640(24635-11016\sqrt{5})$	K
$i\sqrt{2/5}$	$8640(24635-11016\sqrt{5})$	F
$i\sqrt{5}$	$320(1975+884\sqrt{5})$	K
$i/\sqrt{5}$	$320(1975+884\sqrt{5})$	F
$(1+i\sqrt{5})/2$	$320(1975-884\sqrt{5})$	K
$(1+i/\sqrt{5})/2$	$320(1975-884\sqrt{5})$	E, G
$(1+i\sqrt{15})/2$	$-\frac{135}{2}(1415+637\sqrt{5})$	[29]
$i\sqrt{15}$	$\frac{135}{2}(274207975+122629507\sqrt{5})$	E, G
$i/\sqrt{15}$	$\frac{135}{2}(274207975+122629507\sqrt{5})$	F

A : Use the transformation formula $\eta(-1/\tau) = \sqrt{-i\tau}\eta(\tau)$ together with the value for $\eta(-1/\tau_0)$ to find the value of $\eta(\tau_0)$.

B : Use the relationship between $t_2(\tau) = -64/j_{2B}(\tau)$ and $j(\tau)$ (t_2 is given in terms of $\eta(\tau)$ and $\eta(2\tau)$) along with the values of $j(\tau_0)$ and $\eta(\tau_0)$ to find the value of $\eta(2\tau_0)$.

C : Use the action of the Atkin–Lehner involution W_3 on $t_{2,6}(\tau)$ sending $i\sqrt{6}$ to $i/3\sqrt{6}$ along with the value of $\eta(i\sqrt{6})$.

D : Use the modular polynomial $\phi_3 = x^4 + 36x^3 + 270x^2 - xj + 756x + 729$ relating $x = -27t_3(\tau)$ to $j(\tau)$ along with the values of $j(\tau_0)$ and $\eta(\tau_0)$ to find the value of $\eta(3\tau_0)$.

E : Use the modular polynomial relating $j(\tau)$ and $j(2\tau)$.

F : Apply the action of the matrix $S = \begin{pmatrix} 0 & -1 \\ 1 & 0 \end{pmatrix}$.

G : Apply the action of the matrix $T = \begin{pmatrix} 1 & 1 \\ 0 & 1 \end{pmatrix}$.

H : Use the modular polynomial relating $j(\tau)$ and $j(5\tau)$.

I : Use the transformation formula $\eta(\gamma\tau)^{24} = (c\tau + d)^{12}\eta(\tau)^{24}$ for $\gamma \in SL_2(\mathbb{Z})$.

Table 5 η-values

τ	$\eta(\tau)$	Method
$i\sqrt{3}$	$\frac{3^{1/8}}{2^{4/3}\pi}\Gamma(1/3)^{3/2}$	[17]
$(-1+i\sqrt{3})/2$	$\frac{3^{1/8}}{e^{\pi i/24}2\pi}\Gamma(1/3)^{3/2}$	[17]
$i\sqrt{3/2}$	$\frac{1}{2^{5/4}3^{1/4}\pi^{3/4}}(\sqrt{2}+1)^{1/12}\left(\Gamma(\frac{1}{24})\Gamma(\frac{5}{24})\Gamma(\frac{7}{24})\Gamma(\frac{11}{24})\right)^{1/4}$	(66), J
$i\sqrt{2/3}$	$\frac{1}{2^{3/2}\pi^{3/4}}(\sqrt{2}+1)^{1/12}\left(\Gamma(\frac{1}{24})\Gamma(\frac{5}{24})\Gamma(\frac{7}{24})\Gamma(\frac{11}{24})\right)^{1/4}$	A
$i\sqrt{6}$	$\frac{1}{2^{5/4}6^{1/4}\pi^{3/4}}(\sqrt{2}-1)^{1/12}\left(\Gamma(\frac{1}{24})\Gamma(\frac{5}{24})\Gamma(\frac{7}{24})\Gamma(\frac{11}{24})\right)^{1/4}$	(66), J
$i/\sqrt{6}$	$\frac{1}{2^{5/4}\pi^{3/4}}(\sqrt{2}-1)^{1/12}\left(\Gamma(\frac{1}{24})\Gamma(\frac{5}{24})\Gamma(\frac{7}{24})\Gamma(\frac{11}{24})\right)^{1/4}$	A
$2i\sqrt{2/3}$	$(\frac{3}{32}\sqrt{2}-\frac{71}{512}+\frac{21}{256}\sqrt{3}-\frac{27}{512}\sqrt{6})^{1/24}\eta\left(i\sqrt{2/3}\right)$	E, B
$i/3\sqrt{6}$	$6^{1/4}\eta(i\sqrt{6})((-12\sqrt{2}+12)(12\sqrt{2}+17)^{2/3}-18\sqrt{2}+36(12\sqrt{2}+17)^{1/3}-39)^{1/12}$	C
$2i\sqrt{6}$	$\left(\frac{3-2\sqrt{2}}{2359+1668\sqrt{2}+3\sqrt{1236594+874404\sqrt{2}}}\right)^{1/24}\cdot \frac{1}{6\cdot 2^{7/8}\cdot\pi^{3/4}}\left(\Gamma\left(\frac{1}{24}\right)\Gamma\left(\frac{5}{24}\right)\Gamma\left(\frac{7}{24}\right)\Gamma\left(\frac{11}{24}\right)\right)^{1/4}$	B
$(1+i\sqrt{15})/2$	$e^{\pi i/24}\frac{1}{2^{3/4}3^{1/4}5^{1/4}\pi^{3/4}}\left(\frac{1+\sqrt{5}}{2}\right)^{-1/12}\left(\Gamma(\frac{1}{15})\Gamma(\frac{2}{15})\Gamma(\frac{4}{15})\Gamma(\frac{8}{15})\right)^{1/4}$	[27]
$i\sqrt{15}$	$\frac{1}{2^{5/3}3^{1/4}5^{1/4}\pi^{3/4}}(\sqrt{5}-1)^{5/12}\left(\Gamma(\frac{1}{15})\Gamma(\frac{2}{15})\Gamma(\frac{4}{15})\Gamma(\frac{8}{15})\right)^{1/4}$	B, G
$i/\sqrt{15}$	$\frac{1}{2^{5/3}\pi^{3/4}}(\sqrt{5}-1)^{5/12}\left(\Gamma(\frac{1}{15})\Gamma(\frac{2}{15})\Gamma(\frac{4}{15})\Gamma(\frac{8}{15})\right)^{1/4}$	A
$i\sqrt{3/5}$	$\frac{1}{2^{7/4}\pi^{3/4}}(\sqrt{5}-1)^{5/12}(123+55\sqrt{5})^{1/12}\left(\frac{1}{3}\Gamma(\frac{1}{15})\Gamma(\frac{2}{15})\Gamma(\frac{4}{15})\Gamma(\frac{8}{15})\right)^{1/4}$	D
$i/3\sqrt{15}$	$\frac{(3(-377-165\sqrt{5}-b/a^{1/3}+(10a)^{1/3}))^{1/12}}{2(2\pi)^{3/4}}(\sqrt{5}-1)^{5/12}\left(\Gamma(\frac{1}{15})\Gamma(\frac{2}{15})\Gamma(\frac{4}{15})\Gamma(\frac{8}{15})\right)^{1/4}$	D, A
$i\sqrt{5/3}$	$\frac{1}{15}3^{1/4}5^{3/4}\eta(i/3\sqrt{15})$	A
$i\sqrt{5/2}$	$\frac{(\sqrt{5}-1)^{-1/4}}{2^{3/2}5^{1/4}\pi^{5/4}}\left(\Gamma(\frac{1}{40})\Gamma(\frac{7}{40})\Gamma(\frac{9}{40})\Gamma(\frac{11}{40})\Gamma(\frac{13}{40})\Gamma(\frac{19}{40})\Gamma(\frac{23}{40})\Gamma(\frac{37}{40})\right)^{1/4}$	B, J
$i\sqrt{10}$	$\frac{(\sqrt{5}-1)^{1/4}}{2^{9/4}5^{1/4}\pi^{5/4}}\left(\Gamma(\frac{1}{40})\Gamma(\frac{7}{40})\Gamma(\frac{9}{40})\Gamma(\frac{11}{40})\Gamma(\frac{13}{40})\Gamma(\frac{19}{40})\Gamma(\frac{23}{40})\Gamma(\frac{37}{40})\right)^{1/4}$	B, J
$i/\sqrt{10}$	$\frac{(\sqrt{5}-1)^{1/4}}{4\pi^{5/4}}\left(\Gamma(\frac{1}{40})\Gamma(\frac{7}{40})\Gamma(\frac{9}{40})\Gamma(\frac{11}{40})\Gamma(\frac{13}{40})\Gamma(\frac{19}{40})\Gamma(\frac{23}{40})\Gamma(\frac{37}{40})\right)^{1/4}$	A
$i\sqrt{2/5}$	$\frac{(161+72\sqrt{5})^{1/24}}{22^{3/4}(1+\sqrt{5})^{1/4}\pi^{5/4}}\left(\Gamma(\frac{1}{40})\Gamma(\frac{7}{40})\Gamma(\frac{9}{40})\Gamma(\frac{11}{40})\Gamma(\frac{13}{40})\Gamma(\frac{19}{40})\Gamma(\frac{23}{40})\Gamma(\frac{37}{40})\right)^{1/4}$	B
$2i\sqrt{2/5}$	$\frac{\left((161+72\sqrt{5})c\right)^{1/24}}{4\cdot 2^{1/8}(1+\sqrt{5})^{1/4}\pi^{5/4}}\left(\Gamma(\frac{1}{40})\Gamma(\frac{7}{40})\Gamma(\frac{9}{40})\Gamma(\frac{11}{40})\Gamma(\frac{13}{40})\Gamma(\frac{19}{40})\Gamma(\frac{23}{40})\Gamma(\frac{37}{40})\right)^{1/4}$	E, B

J : Use the Chowla–Selberg formula, as in Example 7.1.

K : Use Class Field Theory and the L-functions and Modular Forms Database, as in (64).

In Table 5 we also use the following definitions to preserve space:

$$a = (71639575 + 32038171\sqrt{5} + 77\sqrt{2838511914270 + 1269421119050\sqrt{5}}),$$

$$b = (6\cdot 10^{2/3}(9125 + 4081\sqrt{5})),$$

$$c = (-647 + 288\sqrt{5} + 9\sqrt{2(5145 - 2300\sqrt{5})}).$$

Table 6 E_k-values

τ	$E_4(\tau)$	$E_6(\tau)$
$i\sqrt{2}$	$\frac{5}{2^9\pi^6}\Gamma\left(\frac{1}{8}\right)^4\Gamma\left(\frac{3}{8}\right)^4$	$\frac{7}{2^{13}\pi^{12}}\Gamma\left(\frac{1}{8}\right)^6\Gamma\left(\frac{3}{8}\right)^6$
$i\sqrt{3}$	$\frac{3^2 5}{\sqrt[3]{2}\cdot 2^9\pi^8}\Gamma\left(\frac{1}{3}\right)^{12}$	$\frac{3^3 11}{2^{14}\pi^{12}}\Gamma\left(\frac{1}{3}\right)^{18}$

Table 7 $E_{2,k}$-values

k	τ	$E_{2,k}(\tau)$
2	$i/\sqrt{6}$	$-\frac{1+\sqrt{2}+(-1+\sqrt{2})^{2/3}(-7925+5695\sqrt{2}+4584\sqrt{3}-3282\sqrt{6})^{1/3}}{32(\sqrt{2}-1)^{2/3}(-71+48\sqrt{2}+42\sqrt{3}-27\sqrt{6})^{1/3}\pi^3}\Gamma\left(\frac{1}{24}\right)\Gamma\left(\frac{5}{24}\right)\Gamma\left(\frac{7}{24}\right)\Gamma\left(\frac{11}{24}\right)$

To compute the special values in Table 6, we utilize the relationship between E_4, E_6, and Δ, and in Table 7 we utilize the relationship $E_{2,2}(\tau) = -(\theta_3(2\tau)^4 + \theta_2(2\tau)^4)$.

References

1. G.E. Andrews, R. Askey, R. Roy, R. Roy, R. Askey, *Special Functions*, vol. 71 (Cambridge University, Cambridge, 1999)
2. G. Bauer, Von den Coefficienten der Reihen von Kugelfunctionen einer Variablen. J. Reine Angew. Math. **56**, 101–121 (1859)
3. J.M. Borwein, P.B. Borwein, Pi and the AGM, in *Canadian Mathematical Society Series of Monographs and Advanced Texts* (Wiley, New York, 1987). A study in analytic number theory and computational complexity, A Wiley-Interscience Publication
4. R. Bröker, K. Lauter, A. Sutherland, Modular polynomials via isogeny volcanoes. Math. Comput. **81**(278), 1201–1231 (2012)
5. J.H. Bruinier, G. van der Geer, G. Harder, D. Zagier, *The 1-2-3 of modular forms*. Universitext. (Springer, Berlin, 2008). Lectures from the Summer School on Modular Forms and their Applications held in Nordfjordeid, June 2004, Edited by Kristian Ranestad
6. H.H. Chan, S.H. Chan, Z. Liu, Domb's numbers and Ramanujan–Sato type series for $1/\pi$. Adv. Math. **186**(2), 396–410 (2004)
7. H.H. Chan, S. Cooper, Rational analogues of Ramanujan's series for $1/\pi$. Math. Proc. Cambridge Philos. Soc. **153**(2), 361–383 (2012)
8. H.H. Chan, H. Verrill, The Apéry numbers, the Almkvist-Zudilin numbers and new series for $1/\pi$. Math. Res. Lett. **16**(3), 405–420 (2009)
9. K. Chandrasekharan, Elliptic functions, in *Grundlehren der Mathematischen Wissenschaften [Fundamental Principles of Mathematical Sciences]*, vol. 281 (Springer, Berlin, 1985)
10. R. Chapman, W. Hart, Evaluation of the dedekind eta function. Can. Math. Bull. **49**(1), 21–35 (2006)
11. S.Y. Choi, C.H. Kim, Valence formulas for certain arithmetic groups and their applications. J. Math. Anal. Appl. **420**(1), 447–463 (2014)
12. D.V. Chudnovsky, G.V. Chudnovsky, Approximations and complex multiplication according to Ramanujan, in *Ramanujan revisited (Urbana-Champaign, Ill., 1987)* (Academic Press, Boston, 1988), pp. 375–472

13. J.H. Conway, S.P. Norton, Monstrous moonshine. Bull. Lond. Math. Soc. **11**(3), 308–339 (1979)
14. D.A. Cox, Primes of the form $x^2 + ny^2$, in *A Wiley-Interscience Publication* (Wiley, New York, 1989). Fermat, class field theory and complex multiplication
15. F. Diamond, J. Shurman, A first course in modular forms, in *Graduate Texts in Mathematics*, vol. 228 (Springer, New York, 2005)
16. N. Kaiblinger, Product of two hypergeometric functions with power arguments. J. Math. Anal. Appl. **479**(2), 2236–2255 (2019)
17. W.-C.W. Li, L. Long, F.-T. Tu, Computing special L-values of certain modular forms with complex multiplication, in *SIGMA 14 (2018), 090* (2018)
18. R.S. Maier, On rationally parametrized modular equations. J. Ramanujan Math. Soc. **24**(1), 1–73 (2009)
19. R. Miranda, *Algebraic Curves and Riemann Surfaces*, vol. 5 (American Mathematical Society, New York, 1995)
20. S. Ramanujan, Modular equations and approximations to π [Quart. J. Math. **45**, 350–372 (1914)], in *Collected papers of Srinivasa Ramanujan* (AMS Chelsea Publication, Providence, 2000), pp. 23–39
21. J.H. Silverman, Advanced topics in the arithmetic of elliptic curves, in *Graduate Texts in Mathematics*, vol. 151 (Springer, New York, 1994)
22. J. Sturm, On the congruence of modular forms, in *Number Theory* (Springer, Berlin, 1987), pp. 275–280
23. D. Sutherland, Modular Polynomials. https://math.mit.edu/~drew/ClassicalModPolys.html
24. K. Takeuchi, Arithmetic triangle groups. J. Math. Soc. Jpn. **29**(1), 91–106 (1977)
25. K. Takeuchi, Commensurability classes of arithmetic triangle groups. J. Fac. Sci. Univ. Tokyo Sect. IA Math. **24**(1), 201–212 (1977)
26. The LMFDB Collaboration, The L-functions and modular forms database. http://www.lmfdb.org (2021). [Online; accessed 25 August 2021]
27. A. van der Poorten, K.S. Williams, Values of the Dedekind eta function at quadratic irrationalities. Canad. J. Math. **51**(1), 176–224 (1999)
28. Y. Yang, Schwarzian differential equations and Hecke eigenforms on Shimura curves. Compos. Math. **149**(1), 1–31 (2013)
29. D. Zagier, Traces of singular moduli, in *Motives, Polylogarithms and Hodge Theory, Part I (Irvine, CA, 1998)*. International Press Lecture Series, vol. 3 (International Press, Somerville, MA, 2002), pp. 211–244
30. W. Zudilin, Ramanujan-type formulae for $1/\pi$: a second wind? in *Modular Forms and String Duality*. Fields Institute Communications, vol. 54 (American Mathematical Society, Providence, 2008), pp. 179–188

Mock Theta Functions and Related Combinatorics

Cristina Ballantine, Hannah Burson, Amanda Folsom, Chi-Yun Hsu, Isabella Negrini, and Boya Wen

MSC 2020 11P81, 11P82, 05A17, 05A19

1 Introduction

1.1 Mock Theta Functions

In Ramanujan's last letter to Hardy from 1920, he presented his *mock theta functions,* a collection of 17 curious q-hypergeometric series including

C. Ballantine (✉)
Department of Mathematics and Computer Science, College of the Holy Cross, Worcester, MA, USA
e-mail: cballant@holycross.edu

H. Burson
School of Mathematics, University of Minnesota, Minneapolis, MN, USA
e-mail: hburson@umn.edu

A. Folsom
Department of Mathematics and Statistics, Amherst College, Amherst, MA, USA
e-mail: afolsom@amherst.edu

C.-Y. Hsu
Department of Mathematics, University of California, Los Angeles, Los Angeles, CA, USA

Department of Mathematics and Computer Science, Santa Clara University, Santa Clara, CA, USA
e-mail: chsu6@scu.edu

I. Negrini
Department of Mathematics, University of Toronto, Toronto, ON, Canada
e-mail: isabella.negrini@utoronto.ca

B. Wen
Department of Mathematics, University of Wisconsin-Madison, Madison, WI, USA
e-mail: bwen25@wisc.edu

A. Bucur et al. (eds.), *Research Directions in Number Theory*, Association for Women in Mathematics Series 33, https://doi.org/10.1007/978-3-031-51677-1_4

$$\omega(q) := \sum_{k=0}^{\infty} \frac{q^{2k(k+1)}}{(q;q^2)_{k+1}^2},$$

$$\nu(q) := \sum_{k=0}^{\infty} \frac{q^{k(k+1)}}{(-q;q^2)_{k+1}},$$

$$\phi(q) := \sum_{k=0}^{\infty} \frac{q^{k^2}}{(-q^2;q^2)_k},$$

of the *third order*. Here and throughout, the q-Pochhammer symbol is defined for $n \in \mathbb{N}_0 \cup \{\infty\}$ by

$$(a;q)_n := \prod_{j=0}^{n-1}(1-aq^j) = (1-a)(1-aq)(1-aq^2)\cdots(1-aq^{n-1}),$$

and we assume $|q| < 1$, so that all series converge absolutely. Ramanujan did not define what he meant by the order of a mock theta function, nor did he precisely define a mock theta function. However, we have since been able to extract a definition from his own writing [13] (see also the recent works [21, 25]):

> Suppose there is a function in the Eulerian form and suppose that all or an infinity of points $q = e^{2i\pi m/n}$ are exponential singularities and also suppose that at these points the asymptotic form of the function closes neatly…The question is: is the function taken the sum of two functions one of which is an ordinary theta function and the other a (trivial) function which is $O(1)$ at all the points $e^{2i\pi m/n}$? The answer is it is not necessarily so. When it is not so I call the function Mock θ-function. I have not proved rigorously that it is not necessarily so. But I have constructed a number of examples…

Ramanujan's reference to theta functions, a class of modular forms, and Eulerian forms, which are q-series expressible in terms of *q-hypergeometric* series [17, 20] and similar in shape to $\omega(q)$, $\nu(q)$, and $\phi(q)$, indirectly points back to earlier examples of Eulerian modular forms. For example, Dedekind's η-function is an important modular theta function of weight $1/2$ that can be expressed in terms of a q-hypergeometric series as follows:

$$q^{\frac{1}{24}}\eta^{-1}(\tau) = \sum_{n=0}^{\infty} \frac{q^{n^2}}{(q;q)_n^2}, \tag{1}$$

where $q = e^{2\pi i\tau}$ is the usual modular variable, with τ in the upper half complex plane. Ramanujan's letter on his mock theta functions claimed that mock theta functions behave like (weakly holomorphic) modular forms near roots of unity but are not themselves modular, hence the adjective *mock*.

The precise roles played by the mock theta functions within the theory of modular forms remained unclear in the decades following Ramanujan's death shortly after he wrote his last letter to Hardy. However, the importance of these functions was clear—they have been shown to play meaningful roles in the diverse subjects of combinatorics, q-hypergeometric series, mathematical physics, elliptic curves and traces of singular moduli, Moonshine and representation theory, and more. Within the last 20 years we have also finally understood, thanks to key work by Zwegers, Bruinier–Funke, and others including Bringmann–Ono and Zagier [15], that the mock theta functions turn out to be examples of *mock modular forms*, which are holomorphic parts of *harmonic Maass forms*, modern relatives to ordinary Maass forms and modular forms. This context has also allowed us to make more sense of the notion of the order of a mock theta function. For more background and information on these aspects of the mock theta functions, see, e.g., [15, 16, 18, 27, 28].

Turning to the first application of mock theta functions mentioned above, combinatorics, we recall that Dedekind's modular η-function may also be viewed as the reciprocal of the generating function for integer partitions. That is, (1) may also be written as

$$\prod_{n=1}^{\infty} \frac{1}{1-q^n} = \sum_{n=0}^{\infty} p(n)q^n = 1 + q + 2q^2 + 3q^3 + 5q^4 + 7q^5 + \cdots , \tag{2}$$

where $p(n)$ is the number of partitions of n. That (2) is simultaneously a modular form, and a combinatorial generating function has led to some deep and important results and theory. Namely, Hardy–Ramanujan introduced their famous *Circle Method* in analytic number theory, which combined with the modularity of Dedekind's η-function, led to the following exact formula for the partition numbers [24]

$$p(n) = 2\pi(24n-1)^{-\frac{3}{4}} \sum_{k=1}^{\infty} \frac{A_k(n)}{k} I_{\frac{3}{2}}\left(\frac{\pi\sqrt{24n-1}}{6k}\right),$$

an infinite sum in terms of Kloosterman sums A_k and Bessel functions I_s.

Like the modular η-function, the mock theta functions may also be viewed as combinatorial generating functions. For example, we have

$$q\omega(q) = \sum_{n=1}^{\infty} a_\omega(n)q^n, \qquad \nu(-q) = \sum_{n=0}^{\infty} a_\nu(n)q^n, \qquad \phi(q) = \sum_{n=0}^{\infty} a_\phi(n)q^n,$$

where $a_\omega(n)$ counts the number of partitions of n whose parts, except for one instance of the largest part, form pairs of consecutive non-negative integers [17, (26.84)]; $a_\nu(n)$ counts the number of partitions of n whose even parts are distinct, and if m occurs as a part, then so does every positive even number less than m; and $a_\phi(n) := sc_e(n) - sc_o(n)$, where $sc_{o/e}(n)$ counts the number of self-conjugate partitions λ of n with $L(\lambda)$ odd/even. Here, $L(\lambda)$ is the number of parts of λ minus the side length of its Durfee square. (See, e.g., [1] and Sect. 2 for more background on integer partitions.)

Using the newer theory of mock modular forms, we have results analogous to the celebrated Hardy–Ramanujan–Rademacher exact formula for $p(n)$; for example, due to Garthwaite [19], we have

$$a_\omega(n) = \frac{\pi}{2\sqrt{2}}(3n+2)^{-\frac{1}{4}} \sum_{\substack{k=1 \\ (k,2)=1}}^{\infty} \frac{(-1)^{\frac{k-1}{2}} A_k\left(\frac{n(k+1)}{2} - \frac{3(k^2-1)}{8}\right)}{k} I_{\frac{1}{2}}\left(\frac{\pi\sqrt{3n+2}}{3k}\right).$$

Numerous other papers, some of which we discuss in the sections that follow, have established further meaningful combinatorial results pertaining to the mock theta functions, including congruence properties, asymptotic properties, and more, adding to broader and older theories that rest at the intersection of combinatorics and modular forms.

1.2 Beck-Type Partition Identities

In this article, we seek to add to the growing literature on understanding the combinatorial nature of the mock theta functions. More specifically, we study the total number of parts in certain sets of partitions of n related to the third-order mock theta functions $\omega(q)$, $\nu(q)$, and $\phi(q)$. In general, identities on the number of parts in all partitions of n of a certain type have been of interest in the literature, dating back to work of Beck and Andrews. Their work was motivated by Euler's famous partition identity, which states that for any positive integer n,

$$p(n \mid \text{odd parts}) = p(n \mid \text{distinct parts}),$$

and which may be immediately deduced from the identity

$$\prod_{n=1}^{\infty} \frac{1}{1-q^{2n-1}} = \prod_{n=1}^{\infty}(1+q^n)$$

upon realizing that the "modular" products appearing are generating functions for the partition functions in Euler's identity. Here and throughout, we use the common notation $p(n \mid \textit{conditions})$ to denote the number of partitions of n subject to the given conditions. For example, $p(n \mid \text{odd parts})$ equals the number of partitions of n with odd parts.

While the natural number-of-parts refinement of Euler's identity is not true, namely the number of partitions of n into exactly m odd parts is not in general equal to the number of partitions of n into exactly m distinct parts, Beck conjectured and Andrews proved [2] that the excess in the number of parts in all partitions of n into odd parts over the number of parts in all partitions of n into distinct parts is equal to the number of partitions with only one (possibly repeated) even part and all other parts odd. Andrews additionally showed that this excess is also equal to the number of partitions of n with only one repeated part and all other parts distinct. Andrews provided an analytic proof of this theorem using generating functions, and Yang [26] and Ballantine–Bielak [8] later independently provided combinatorial proofs.

Since Beck made the first conjecture of this type, combinatorial identities on the excess between the number of parts in all partitions of n arising from a partition identity like Euler's are now fairly commonly referred to as "Beck-type identities." In the recent past, a number of other interesting Beck-type companions to other important identities have been established—see, e.g., [4, 9, 10, 22, 26].

Here, we establish Beck-type identities associated to the third-order mock theta functions $\omega(q)$, $\nu(q)$, and $\phi(q)$ in Theorems 3.2, 4.2, and 5.1, respectively. Our results may be viewed as Beck-type companion identities to partition identities for the third-order mock theta functions $\omega(q)$, $\nu(q)$, and $\phi(q)$ due to Andrews, Dixit and Yee in [5]. We devote Sect. 2 to preliminaries on partitions and state and prove our main results on $\omega(q)$, $\nu(q)$, and $\phi(q)$ in Sects. 3, 4, and 5, respectively. As a Corollary to our main results, we also establish mock theta pentagonal-number-theorem-type results in Theorem 4.3 and Corollary 4.4. Generally speaking, our proofs are both analytic and combinatorial in nature and involve mock theta generating functions and combinatorial bijections.

2 Preliminaries on Partitions

Let $n \in \mathbb{N}_0$. A *partition* of n, denoted $\lambda = (\lambda_1, \lambda_2, \ldots, \lambda_j)$, is a non-increasing sequence of positive integers $\lambda_1 \geq \lambda_2 \geq \cdots \geq \lambda_j$ called *parts* that add up to n. We refer to n as the *size* of λ. The *length* of λ is the number of parts of λ, denoted by $\ell(\lambda)$. We denote by $\ell_o(\lambda)$ and $\ell_e(\lambda)$ the number of odd, respectively even, parts of λ. For convenience, we abuse notation and use λ to denote either the multiset of its parts or the non-increasing sequence of parts. We write $a \in \lambda$ to mean the positive integer a is a part of λ. As mentioned in the introduction, we denote by $p(n)$ the number of partitions of n. The empty partition is the only partition of size 0; thus, $p(0) = 1$. We write $|\lambda|$ for the size of λ and $\lambda \vdash n$ to mean that λ is a partition of size n. For a pair of partitions (λ, μ), we also write $(\lambda, \mu) \vdash n$ to mean $|\lambda| + |\mu| = n$. We use the convention that $\lambda_k = 0$ for all $k > \ell(\lambda)$. When convenient, we will also use the exponential notation for parts in a partition: The exponent of a part is the multiplicity of the part in the partition. This notation will be used mostly for rectangular partitions. We write (a^b) for the partition consisting of b parts equal to a. Further, we denote by calligraphy style capital letters the set of partitions enumerated by the function denoted by the same letter. For example, we denote by $q_o(n)$ the number of partitions of n into distinct odd parts and by $\mathcal{Q}_o(n)$ the set of partitions of n into distinct odd parts. Moreover, when the size of the partitions is not explicitly stated in the notation of a set, we mean the set

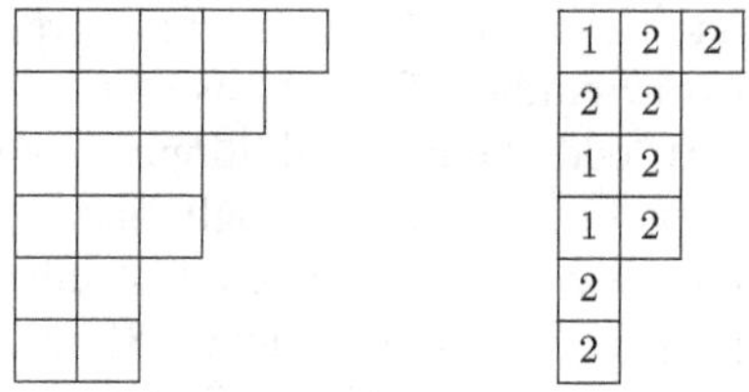

Fig. 1 A Ferrers diagram and 2-modular Ferrers diagram

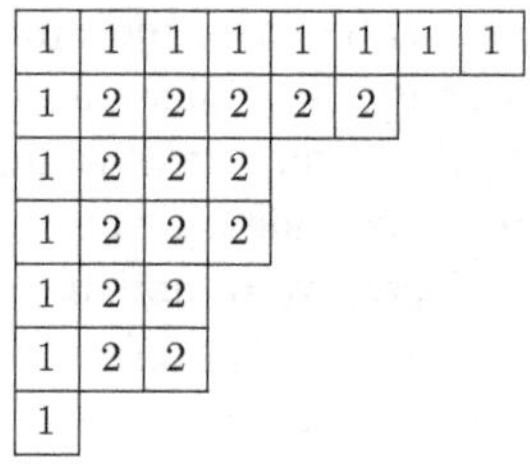

Fig. 2 An odd Ferrers diagram

of all partitions with the properties implied by the notation. For example, $\mathcal{Q}_o = \bigcup_{n\geq 0} \mathcal{Q}_o(n)$.

The *Ferrers diagram* of a partition $\lambda = (\lambda_1, \lambda_2, \ldots, \lambda_j)$ is an array of left justified boxes such that the ith row from the top contains λ_i boxes. In the literature, these are also referred to as *Young diagrams*. We abuse notation and use λ to mean a partition or its Ferrers diagram. The 2*-modular Ferrers diagram* of λ is a Ferrers diagram in which row i has $\lceil \frac{\lambda_i}{2} \rceil$ boxes, all but the first filled with 2. The first box of row i is filled with 2, respectively 1, if λ_i is even, respectively odd.

Example 1 The Ferrers diagram and the 2-modular Ferrers diagram of $\lambda = (5, 4, 3, 3, 2, 2)$ are shown in Fig. 1.

Given a partition λ, its *conjugate* λ' is the partition for which the rows in its Ferrers diagram are precisely the columns in the Ferrers diagram of λ. For example, the conjugate of $\lambda = (5, 4, 3, 3, 2, 2)$ is $\lambda' = (6, 6, 4, 2, 1)$. A partition is called *self-conjugate* if it is equal to its conjugate.

The Durfee square of a partition λ is the largest square that fits inside the Ferrers diagram of λ, i.e., the partition (a^a), where a is such that $\lambda_a \geq a$ and $\lambda_{a+1} \leq a$. For example, the Durfee square of $\lambda = (5, 4, 3, 3, 2, 2)$ is $(3^3) = (3, 3, 3)$.

For more details on partitions, we refer the reader to [1].

An *odd Ferrers diagram* F is a Ferrers diagram filled with 1 and 2 such that the first row is filled with 1 and the remaining rows form the 2-modular Ferrers diagram of a partition λ with all parts odd. If the first row has length k, we identify the odd Ferrers diagram F with the pair (k, λ). The *size* of an odd Ferrers diagram F is the sum of all entries in the boxes of diagram and is denoted by $|F|$.

Example 2 Figure 2 shows the odd Ferrers diagram of size 44 with 7 rows corresponding to the pair (k, λ) with $k = 8$ and $\lambda = (11, 7, 7, 5, 5, 1)$.

The rank of a partition λ, denoted $r(\lambda)$, is defined as $r(\lambda) = \lambda_1 - \ell(\lambda)$, or equivalently, the number of columns minus the number of rows in its Ferrers diagram. In [11], the M_2-rank of a partition is defined as the number of columns minus the number of rows in its 2-modular diagram. The rank of an odd Ferrers diagram $F = (k, \lambda)$, denoted rank(F), is defined as the number of columns minus the number of rows of F, or equivalently, rank$(F) = k - \ell(\lambda) - 1$.

3 The Mock Theta Function ω

Recall from Sect. 1 that Ramanujan's third-order mock theta function ω is defined by

$$\omega(q) := \sum_{k=0}^{\infty} \frac{q^{2k(k+1)}}{(q;q^2)_{k+1}^2}.$$

It is known [17, (26.84)] that

$$q\omega(q) = A_\omega(q) := \sum_{k=1}^{\infty} \frac{q^k}{(q;q^2)_k} = \sum_{n=1}^{\infty} a_\omega(n) q^n,$$

where $a_\omega(n)$ counts the number of partitions of n whose parts, except for one instance of the largest part, form pairs of consecutive non-negative integers. We are allowing pairs of consecutive integers to be (0, 1), but we are not considering 0 as a part of the partition. There is also the (highly non-trivial) identity by Andrews–Dixit–Yee [5]:

$$q\omega(q) = B_\omega(q) := \sum_{k=1}^{\infty} \frac{q^k}{(q^k;q)_{k+1}(q^{2k+2};q^2)_\infty} = \sum_{n=1}^{\infty} b_\omega(n) q^n,$$

where $b_\omega(n)$ counts the number of partitions of n such that all odd parts are less than twice the smallest part. Hence, $a_\omega(n) = b_\omega(n)$.

We define two-variable generalizations of $A_\omega(q)$ and $B_\omega(q)$ as follows. Let

$$A_\omega(z;q) := \sum_{k=1}^{\infty} \frac{zq^k}{(1-zq)(z^2q^3;q^2)_{k-1}} = \sum_{m=1}^{\infty}\sum_{n=1}^{\infty} a_\omega(m,n) z^m q^n, \tag{3}$$

where $a_\omega(m,n)$ counts the number of partitions of n with m parts, which except for one instance of the largest part form pairs of consecutive non-negative integers. To see this, one can rewrite the denominator of the kth summand as

$$(1-zq^{0+1})(1-z^2q^{1+2})(1-z^2q^{2+3})\cdots(1-z^2q^{(k-1)+k}),$$

and use the same convention as noted above for the combinatorial interpretation of $a_\omega(n)$ for which (0, 1) is an allowed pair of consecutive non-negative integers but for which 0 is not considered a part of the partition. Let

$$B_\omega(z;q) := \sum_{k=1}^{\infty} \frac{zq^k}{(zq^k;q)_{k+1}(zq^{2k+2};q^2)_\infty} = \sum_{m=1}^{\infty}\sum_{n=1}^{\infty} b_\omega(m,n) z^m q^n, \tag{4}$$

where $b_\omega(m,n)$ counts the number of partitions of n with m parts, whose odd parts are less than twice the smallest part. In particular, we have $A_\omega(1;q) = A_\omega(q)$, and $B_\omega(1;q) = B_\omega(q)$.

Following the notation convention introduced in Sect. 2, $\mathcal{A}_\omega(n)$ is the set of partitions of n whose parts, except for one instance of the largest part, form pairs of consecutive non-negative integers. We denote by $\mathcal{A}_{\omega,2}(n)$ the set of odd Ferrers diagrams of size n, and then $a_{\omega,2}(n) = |\mathcal{A}_{\omega,2}(n)|$.

We next define two generating functions, $A_{\omega,2}$ and $\widetilde{A}_{\omega,2}$, for odd Ferrers diagrams, which we later show are related to A_ω and B_ω. Namely, we let

$$A_{\omega,2}(z;q) := \sum_{k=1}^{\infty} \frac{zq^k}{(zq;q^2)_k} = \sum_{m=1}^{\infty}\sum_{n=1}^{\infty} a_{\omega,2}(m,n) z^m q^n, \tag{5}$$

where $a_{\omega,2}(m,n)$ counts the number of odd Ferrers diagrams of size n with m rows. We note that this interpretation was introduced by Andrews in [3]. We also let

$$\widetilde{A}_{\omega,2}(z;q) := \sum_{k=1}^{\infty} \frac{z^k q^k}{(q;q^2)_k} = \sum_{m=1}^{\infty}\sum_{n=1}^{\infty} \widetilde{a}_{\omega,2}(m,n) z^m q^n, \tag{6}$$

where $\widetilde{a}_{\omega,2}(m,n)$ counts the number of odd Ferrers diagrams of size n with m columns. The combinatorial interpretation of $\widetilde{A}_{\omega,2}(z;q)$ was first described by Li and Yang in [23, (2.22)].

Lemma 3.1 *There is an explicit bijection $\mathcal{A}_\omega(n) \xrightarrow{\sim} \mathcal{A}_{\omega,2}(n)$. Moreover, if $\mu \mapsto F_\mu$ under this bijection, then the number of parts of μ is equal to the number of rows of F_μ plus the number of rows of F_μ containing at least one 2, i.e.,*

$$A_\omega(z;q) = A_{\omega,2}(z^2;q) \cdot \frac{1-z^2q}{z(1-zq)}. \tag{7}$$

Proof Start with $\mu \in \mathcal{A}_\omega(n)$, remove one instance of the largest part μ_1, and merge the (consecutive) pairs of parts of the remaining partition to obtain a partition λ into odd parts. Then the corresponding odd Ferrers diagram is $F_\mu = (\mu_1, \lambda)$. This transformation is invertible: Given $(k, \lambda) \in \mathcal{A}_{\omega,2}(n)$, each part of λ is odd and, hence, the sum of a pair of consecutive non-negative integers. The corresponding partition has parts k and all pairs of parts obtained by splitting the parts of λ into consecutive integers. The connection between the number of parts of μ and the number of rows of F_μ is clear from this explicit bijection. □

Theorem 3.2 *The excess of the number of parts in all partitions in $\mathcal{A}_\omega(n)$ over the number of parts in all partitions in $\mathcal{B}_\omega(n)$ equals the number of rows containing at least one 2 in all odd Ferrers diagrams $F = (k, \lambda)$ of size n, which is the same as the number of parts greater than 1 in λ.*

We will provide four proofs of this theorem.

We first introduce some useful identities. From Equation (8) of [6], we have

$$B_\omega(z; q) = \widetilde{A}_{\omega,2}(z; q). \tag{8}$$

Moreover, from Equation (16) of [6], we have

$$\widetilde{A}_{\omega,2}(z; q) = A_{\omega,2}(z; q). \tag{9}$$

(This can also be seen from the fact that conjugation provides a bijection $\widetilde{\mathcal{A}}_{\omega,2}(m, n) \xrightarrow{\sim} \mathcal{A}_{\omega,2}(m, n)$ [23, p.539].) From (8) and (9), we have

$$B_\omega(z; q) = A_{\omega,2}(z; q). \tag{10}$$

By Lemma 3.1 and Theorem 3.2, or, after differentiating (10) at $z = 1$, we have the following result.

Corollary 3.3 *The total number of parts in all partitions in $\mathcal{B}_\omega(n)$ equals the total number of rows in all odd Ferrers diagrams of size n.*

All four proofs of Theorem 3.2 make use of the fact that

$$\left.\frac{\partial(A_\omega(z; q) - B_\omega(z; q))}{\partial z}\right|_{z=1} \tag{11}$$

is the generating function for the excess of the number of parts in all partitions in $\mathcal{A}_\omega(n)$ over the number of parts in all partitions in $\mathcal{B}_\omega(n)$.

First Proof We compute the derivative difference (11), using (7), (8), and (9):

$$\begin{aligned}
\left.\frac{\partial(A_\omega(z; q) - B_\omega(z; q))}{\partial z}\right|_{z=1} &= \left.\frac{\partial}{\partial z}\right|_{z=1} \left(\frac{1 - z^2 q}{z(1 - zq)} \cdot \widetilde{A}_{\omega,2}(z^2; q) - \widetilde{A}_{\omega,2}(z; q)\right) \\
&= \left.\frac{\partial \widetilde{A}_{\omega,2}(z; q)}{\partial z}\right|_{z=1} - \frac{1}{1-q} A_\omega(q) \\
&= \sum_{k=1}^{\infty} \frac{kq^k}{(q; q^2)_k} - \frac{1}{1-q} \sum_{k=1}^{\infty} \frac{q^k}{(q; q^2)_k}.
\end{aligned}$$

The second term $\frac{1}{1-q} \sum_{k=1}^{\infty} \frac{q^k}{(q;q^2)_k}$ is the generating function for the number of pairs $(F, (1^b)) \vdash n$, where F is an odd Ferrers diagram and $b \geq 0$ is an integer.

By mapping a pair $(F, (1^b))$ to an odd Ferrers diagram with at least b rows of size 1 and coloring the final b rows of size 1, we can see that $\frac{1}{1-q}\sum_{k=1}^{\infty}\frac{q^k}{(q;q^2)_k}$ is also the generating function for the number of odd Ferrers diagrams $F = (k, \lambda)$ weighted by $m_\lambda(1) + 1$, where $m_\lambda(1)$ is the number of parts equal to 1 in λ.

Hence, $\left.\frac{\partial(A_\omega(z;q)-B_\omega(z;q))}{\partial z}\right|_{z=1}$ is the generating function for the number of odd Ferrers diagrams $F = (k, \lambda)$ weighted by $k - (m_\lambda(1) + 1)$.

Note that conjugation provides a bijection between odd Ferrers diagrams of size n with m rows and odd Ferrers diagrams of size n with m columns. Hence for a conjugate pair $F = (k, \lambda)$ and $F' = (j, \mu)$, we have

$$\begin{aligned}
&k - (m_\lambda(1) + 1) + j - (m_\mu(1) + 1) = j - (m_\lambda(1) + 1) + k - (m_\mu(1) + 1)\\
=&(\ell(\lambda) + 1) - (m_\lambda(1) + 1) + (\ell(\mu) + 1) - (m_\mu(1) + 1).
\end{aligned}$$

Therefore, summing over all odd Ferrers diagrams of size n, the generating function stays the same if we replace the weight by $(\ell(\lambda) + 1) - (m_\lambda(1) + 1)$, which is the number of rows containing at least one 2 in F. □

Second Proof We compute the derivative difference (11), using (7) and (10):

$$\begin{aligned}
\left.\frac{\partial(A_\omega(z;q) - B_\omega(z;q))}{\partial z}\right|_{z=1} &= \left.\frac{\partial}{\partial z}\right|_{z=1}\left(\frac{1-z^2q}{z(1-zq)}\cdot A_{\omega,2}(z^2;q) - A_{\omega,2}(z;q)\right)\\
&= \left.\frac{\partial A_{\omega,2}(z;q)}{\partial z}\right|_{z=1} - \frac{1}{1-q}A_\omega(q).
\end{aligned}$$

We have seen in the first proof that the second term $\frac{1}{1-q}A_\omega(q)$ is the generating function for the number of odd Ferrers diagrams $F = (k, \lambda)$ weighted by $m_\lambda(1) + 1$. Hence, $\left.\frac{\partial(A_\omega(z;q)-B_\omega(z;q))}{\partial z}\right|_{z=1}$ is the generating function for the number of rows containing at least one 2 in all odd Ferrers diagrams of size n. □

Third Proof We compute the derivative difference (11), using (10):

$$\left.\frac{\partial(A_\omega(z;q) - B_\omega(z;q))}{\partial z}\right|_{z=1} = \left.\frac{\partial}{\partial z}\right|_{z=1}\left(A_\omega(z;q) - A_{\omega,2}(z;q)\right) \tag{12}$$

$$= \sum_{k=1}^{\infty}\frac{q^k}{(q;q^2)_k}\left(\sum_{j=1}^{k-1}\frac{q^{2j+1}}{1-q^{2j+1}}\right). \tag{13}$$

This is the generating function for the number of pairs $(F, ((2j + 1)^b)) \vdash n$, where F is an odd Ferrers diagram and $j, b \geq 1$ are integers. For each pair $(F, ((2j + 1)^b)) \vdash n$, we insert b copies of $(2j + 1)$ as 2-modular rows into F and color the final b rows of size $(2j + 1)$ to obtain a colored odd Ferrers diagram.

The number of such colored odd Ferrers diagrams of size n is equal to the number of rows containing at least one 2 in all odd Ferrers diagrams of size n. □

Fourth Proof From (12), we have

$$\sum_{m=1}^{\infty}(ma_{\omega}(m,n) - mb_{\omega}(m,n)) = \sum_{m=1}^{\infty}(ma_{\omega}(m,n) - ma_{\omega,2}(m,n)), \tag{14}$$

for each $n \geq 1$. The left-hand side of (12) is the excess in the statement of Theorem 3.2, whereas the right-hand side is the excess of the number of parts in all partitions in $\mathcal{A}_{\omega}(n)$ over the number of rows in all odd Ferrers diagrams in $\mathcal{A}_{\omega,2}(n)$. By Lemma 3.1, $\mathcal{A}_{\omega}(n) \xrightarrow{\sim} \mathcal{A}_{\omega,2}(n)$, and the excess is precisely the number of rows containing at least one 2 in all odd Ferrers diagrams of size n. □

From the third proof of Theorem 3.2, we obtain new interpretations of the derivative difference (12) in Corollaries 3.4 and 3.5. These are analogous to the original Beck identity that can be reinterpreted as follows. The excess of the total number of parts in all partitions of n into distinct parts over the total number of parts in all partitions of n into odd parts equals the number of pairs $(\xi, \eta) \vdash n$, where ξ is a partition into odd parts and η is a rectangular partition into equal even parts. This is also the number of pairs $(\xi, \eta) \vdash n$, where ξ is a partition into distinct parts and η is a rectangular partition with at least two parts.

Corollary 3.4 *The excess of the total number of parts in all partitions in $\mathcal{A}_{\omega}(n)$ over the total number of parts in all odd Ferrers diagrams of size n equals the number of pairs $(\xi, \eta) \vdash n$, where ξ is an odd Ferrers diagram and η is a rectangular partition into odd parts of size at least* 3.

Corollary 3.5 *The excess of the number of parts in all partitions in $\mathcal{A}_{\omega}(n)$ over the number of parts in all partitions in $\mathcal{B}_{\omega}(n)$ equals the number of pairs $(\lambda, \eta) \vdash n$, where $\lambda \in \mathcal{A}_{\omega}$ and η is a rectangular partition into odd parts of size at least* 3.

4 The Mock Theta Function ν

Recall from Sect. 1 the mock theta function

$$\nu(q) := \sum_{k=0}^{\infty} \frac{q^{k(k+1)}}{(-q;q^2)_{k+1}}.$$

We write

$$\nu(-q) = A_{\nu}(q) = \sum_{n=0}^{\infty} a_{\nu}(n)q^n. \tag{15}$$

Since $k(k+1) = 2+4+\cdots+2k$, $a_\nu(n)$ counts the number of partitions of n whose even parts are distinct, and if m occurs as a part, then so does every positive even number less than m.

Let

$$A_{\nu,2}(q) := \sum_{k=0}^{\infty}(-q;q^2)_k q^k =: \sum_{n=0}^{\infty} a_{\nu,2}(n)q^n.$$

We recall [5, (44)] that gives the identity

$$\sum_{m=0}^{\infty} \frac{q^{m^2}x^m}{(y;q^2)_{m+1}} = \sum_{m=0}^{\infty}(-xq/y;q^2)_m\, y^m.$$

Letting $x = y = q$, we have

$$\nu(-q) = A_{\nu,2}(q).$$

Note that $A_{\nu,2}(q)$ is the generating function for the number of odd Ferrers diagrams (k, λ) where the partition λ has distinct parts.

We also define

$$B_\nu(q) := \sum_{k=0}^{\infty} q^k(-q^{k+1};q)_k(-q^{2k+2};q^2)_\infty = \sum_{n=0}^{\infty} b_\nu(n)q^n,$$

where $b_\nu(n)$ counts the number of partitions of n into distinct parts, in which each odd part is less than twice the smallest part, and zero can be a part (note that this is different from our usual convention). For example, $(6, 4, 3, 2)$ and $(6, 4, 3, 2, 0)$ are counted as different partitions, the former from the term $q^k(-q^{k+1};q)_k(-q^{2k+2};q^2)_\infty$ for $k = 2$, while the latter from the term for $k = 0$. Then, as stated in [5, Theorem 4.1], we have the identity

$$\nu(-q) = B_\nu(q). \tag{16}$$

Lemma 4.1 *There is an explicit bijection* $\mathcal{A}_{\nu,2}(n) \xrightarrow{\sim} \mathcal{A}_\nu(n)$. *Moreover, if* $(k, \lambda) \mapsto \pi$ *under this bijection, then* $\ell(\pi) = k$ *and* $\ell(\lambda)$ *is the number of even parts in* π.

Proof We adapt the bijection in [23, Theorem 1.3]. We start with an odd Ferrers diagram $F = (k, \lambda) \in \mathcal{A}_{\nu,2}(n)$, where λ has distinct parts and length ℓ. We will associate to F a partition π in $\mathcal{A}_\nu(n)$. Consider the subdiagram $T = (\ell, (2\ell-1, 2\ell-3, \ldots, 3, 1))$ of F. We map T to the partition $\varepsilon = (2\ell, 2\ell-2, \ldots, 4, 2)$. We remove T from F and shift all remaining boxes to the left to obtain a diagram R. The conjugate of R is the 2-modular diagram of a partition ρ with odd parts. Define $\pi := \varepsilon \cup \rho$.

Fig. 3 A decomposition of an odd Ferrers diagram as in Lemma 4.1

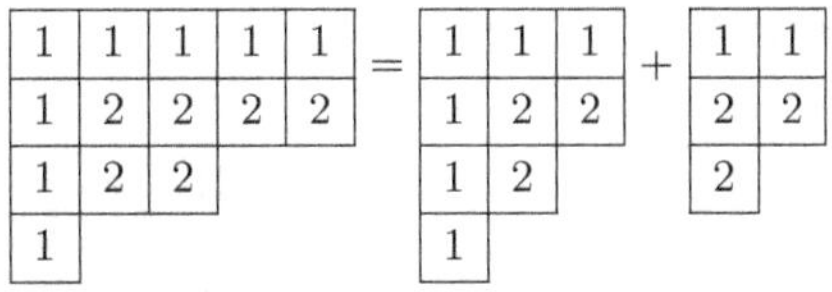

From the procedure above, we see that π has k parts and the number of parts of λ is equal to the number of even parts in π. □

Example 3 Let $F = (k, \lambda) = (5, (9, 5, 1))$. As in Lemma 4.1, we decompose $F = T + R$ as shown in Fig. 3.

Then, $\varepsilon = (6, 4, 2)$, $\rho = (5, 3)$, and $\pi = (6, 5, 4, 3, 2)$.

As in Sect. 3, we introduce two variable generalizations of $A_\nu(q)$, $A_{\nu,2}(q)$, and $B_\nu(q)$ in which the exponent of z keeps track of the number of parts in partitions.

Let

$$A_\nu(z; q) := \sum_{k=0}^{\infty} \frac{z^k q^{k^2+k}}{(zq; q^2)_{k+1}} =: \sum_{m=0}^{\infty}\sum_{n=0}^{\infty} a_\nu(m, n) z^m q^n. \tag{17}$$

Since $z^k q^{k^2+k} = zq^2 \cdot zq^4 \cdots zq^{2k}$, we find that $a_\nu(m, n)$ counts the number of partitions in $\mathcal{A}_\nu(n)$ with m parts.

Let

$$A_{\nu,2}(z; q) := \sum_{k=0}^{\infty} (-zq; q^2)_k zq^k = \sum_{m=0}^{\infty}\sum_{n=0}^{\infty} a_{\nu,2}(m, n) z^m q^n, \tag{18}$$

where $a_{\nu,2}(m, n)$ counts the number of odd Ferrers diagrams (k, λ) in $\mathcal{A}_{\nu,2}(n)$ with m rows.

Let

$$B_\nu(z; q) := \sum_{k=0}^{\infty} zq^k(-zq^{k+1}; q)_k(-zq^{2k+2}; q^2)_\infty = \sum_{m=0}^{\infty}\sum_{n=0}^{\infty} b_\nu(m, n) z^m q^n, \tag{19}$$

where $b_\nu(m, n)$ counts the number of partitions in $\mathcal{B}_\nu(n)$ with m parts. Recall that partitions in $\mathcal{B}_\nu(n)$ can have 0 as a part, which we do count in the number of parts; for example, the partition (6, 4, 2) has three parts, while (6, 4, 2, 0) has four parts.

Theorem 4.2 *The excess of the total number of parts in all partitions in $\mathcal{A}_\nu(n)$ over the total number of parts in all partitions in $\mathcal{B}_\nu(n)$ equals the sum of the number of odd parts minus* 1 *over all partitions in $\mathcal{A}_\nu(n)$, or equivalently the sum of ranks over all odd Ferrers diagrams in $\mathcal{A}_{\nu,2}(n)$. If $n \geq 1$, the excess is non-negative.*

We provide two proofs of this theorem.

Proof 1 We have

$$\begin{aligned}
&\frac{\partial}{\partial z}\left(A_\nu(z;q)-B_\nu(z;q)\right)\Big|_{z=1}\\
&=\frac{\partial}{\partial z}\left(\sum_{k=0}^{\infty}\frac{z^kq^{k^2+k}}{(zq;q^2)_{k+1}}-\sum_{k=0}^{\infty}zq^k(-zq^{k+1};q)_k(-zq^{2k+2};q^2)_\infty\right)\Bigg|_{z=1}\\
&=\frac{\partial}{\partial z}\left(\sum_{k=0}^{\infty}\frac{z^kq^{k^2+k}}{(zq;q^2)_{k+1}}-\sum_{k=0}^{\infty}\frac{z^{k+1}q^{k^2+k}}{(q;q^2)_{k+1}}\right)\Bigg|_{z=1}\\
&=\sum_{k=0}^{\infty}\frac{q^{k^2+k}}{(q;q^2)_{k+1}}\left(\left(\sum_{j=0}^{k}\frac{q^{2j+1}}{1-q^{2j+1}}\right)-1\right)\\
&=:\sum_{n=0}^{\infty}c(n)q^n,
\end{aligned}$$

where we use [6, (7)] in the second line above. Note that $c(n)$ counts the number of odd parts in all partitions in $\mathcal{A}_\nu(n)$ minus the number of partitions in $\mathcal{A}_\nu(n)$. Rephrased, we have

$$\sum_{n=0}^{\infty}c(n)q^n=\sum_{n=0}^{\infty}\sum_{\pi\in\mathcal{A}_\nu(n)}(\ell_o(\pi)-1)q^n.$$

To show that $c(n)\geq 0$ for $n\geq 1$, notice that the only partition π with $\ell_o(\pi)-1<0$ is $\pi=(2m,2m-2,\ldots,2)$. For this π, we can split the largest part $2m$ into two odd parts, namely $2m-1$ and 1, to obtain $\widetilde{\pi}=(2m-1,2m-2,\ldots,2,1)$, another partition of the same size in $\mathcal{A}_\nu$, such that $(\ell_o(\pi)-1)+(\ell_o(\widetilde{\pi})-1)=-1+1=0$. All the other partitions in $\mathcal{A}_\nu$ have at least one odd part. Therefore, $c(n)=\sum_{\pi\in\mathcal{A}_\nu(n)}(\ell_o(\pi)-1)\geq 0$. □

Proof 2 From Lemma 4.1, if the partition $\pi\in\mathcal{A}_\nu(n)$ corresponds to the odd Ferrers diagram $F=(k,\lambda)\in\mathcal{A}_{\nu,2}(n)$, then the number of parts of π is k. In terms of generating functions, we have the identity

$$A_\nu(z;q)=\sum_{k=0}^{\infty}\frac{z^kq^{k^2+k}}{(zq;q^2)_{k+1}}=\sum_{k=0}^{\infty}(-q;q^2)_kz^kq^k.$$

Thus, $\frac{\partial}{\partial z}\Big|_{z=1}A_\nu(z;q)$ is the generating function for the total number of columns in all odd Ferrers diagrams in $\mathcal{A}_{\nu,2}(n)$.

On the other hand, from [6, Theorem 1], we have

$$B_\nu(z;q) = \sum_{k=0}^{\infty} \frac{z^{k+1}q^{k^2+k}}{(q;q^2)_{k+1}}.$$

Again from Lemma 4.1, if the partition $\pi \in \mathcal{A}_\nu(n)$ corresponds to the odd Ferrers diagram $F = (k, \lambda) \in \mathcal{A}_{\nu,2}(n)$, then the number of even parts in π is equal to $\ell(\lambda)$. In terms of generating functions, we have

$$\sum_{k=0}^{\infty} \frac{z^{k+1}q^{k^2+k}}{(q;q^2)_{k+1}} = \sum_{k=0}^{\infty}(-zq;q^2)zq^k = A_{\nu,2}(z;q).$$

(This is also [6, Theorem 2].) Hence, $B_\nu(z;q) = A_{\nu,2}(z;q)$ and so $\frac{\partial}{\partial z}\Big|_{z=1} B_\nu(z;q)$ is the generating function for the total number of rows in all odd Ferrers diagrams in $\mathcal{A}_{\nu,2}(n)$.

Combining these, we conclude that $\frac{\partial}{\partial z}\Big|_{z=1} (A_\nu(z;q) - B_\nu(z;q))$ is the generating function for the sum of ranks of all odd Ferrers diagrams in $\mathcal{A}_{\nu,2}(n)$.

Given an odd Ferrers diagram $F = (m, \lambda) \in \mathcal{A}_{\nu,2}(n)$, we have $m \geq \ell(\lambda)$ since the parts of λ are distinct. Then, $\text{rank}(F) = m - \ell(\lambda) - 1 \geq -1$ and $\text{rank}(F) = -1$ if and only if $m = \ell(\lambda)$, in which case $F = (m, (2m-1, 2m-3, \ldots, 1))$. Hence, there is at most one odd Ferrers diagram with rank -1 in $\mathcal{A}_{\nu,2}(n)$. If $\text{rank}(F) = -1$, the conjugate of F is $F' = (m+1, (2m-1, 2m-3, \ldots, 3)) \in \mathcal{A}_{\nu,2}(n)$, $\text{rank}(F') = (m+1) - m = 1$, and thus $\text{rank}(F) + \text{rank}(F') = 0$. Since all other odd Ferrers diagrams in $\mathcal{A}_{\nu,2}(n)$ have non-negative rank, it follows that $c(n) \geq 0$. □

We end this section by investigating the parity of $c(n)$. To this end, we first prove a result similar to Euler's Pentagonal Number Theorem. Let $a_\nu^e(n)$ (resp., $a_\nu^o(n)$) be the number of partitions in $\mathcal{A}_\nu$ with an even (resp., odd) number of parts.

Theorem 4.3 *For any non-negative integer n, we have*

$$a_\nu^e(n) = a_\nu^o(n) + e(n),$$

where

$$e(n) = \begin{cases} 1, & \text{if } n = 3j^2 + 2j \text{ for some } j \geq 0, \\ -1, & \text{if } n = 3j^2 + 4j + 1 \text{ for some } j \geq 0, \\ 0, & \text{otherwise.} \end{cases}$$

Proof We write a partition $\pi \in A_\nu(n)$ as $\pi = (\pi^e, \pi^o)$, where π^e (resp., π^o) is the partition consisting of the even (resp., odd) parts of π. As usual, the largest part of π^e is π_1^e. We denote by π_s^o the smallest part of π^o and by $m^o(\pi)$ the multiplicity of $\pi_1^e + 1$ in π^o. We have $m^o(\pi) \geq 0$.

Let $\tilde{A}_\nu(n) = \{\pi \in A_\nu(n) \mid \pi^o = (\pi_1^e + 1)^{m^o(\pi)},\ \text{with } m^o(\pi) \in \{\frac{\pi_1^e}{2}, \frac{\pi_1^e}{2} + 1\}\}$. Then, $|\tilde{A}_\nu(n)| = 0$ or 1.

We define an involution on $A_\nu(n) \setminus \tilde{A}_\nu(n)$ as follows:

(i) If $\pi_s^o \geq 2m^o(\pi) + 1$, remove π_1^e from π^e and the last two columns (of length $m^o(\pi)$) from π^o, and add parts $\pi_1^e - 1$ and $2m^o(\pi) + 1$ to π^o.

(ii) If $\pi_s^o < 2m^o(\pi) + 1$, remove from π^o one part equal to $\pi_1^e + 1$ (largest part) and one part equal to π_s^o, and add a part equal to $\pi_1^e + 2$ to π^e and two columns equal to $\frac{\pi_s^o - 1}{2}$.

Note that the transformations in (i) and (ii) are inverses of each other.

We have $|\tilde{A}_\nu(n)| = 1$ if and only if $n = 3j^2 + 2j$ or $3j^2 + 4j + 1$. Moreover, for $\pi \in \tilde{A}_\nu(n)$, $j = \ell(\pi^e) = \ell_e(\pi)$, which completely determines the parity of $\ell(\pi)$.

□

Corollary 4.4 *Let $n \in \mathbb{N}$. Then $c(n)$ is odd if and only if n is eight times a generalized pentagonal number.*

Proof We have

$$c(n) = \sum_{\pi \in A_\nu(n)} (\ell_o(\pi) - 1).$$

With the notation in the proof of Theorem 4.3, we have $\ell_o(\pi) = \ell(\pi^o) \equiv n \pmod 2$ because the number of parts in a partition with odd parts has the same parity as its size. Therefore, if n is odd, $\ell_o(\pi) - 1$ is even for every $\pi \in A_\nu(n)$ and $c(n)$ is even.

If n is even, $c(n) \equiv |A_\nu(n)| \pmod 2$. From Theorem 4.3, it follows that $|A_\nu(n)| \equiv 1 \pmod 2$ if and only if $n = 3j^2 + 2j$ or $3j^2 + 4j + 1$ for some $j \geq 0$. Since n is even, if $n = 3j^2 + 2j$, then j must be even, and if $n = 3j^2 + 4j + 1$, then j must be odd. Therefore, $|A_\nu(n)| \equiv 1 \pmod 2$ if and only if n is eight times a generalized pentagonal number. □

Remark 1 Theorem 4.3 also follows from setting $a = 1$ in Entry 3.7 of [14]. Using Lemma 4.1, Theorem 4.3 can be adapted to a pentagonal number theorem for $A_{\nu,2}(n)$. Then, [23, Section 3.2] leads to a combinatorial proof of [5, Theorem 5.4].

5 The Mock Theta Function ϕ

Recall from Sect. 1 that the third-order mock theta function ϕ is defined by

$$\phi(q) := \sum_{n=0}^{\infty} \frac{q^{n^2}}{(-q^2, q^2)_n} = \sum_{n=0}^{\infty} (sc_e(n) - sc_o(n))q^n, \tag{20}$$

where $sc_e(n)$ (resp., $sc_o(n)$) counts the number of self-conjugate partitions λ of n with $L(\lambda)$ even (resp., odd). Here, $L(\lambda)$ is the number of parts of λ minus the side length of its Durfee square. From [5, Proof of Theorem 4.2], we have

$$\phi(q) = 1 + \sum_{n=0}^{\infty}(-1)^n q^{2n+1}(q;q^2)_n.$$

We first define the following generalization of $\phi(q)$:

$$B_\phi(z;q) := 1 + \sum_{n=0}^{\infty} z^n q^{2n+1}(q;q^2)_n = \sum_{n=0}^{\infty}\sum_{m=0}^{\infty} b_\phi(m,n) z^m q^n,$$

where $b_\phi(0,0) := 1$, and for $(m,n) \neq (0,0)$, $b_\phi(m,n)$ equals the difference between the number of partitions of n into distinct odd parts with largest part $2m+1$ and an odd number of parts, and the number of such partitions with an even number of parts. Note that $B_\phi(-1,q) = \phi(q)$ and that this gives rise to a different combinatorial interpretation for the coefficients of ϕ than the one given in (20). Namely, the coefficient of q^n in the q-series expansion for $\phi(q)$ also equals $do_e(n) - do_o(n)$, where $do_e(n)$ (resp., $do_o(n)$) counts the number of partitions of n into distinct odd parts with M_2-rank even (resp., odd).

Next we define another bivariate function, which we later explain is related to ϕ when $z = 1$ (see (22)):

$$A_\phi(z;q) := q\sum_{n=0}^{\infty} zq^n(-zq^{n+1};q)_n(-zq^{2n+1};q^2)_\infty = \sum_{n=0}^{\infty}\sum_{m=0}^{\infty} a_\phi(m,n) z^m q^{n+1}, \tag{21}$$

where $a_\phi(m,n)$ is the number of partitions of n into distinct parts, with m parts, such that each even part is at most twice the smallest part. The function $A_\phi(z;q)$ is related to $\phi(q)$ by the following identity:

$$A_\phi(1;q) = 1 - \phi(q) + 2(-q;q^2)_\infty \sum_{n=1}^{\infty} q^{n^2}. \tag{22}$$

Using Jacobi's triple product identity [1, (2.2.10) with $z = 1$], identity (22) is essentially [5, Theorem 4.2] with some minor typographical errors corrected. Unlike the mock theta functions $\omega(q)$ and $\nu(-q)$ studied in Sects. 3 and 4, the q-series coefficients of $\phi(q)$ are not uniformly non-negative, e.g.,

$$\begin{aligned}\phi(q) = 1 &+ q - q^3 + q^4 + q^5 - q^6 - q^7 + 2q^9 - 2q^{11} + q^{12} + q^{13} - q^{14}\\ &- 2q^{15} + q^{16} + O(q^{17}).\end{aligned}$$

However, the authors of [5] present (22) for $\phi(q)$ as a companion identity to their similar result [5, Theorem 4.1] (see (16)), which shows that the mock theta function $\nu(-q)$ is equal to the generating function for partitions into distinct parts, in which each odd part is less than twice the smallest part. Identity (22) similarly relates the mock theta function $\phi(q)$ to the generating function for partitions into distinct parts in which each even part is at most twice the smallest part, but up to a theta function. Indeed, it is identity (22) that leads to our "Beck-type" Theorem 5.1 for the mock theta function $\phi(q)$ below. To state it, we introduce the functions

$$F_1(q) := F_3(q)\Big(1 + 2\sum_{n=1}^{\infty} q^{n^2}\Big) + 2(-q;q^2)_\infty \sum_{n=1}^{\infty} q^{n^2}, \tag{23}$$

$$F_2(q) := (-q;q^2)_\infty \sum_{m=1}^{\infty} \frac{q^{2m-1}}{1+q^{2m-1}}, \tag{24}$$

$$F_3(q) := (-q;q^2)_\infty \sum_{m=1}^{\infty} \frac{q^{2m}}{1+q^{2m}}. \tag{25}$$

The functions $F_2(q)$ and $F_3(q)$, including their combinatorial interpretations, are studied in [9].

In what follows, we use the notation $G(q) \succeq_S 0$, where $S \subseteq \mathbb{N}$, to mean that when expanded as a q-series, the coefficients of $G(q)$ are non-negative, with the exception of the coefficients of q^n for $n \in S$. When $S = \emptyset$, we simply use the notation $\succeq 0$.

Theorem 5.1 *We have*

$$\frac{\partial}{\partial z}\Big|_{z=1} (A_\phi(z;q) + B_\phi(-z^{-1};q)) = F_1(q) - F_2(q). \tag{26}$$

Moreover, we have

$$\frac{\partial}{\partial z}\Big|_{z=1} (A_\phi(z;q) + B_\phi(-z^{-1};q)) \succeq 0.$$

Remark 2 A combinatorial interpretation of (26) in Theorem 5.1 can be deduced from the combinatorial definitions of $a_\phi(m,n)$ and $b_\phi(m,n)$ provided above, together with combinatorial interpretations of the q-series coefficients in the functions $F_2(q)$ and $F_3(q)$ provided in [9], and the definition of $F_1(q)$. While this combinatorial interpretation involves partition differences, Theorem 5.1 establishes the non-negativity of the q-series coefficients of (26). On the other hand, it is of interest to find another proof of this fact by finding a different and *manifestly* positive combinatorial interpretation of the q-series coefficients of $F_1(q) - F_2(q)$ (i.e., one that does not involve a combinatorial difference). We leave this as an open problem.

5.1 Proof of Theorem 5.1

In this section, we prove Theorem 5.1, assuming the truth of Proposition 5.2 and Proposition 5.4 stated below. We provide a combinatorial proof of Proposition 5.2 in Sect. 5.2 and provide both combinatorial and analytic proofs of Proposition 5.4 in Sect. 5.3.

Proposition 5.2 *We have*

$$2F_3(q) - F_2(q) \succeq_S 0,$$

where $S := \{1, 4, 8, 16\}$. *Moreover, the coefficients of* $2F_3(q) - F_2(q)$ *are at least 4, with the exception of the coefficients of* q^n *with* n *in the set* $U := \{1, 2, 3, 4, 5, 8, 9, 12, 13, 16, 17\}$.

Corollary 5.3 *We have*

$$(2F_3(q) - F_2(q)) \sum_{n=1}^{\infty} q^{n^2} \succeq_T 0, \tag{27}$$

where $T := \{2, 5, 9, 13, 17\}$.

Proof of Corollary 5.3 Writing $2F_3(q) - F_2(q) =: \sum_{n=1}^{\infty} a_n q^n$, we have the coefficient of q^n in $(2F_3(q) - F_2(q)) \sum_{j=1}^{\infty} q^{j^2}$ is equal to

$$\sum_{\substack{k=1 \\ k+m^2=n,\ 1\le m^2\le n-1}}^{n-1} a_k. \tag{28}$$

By Proposition 5.2, we have $a_k \geq 0$ for any $k \notin S = \{1, 4, 8, 16\}$, and $a_k \geq 4$ for any $k \notin U$.

Let $n \notin T$ be a positive integer. We can directly compute the q-series in (27) to $O(q^{65})$ to see the coefficients are non-negative. (Here and throughout we use standard "big-O," also called "big-oh," notation; see, e.g., [7].) Now assume that $n > 65$. To prove that (28) is non-negative, it suffices to show that for any $k \in S$ with $k + m^2 = n$, there is another $k' \neq k$ with $k' + m'^2 = n$ such that $a_{k'} + a_k \geq 0$. Because $n > 65$, we must have $m \geq 2$. Let $k' := k + 2m - 1$. Then $1 \leq k' \leq n - 1$, $k' > k$, and $k' + (m-1)^2 = n$, where $1 \leq (m-1)^2 \leq n - 1$. Note that $k' \notin U$. This is because for $k \in S$ and $k' \in U$, we have $n = k + m^2 = k + (\frac{k'-k+1}{2})^2 \leq 65$. By direct calculation, we find that $\min\{a_k\}_{k\in S} = -4$, and hence, $a_k + a_{k'} \geq 0$. □

Proposition 5.4 *Let* $F_2(q) =: \sum_{n=1}^{\infty} b_n q^n$. *For* $n \geq 9$, *we have*

$$b_n \leq b_{n-1} + b_{n-4}.$$

Corollary 5.5 *We have*

$$F_2(q)\sum_{n=1}^{\infty} q^{n^2} - F_2(q) \succeq_V 0, \tag{29}$$

where $V := \{1, 3, 4, 6, 8\}$.

Proof of Corollary 5.5 To prove (29), we show

$$\sum_{1\leq m^2\leq n-1} b_{n-m^2} \geq b_n \tag{30}$$

for $n \in \mathbb{N} \setminus V$. By Proposition 5.4, and the non-negativity of b_n, for $n \geq 9$, we have

$$\sum_{1\leq m^2\leq n-1} b_{n-m^2} \geq b_{n-1} + b_{n-4} \geq b_n.$$

For $n \in \{2, 5, 7\}$, the inequality (30) can be verified directly. □

Proof of Theorem 5.1

First, by straightforward manipulations, we find that [23, (2.4)] leads to the identity

$$B_\phi(-z^{-1}; q) = 1 + \sum_{n=0}^{\infty} \frac{z^{-n} q^{(n+1)^2}}{(-z^{-1}q^2; q^2)_{n+1}}. \tag{31}$$

We recall [12, Theorem 6.11], which gives the interesting identity

$$\begin{aligned}
&\sum_{n=0}^{\infty} q^n(-zq^n; q)_{n+1}(-zq^{2n+2}; q^2)_\infty \\
&\quad = -\frac{q}{z}\nu\left(\frac{q^2}{z}, -\frac{q^2}{z}; -q\right) + \frac{1}{q}(-z; q^2)_\infty \\
&\quad \times \left(-1 + \frac{(-q; q)_\infty (q^2; q^2)_\infty (-q^2/z; q^2)_\infty}{(-q^3/z; q^2)_\infty}\right),
\end{aligned} \tag{32}$$

where the function $\nu(\alpha, z; q)$ is defined in [12, (1.18)] by

$$\nu(\alpha, z; q) := \sum_{n=0}^{\infty} \frac{\alpha^n q^{n^2+n}}{(-zq; q^2)_{n+1}}.$$

We let $z \mapsto zq$ in (32) and multiply the resulting identity by zq. Using this and (31), we find that

$$A_\phi(z;q) + B_\phi(-z^{-1};q) = D_\phi(z;q),$$

where

$$\begin{aligned} D_\phi(z;q) &:= 1 + z(-zq;q^2)_\infty \left(-1 + \frac{(-q;q)_\infty (q^2;q^2)_\infty (-q/z;q^2)_\infty}{(-q^2/z;q^2)_\infty}\right) \\ &= 1 - z(-zq;q^2)_\infty + z\frac{(-q;q)_\infty}{(-q^2/z;q^2)_\infty}\left(1 + \sum_{n=1}^{\infty}(z^n + z^{-n})q^{n^2}\right). \end{aligned}$$

Above, we have also used the Jacobi triple product [1, (2.2.10)]. Thus, the derivative difference on the left-hand side of (26) equals $\frac{\partial}{\partial z}\big|_{z=1} D_\phi(z;q)$. After a direct calculation using the definition of $D_\phi(z;q)$ and some simplification, we obtain that this derivative difference equals $F_1(q) - F_2(q)$.

To prove the second assertion of the theorem, it now suffices to show that $F_1(q) - F_2(q) \succeq 0$. From [9], we have

$$F_3(q) \succeq 0.$$

From this and (23), it is not difficult to see that

$$F_1(q) - 2F_3(q)\sum_{n=1}^{\infty} q^{n^2} \succeq 0. \tag{33}$$

Thus, we have from (27), (29), and (33) that $F_1(q) - F_2(q) \succeq_W 0$ for some explicit, finite, set W. The proof is complete after a direct calculation of the q-series for $F_1(q) - F_2(q)$ up to $O(q^{n_W})$, where $n_W := \max W$, which reveals that, in fact, $F_1(q) - F_2(q) \succeq 0$ as claimed.

5.2 *Proof of Proposition 5.2*

Setting $r = 1$ and $\ell = 1$ in [9, Section 5.2] shows that $F_2(q)$ is the generating function for $|A(n)|$, where

$$A(n) := \{(\lambda, (a)) \vdash n \mid \lambda \in \mathcal{Q}_o, a \text{ odd}, a \notin \lambda\}.$$

Similarly, setting $r = 1$, $\ell = 2$ in [9, Section 5.2] shows that $F_3(q)$ is the generating function for $|B(n)| - \varepsilon(n)$, where

$$\varepsilon(n) := \begin{cases} 1, & \text{if } n \equiv 0 \pmod 4, \\ 0, & \text{else,} \end{cases}$$

and

$$B(n) := \{(\lambda, (c^d)) \vdash n \mid \lambda \in \mathcal{Q}_o, c \text{ even}, d \text{ odd}, \lambda_1 - \lambda_2 \leq c, \text{ and } \lambda \neq \mu(c) \text{ if } c \geq 4\}.$$

Here, $\mu(c)$ is defined for even $c \geq 4$ to be the partition

$$\mu(c) = \begin{cases} (\frac{c}{2}+1, \frac{c}{2}-1), & \text{if } c \equiv 0 \pmod 4, \\ (\frac{c}{2}+2, \frac{c}{2}-2), & \text{if } c \equiv 2 \pmod 4. \end{cases}$$

Thus, $\mu(c)$ is a partition of c into two distinct odd parts with smallest possible difference between the parts. We remark that, if $n \not\equiv 0 \pmod 4$, c even, and d odd, then $(\lambda, (c^d)) \neq (\mu(c), (c^d))$ vacuously.

We will show that $2(|B(n)| - \varepsilon(n)) - |A(n)| \geq 4$ for all $n \notin S$. For $n < 53$, this can be verified directly. For the remainder of the proof, let $n \geq 53$.

Roadmap of the Proof Since the proof is intricate, we begin by providing a roadmap for the benefit of the reader. Ultimately, we show that $2|B(n)| \geq |A(n)|+6$, which is sufficient to prove the proposition. To do this, we establish relevant injections. To describe them, we denote by $B'(n)$ the multiset whose elements are precisely those of $B(n)$, each appearing with multiplicity 2. Equivalently, $B'(n)$ is the disjoint union of two copies of $B(n)$. We also let

$$U(n) := \{(\lambda, (c^d)) \vdash n \mid \lambda \text{ has odd parts}, c \text{ even}, d \text{ odd}, \lambda_1 - \lambda_2 \leq c\}.$$

Note that $B(n) \subseteq U(n)$. Finally, we let $U'(n)$ be the multiset whose elements are precisely those of $U(n)$, each appearing with multiplicity 2. In the proof of the proposition below, we define an injection ψ from $A(n)$ to $U'(n)$ as a composition of two mappings Ψ_1 (Step 1) and Ψ_2 (Step 2). At the end of Step 2, we describe the image of ψ. Then we define an injection ζ (Step 3)

$$\psi(A(n)) \setminus B'(n) \to B'(n) \setminus \psi(A(n))$$

and show that

$$|B'(n) \setminus \psi(A(n))| \geq |\psi(A(n)) \setminus B'(n)| + 6$$

by explicitly listing six elements in $B'(n) \setminus \psi(A(n))$ that are not in the image of ζ.

This shows that $|B'(n)| \geq |\psi(A(n))| + 6$, which is equivalent to $2|B(n)| \geq |A(n)|+6$, and implies $2(|B(n)| - \varepsilon(n)) \geq |A(n)|+4$ as desired. The details of the proof of the proposition begin now.

Let

$$A_1(n) = \{(\lambda, (a)) \in A(n) \mid 1 \in \lambda\},$$
$$A_2(n) = \{(\lambda, (a)) \in A(n) \mid 1 \notin \lambda, a \neq 1\},$$
$$A_3(n) = \{(\lambda, (a)) \in A(n) \mid 1 \notin \lambda, a = 1\}.$$

Then, $A(n) = A_1(n) \sqcup A_2(n) \sqcup A_3(n)$. We also define

$$C_1(n) = \{(\eta, (c)) \vdash n \mid c \text{ even}, c \geq 4, \eta \in \mathcal{Q}_o, 1 \notin \eta, c-1 \notin \eta\},$$
$$C_2(n) = \{(\eta, (c)) \vdash n \mid c \text{ even}, c \geq 2, \eta \in \mathcal{Q}_o, 1 \in \eta, c+1 \notin \eta\},$$

and set $C(n) = C_1(n) \sqcup C_2(n)$.

We define $\psi : A(n) \to U'(n)$ as $\psi = \Psi_2 \circ \Psi_1$, where $\Psi_1 : A(n) \to C(n)$ and $\Psi_2 : C(n) \to U'(n)$ are defined in the steps below.

Step 1: $\Psi_1 : A(n) \to C(n)$. Given $(\lambda, (a)) = ((\lambda_1, \lambda_2, \ldots, \lambda_{\ell(\lambda)}), (a)) \in A(n)$, we define

$$\Psi_1(\lambda, (a)) = (\eta, (c)) = \begin{cases} ((\lambda \setminus (1), (a+1)), & \text{if } (\lambda, (a)) \in A_1(n), \\ (\lambda \cup (1), (a-1)), & \text{if } (\lambda, (a)) \in A_2(n), \\ (\lambda \setminus (\lambda_1), (\lambda_1 + 1)), & \text{if } (\lambda, (a)) \in A_3(n). \end{cases}$$

Then Ψ_1 induces bijections $\Psi_1 : A_1(n) \to C_1(n)$ and $\Psi_1 : A_2(n) \to C_2(n)$, and the injection $\Psi_1 : A_3(n) \to C_1(n)$ whose image is

$$C_1'(n) = \{(\eta, (c)) \in C_1(n) \mid \eta_1 \leq c - 3\}.$$

We note that for $n \geq 4$ even, the pair $(\emptyset, (n))$ belongs to $C_1'(n)$ and thus also to $C_1(n)$.

Step 2: $\Psi_2 : C(n) \to U'(n)$. Let $(\eta, (c)) \in C(n)$. If $\eta \neq \emptyset$, write $\eta_1 - \eta_2 = q_\eta c + r_\eta$ with $q_\eta, r_\eta \in \mathbb{Z}, q_\eta \geq 0, 0 < r_\eta \leq c$. We use the convention that $\eta_j = 0$ for all $j > \ell(\eta)$. If $\eta = \emptyset$, let $q_\eta = r_\eta = 0$.

For $i = 1, 2$, we write $C_i = C_{i,e} \sqcup C_{i,o}$, where

$$C_{i,e}(n) = \{(\eta, (c)) \in C_i(n) \mid q_\eta \text{ even}\},$$
$$C_{i,o}(n) = \{(\eta, (c)) \in C_i(n) \mid q_\eta \text{ odd}\}.$$

Moreover, if $(\eta, c) \in C_1'(n)$, then $q_\eta = 0$ and thus $C_1'(n) \subseteq C_{1,e}(n)$.

Set $\eta - (q_\eta c) := (\eta_1 - q_\eta c, \eta_2, \ldots, \eta_{\ell(\eta)})$. Note that $\eta - (q_\eta c) \in \mathcal{Q}_0$. For simplicity, we write q for q_η.

For $(\eta, (c)) \in C(n)$, we define

$$\Psi_2(\eta, (c)) = (\xi, (c^d))$$

$$= \begin{cases} (\eta - (qc), (c^{q+1})), & \text{if } (\eta, (c)) \in C_{1,e}(n) \cup C_{2,e}(n), \\ ((\eta - (qc)) \cup (c-1) \cup (1), (c^q)), & \text{if } (\eta, (c)) \in C_{1,o}(n), \\ ((\eta - (qc)) \setminus (1) \cup (c+1), (c^q)), & \text{if } (\eta, (c)) \in C_{2,o}(n). \end{cases}$$

All pairs obtained satisfy $\xi_1 - \xi_2 \leq c$.

We determine the image of the relevant subsets of $C(n)$ under Ψ_2. Notice that Ψ_2 is an injection on each of the relevant subsets. We have

$$\Psi_2(C_{1,e}(n))$$

$$= \left\{ (\xi, (c^d)) \in U(n) \;\middle|\; \begin{array}{l} \xi \in \mathcal{Q}_o, c \geq 4, \text{ and} \\ ((1 \notin \xi, c-1 \notin \xi) \text{ or} \\ \qquad (1 \notin \xi, \xi_1 = c-1, d > 1) \text{ or } (\xi = (1), d > 1)) \end{array} \right\},$$

$$\Psi_2(C_{2,e}(n))$$

$$= \left\{ (\xi, (c^d)) \in U(n) \;\middle|\; \begin{array}{l} \xi \in \mathcal{Q}_o, 1 \in \xi, \text{ and } (c+1 \notin \xi \text{ or } (\xi_1 = c+1, d > 1)), \\ \qquad \text{and if } \ell(\xi) = 1 \text{ then } d = 1 \end{array} \right\},$$

$$\Psi_2(C_{1,o}(n))$$

$$= \left\{ (\xi, (c^d)) \in U(n) \;\middle|\; \begin{array}{l} 1 \in \xi, c \geq 4, c-1 \in \xi \text{ and } \ell(\xi) \geq 3 \text{ and} \\ \xi \notin \mathcal{Q}_o \implies (\xi_1 = \xi_2 = c-1 \text{ and } \xi \setminus (c-1) \in \mathcal{Q}_o) \text{ or} \\ \qquad \xi = (c-1, 1, 1) \end{array} \right\},$$

$$\Psi_2(C_{2,o}(n))$$

$$= \left\{ (\xi, (c^d)) \in U(n) \;\middle|\; \begin{array}{l} 1 \notin \xi \text{ and } c+1 \in \xi \text{ and } \ell(\xi) \geq 2 \text{ and} \\ \xi \notin \mathcal{Q}_o \implies \xi_1 = \xi_2 = c+1 \text{ and } \xi \setminus (c+1) \in \mathcal{Q}_o \end{array} \right\}.$$

Moreover, Ψ_2 is the identity on $C_1'(n)$, so

$$\Psi_2(C_1'(n)) = C_1'(n) = \{(\xi, (c)) \in U(n) \mid \xi \in \mathcal{Q}_o, c \geq 4, 1 \notin \xi, \xi_1 \leq c-3\}.$$

Thus, $\psi(A(n))$ is the multiset

$$\psi(A(n)) = \Psi_2(C_{1,e}(n)) \sqcup \Psi_2(C_{2,e}(n)) \sqcup \Psi_2(C_{1,o}(n)) \sqcup \Psi_2(C_{2,o}(n)) \sqcup \Psi_2(C_1'(n)).$$

To find the pairs $(\xi, (c^d)) \in \psi(A(n))$ occurring with multiplicity 2, we determine the mutual intersections of the images under Ψ_2 of the different subsets of $C(n)$.

We have

$$\Psi_2(C_{1,e}(n)) \cap \Psi_2(C_{2,o}(n)) =$$
$$\{(\xi, (c^d)) \in U(n) \mid \xi \in \mathcal{Q}_o, 1 \notin \xi, c \geq 4,\ c+1 \in \xi, c-1 \notin \xi \text{ and } \ell(\xi) \geq 2\},$$

$$\Psi_2(C_{1,e}(n)) \cap \Psi_2(C_1'(n))$$
$$= \Psi_2(C_1'(n)) = \{(\xi, (c)) \in U(n) \mid \xi \in \mathcal{Q}_o, c \geq 4, 1 \notin \xi, \xi_1 \leq c-3\},$$

$$\Psi_2(C_{2,e}(n)) \cap \Psi_2(C_{1,o}(n)) =$$
$$\left\{(\xi, (c^d)) \in U(n) \middle| \begin{array}{c} \xi \in \mathcal{Q}_o, 1 \in \xi, c \geq 4, c-1 \in \xi, \ell(\xi) \geq 3, \text{ and} \\ ((c+1 \notin \xi) \text{ or } (\xi_1 = c+1, d > 1)) \end{array}\right\}.$$

One can easily verify that all other mutual intersections are empty. Therefore, the mapping ψ is a multiset injection. Let $\mathcal{S}(n)$ be the union of the three sets above, i.e., $\mathcal{S}(n)$ is the (disjoint) set of pairs $(\xi, (c^d)) \in U(n)$ with $\xi \in \mathcal{Q}_o$, $c \geq 4$ and:

- If $1 \notin \xi$, then $(c+1 \in \xi, c-1 \notin \xi$ and $\ell(\xi) \geq 2)$ or $(d = 1$ and $\xi_1 \leq c-3)$.
- If $1 \in \xi$, then $c-1 \in \xi, \ell(\xi) \geq 3,$ and $((c+1 \notin \xi)$ or $(\xi_1 = c+1, d > 1))$.

Thus, the elements of $\psi(A(n))$ occurring twice are precisely the elements of $\mathcal{S}(n)$. We denote by $\mathcal{S}'(n)$ the multiset whose elements are precisely those of $\mathcal{S}(n)$, each appearing with multiplicity 2. Then, every element in $\psi(A(n)) \setminus \mathcal{S}'(n)$ has multiplicity 1.

Step 3: Recall that our goal is to prove that $|B'(n)| \geq |\psi(A(n))| + 6$. To this end, we now show that $|B'(n) \setminus \psi(A(n))| \geq |\psi(A(n)) \setminus B'(n)| + 6$.

As a set, $\psi(A(n)) \setminus B'(n)$ consists of *precisely* the pairs $(\xi, (c^d)) \in U'(n)$ satisfying one of the following conditions:

(i) $\xi = \mu(c)$.
(ii) $\xi_1 = \xi_2 = c-1, \xi \setminus (c-1) \in \mathcal{Q}_o$, and $1 \in \xi$ and $c \geq 4$.
(iii) $\xi = (c-1, 1, 1), c \geq 4$.
(iv) $\xi_1 = \xi_2 = c+1, \xi \setminus (c+1) \in \mathcal{Q}_o$ and $1 \notin \xi$.

Since $B(n) = \{(\lambda, (c^d)) \in U(n) \mid \lambda \text{ has distinct parts}, \lambda \neq \mu(c)\}$, it is clear these pairs do not belong to $B'(n)$ and all the pairs in $\psi(A(n))$ that are not in $B'(n)$ must satisfy one of (i)–(iv).

We define a multiset injection $\zeta : \psi(A(n)) \setminus B'(n) \to B'(n) \setminus \psi(A(n))$. In the process, we also clarify that the pairs satisfying (i)–(iv) above belong to $\psi(A(n))$ and discuss their multiplicities in $\psi(A(n)) \setminus B'(n)$. We note that $B'(n) \setminus \psi(A(n))$ consists of one copy of each pair in $B(n) \setminus \mathcal{S}(n)$ and an additional copy of each pair in $B(n) \setminus \psi(A(n))$. By inspection, we see that the pairs $(\xi, (c^d)) \in B(n)$ satisfying one of the conditions below are *not* in $\psi(A(n))$:

- $1 \in \xi,\ c+1 \in \xi,\ c-1 \notin \xi,\ \xi_1 > c+1$;
- $c = 2,\ 1 \in \xi,\ 3 \in \xi,\ \xi_1 > 3$;
- $1 \notin \xi, c \geq 4,\ c-1 \in \xi,\ c+1 \notin \xi,\ \xi_1 > c+1$.

The list above is not exhaustive, but it is sufficient for our purposes. We denote by $\mathcal{T}(n)$ the set containing each pair $(\xi, (c^d)) \vdash n$ satisfying the conditions above and by $\mathcal{T}'(n)$ the multiset whose elements are precisely those of $\mathcal{T}(n)$, each appearing with multiplicity 2.

Let $(\xi, (c^d)) \in \psi(A(n)) \setminus B'(n)$. We note that when pairs are mapped by ζ to $\mathcal{T}'(n)$ the assignment is *ad hoc*. Moreover, the pairs in $\mathcal{T}'(n)$ that are in the image of ζ are of the form $(\mu(n-k) \cup \alpha, (x^y))$, where α is a partition with small parts and $|\alpha| + xy = k$. Since $n \geq 53$, in all such pairs the parts of $\mu(n-k)$ are larger than the parts of α. This is not a necessary condition, but it allows for an easy check that the map ζ is indeed an injection. In what follows, the congruence conditions on n are imposed by the requirement that, if $(\xi, (c^d)) \in \psi(A(n)) \setminus B'(n)$, then c is even and d is odd.

Case (i) $\xi = \mu(c)$ Since c is even and d is odd, if $(\mu(c), (c^d)) \vdash n$, then $n \equiv 0 \pmod 4$.

If $d = 1$, since $n \geq 53$, it follows that $c \geq 28$. Then $(\mu(c), (c)) \in \Psi_2(C_{1,e}(n)) \cap \Psi_2(C_1'(n))$. We define

$$\zeta(\mu(n/2), (n/2)) := (\mu(n-10) \cup (5, 1), (4)) \in \mathcal{T}'(n).$$

Notice that we are mapping each copy of $(\mu(n/2), (n/2))$ to one of two copies of $(\mu(n-10) \cup (5, 1), (4))$. As we will see below, this is the only pair that occurs twice in $\psi(A(n)) \setminus B'(n)$.

If $d > 1$, $c = 4$, then $(\mu(4), (4^d)) = ((3, 1), (4^{(n-4)/4})) \in \Psi_2(C_{2,e}(n)) \setminus \mathcal{S}(n)$. Thus, $n \equiv 0 \pmod 8$, and we define

$$\zeta((3, 1), (4^{(n-4)/4})) := (\mu(n-14) \cup (7, 1), (6)) \in \mathcal{T}'(n).$$

If $d > 1$, $c = 6$, then $(\mu(6), (6^d)) = ((5, 1), (6^{(n-6)/6}) \in \Psi_2(C_{2,e}(n)) \setminus \mathcal{S}(n)$. Thus, $n \equiv 0 \pmod{12}$, and we define

$$\zeta((5, 1), (6^{(n-6)/6})) := (\mu(n-14) \cup (7, 1), (6)) \in \mathcal{T}'(n).$$

If $d > 1$, $c \geq 8$, then $(\mu(c), (c^d)) \in \Psi_2(C_{1,e}(n)) \setminus \mathcal{S}(n)$. We define

$$\zeta(\mu(c), (c^d)) := (\mu(3c), (c^{d-2})).$$

The parts of $\mu(3c)$ are larger than $c+1$ and thus $(\mu(3c), (c^{d-2})) \notin \mathcal{S}(n)$. Therefore, $(\mu(3c), (c^{d-2})) \in B'(n) \setminus \psi(A(n))$. Moreover, $(\mu(3c), (c^{d-2})) \notin \mathcal{T}(n)$.

Case (ii) If $\xi_1 = \xi_2 = c-1$, $\xi \setminus (c-1) \in \mathcal{Q}_o$, $1 \in \xi$, and $c \geq 4$, then $(\xi, (c^d)) \in \Psi_2(C_{1,o}(n)) \backslash \mathcal{S}(n)$.

To define ζ, we consider two subcases:

(I) $\xi_3 < c-3$. We define

$$\zeta(\xi, (c^d)) := (\xi \setminus (c-1, c-1) \cup \mu(2c-2), (c^d)) =: (\nu, (c^d)).$$

Since $\ell(\xi) \geq 3$, $1 \in \xi$, and $\mu(2c-2) = (c+1, c-3)$, we must have $c > 4$. Since $1 \in \nu$ and $c-1 \notin \nu$, it follows that $(\nu, (c^d)) \notin \mathcal{S}(n)$. Since $c+1 \in \nu$ is the largest part, it follows that $(\nu, (c^d)) \notin \mathcal{T}(n)$.

(II) $\xi_3 = c-3$. If $c = 4, 6, 8, 10$, we define $\zeta(\xi, (c^d)) \in \mathcal{T}'(n)$ in an *ad hoc* manner by the table below:

n	$(\xi, (c^d))$	$\zeta(\xi, (c^d))$
3 (mod 8)	$((3,3,1), (4^{(n-7)/4}))$	$(\mu(n-11) \cup (5,3,1), (2))$
8 (mod 12)	$((5,5,3,1), (6^{(n-14)/6}))$	$(\mu(n-18) \cup (9,1), (8))$
12 (mod 16)	$((7,7,5,1), (8^{(n-20)/8}))$	$(\mu(n-26) \cup (13,1), (12))$
15 (mod 16)	$((7,7,5,3,1), (8^{(n-23)/8}))$	$(\mu(n-11) \cup (5,3,1), (2))$
16 (mod 20)	$((9,9,7,1), (10^{(n-26)/10}))$	$(\mu(n-26) \cup (13,1), (12))$
4 (mod 20)	$((9,9,7,5,3,1), (10^{(n-34)/10}))$	$(\mu(n-26) \cup (13,1), (12))$
1 (mod 20)	$((9,9,7,5,1), (10^{(n-31)/10}))$	$(\mu(n-29) \cup (7,3,1), (2^9))$
19 (mod 20)	$((9,9,7,3,1), (10^{(n-29)/10}))$	$(\mu(n-27) \cup (5,3,1), (2^9))$

.

Note that the congruence conditions on n imply that no pair in the right-hand column occurs more than twice.

For the remaining pairs, i.e., $c \geq 12$, we define

$$\zeta(\xi, (c^d)) := (\xi \setminus (\xi_1, \xi_2, \xi_3, 1) \cup \mu(3c-4), (c^d)) =: (\nu, (c^d)).$$

Since $c \geq 12$, $\mu(3c-4)$ has parts greater than $c+1$. Thus $1, c+1 \notin \nu$, and therefore, $(\nu, (c^d)) \notin \mathcal{S}(n)$. Since $1, c-1 \notin \nu$, we have $(\nu, (c^d)) \notin \mathcal{T}(n)$.

Case (iii) If $\xi = (c-1, 1, 1)$, $c \geq 4$, then $(\xi, (c^d)) \in \Psi_2(C_{1,o}(n)) \backslash \mathcal{S}(n)$.

If $d = 1$, we define

$$\zeta(((n-3)/2, 1, 1), ((n-1)/2)) := (\mu(n-13) \cup (7,3,1), (2)) \in \mathcal{T}'(n).$$

If $d = 3$, we define

$$\zeta(((n-5)/4, 1, 1), ((n-1)/4)^3) := (\mu(n-13) \cup (7,3,1), (2)) \in \mathcal{T}'(n).$$

If $d > 3$, we define

$$\zeta((c-1,1,1),(c^d)) = (\mu(5c) \cup (1), (c^{d-4})).$$

Since $c \geq 4$, the parts of $\mu(5c)$ are larger than $c+1$ and thus $(\mu(5c) \cup (1), (c^{d-2}))$ is neither in $S(n)$ nor in $\mathcal{T}(n)$.

Case (iv) If $\xi_1 = \xi_2 = c+1$, $\xi \setminus (c+1) \in \mathcal{Q}_o$, and $1 \notin \xi$, then $(\xi, (c^d)) \in \Psi_2(C_{2,o}(n)) \setminus \mathcal{S}(n)$.

To define ζ, we consider two subcases:

(I) $\xi_3 < c-1$. We define

$$\zeta((3,3),(2^{(n-6)/2})) := (\mu(n-18) \cup (9,1), (8)) \in \mathcal{T}'(n),$$

and if $(\xi, (c^d)) \neq ((3,3),(2^{(n-6)/2}))$, we define

$$\zeta(\xi,(c^d)) := (\xi \setminus (c+1,c+1) \cup \mu(2c+2), (c^d)) =: (\nu, (c^d)).$$

Since $(\xi, (c^d)) \neq ((3,3),(2^{(n-6)/2}))$, we must have $c > 2$. Moreover, $\mu(2c+2) = (c+3, c-1)$, and these are the largest parts in ν. Since $1, c+1 \notin \nu$ and the largest part in ν is $c+3$, it follows that $(\nu, (c^d)) \notin \mathcal{S}(n)$. In fact, $(\nu, (c^d)) \in \mathcal{T}'(n)$. Furthermore, since $1 \notin \nu$, these pairs do not occur in the images of ζ in cases (i)–(iii) above.

(II) If $\xi_3 = c-1$, we define

$$\zeta(\xi,(c^d)) = (\xi \setminus (\xi_1,\xi_2,\xi_3) \cup \mu(3c) \cup (1), (c^d)) =: (\nu, (c^d)).$$

Since $1 \notin \xi$, we have $c \geq 4$. Then, the parts of $\mu(3c)$ are at least $c+1$, and they are the largest parts in ν. Thus $1 \in \nu$ and $c-1 \notin \nu$, and therefore, $(\nu, (c^d)) \notin \mathcal{S}(n)$. Moreover, $(\nu, (c^d)) \notin \mathcal{T}(n)$.

To finish the proof, for each $n \geq 53$, we display six pairs in $B'(n)$ that do not occur in the image of ζ. These pairs belong to $\mathcal{T}'(n)$.

If $n \equiv 3 \pmod 4$, two copies of each of the following pairs:

$$(\mu(n-15) \cup (5,3,1), (2^3)),$$

$$(\mu(n-19) \cup (5,3,1), (2^5)),$$

$$(\mu(n-23) \cup (5,3,1), (2^7)).$$

If $n \equiv 1 \pmod 4$, two copies of each of the following pairs:

$$(\mu(n-17) \cup (7,3,1), (2^3)),$$

$$(\mu(n-21) \cup (7,3,1), (2^5)),$$

$$(\mu(n-25) \cup (7,3,1), (2^7)).$$

If n is even, two copies of each of the following pairs:

$$(\mu(n-14) \cup (7,3), (4)),$$
$$(\mu(n-20) \cup (9,5), (6)),$$
$$(\mu(n-18) \cup (11,7), (8)).$$

Notice that the last three pairs are the only pairs $(\nu, (c^d))$ such that $1 \notin \nu$. Pairs in $\mathcal{T}'(n)$ with $1 \notin \nu$ occurred in the image of ζ in case (iv) subcase (I). However, in that case $c+3$, $c-1$ are the largest parts of ν, while here $c+3$, $c+1$ are the smallest parts of ν and, since $n \geq 53$, there are larger parts in ν.

Thus, $|B'(n)| \geq |\psi(A(n)| + 6$. □

5.3 *Proof of Proposition 5.4*

We provide two different proofs of Proposition 5.4 below, one of which is combinatorial in nature (see Sect. 5.3), the other of which is analytic (see Sect. 5.3).

Combinatorial Proof of Proposition 5.4

As shown in [9], if $F_2(q) = \sum_{n=1}^{\infty} b_n q^n$, then b_n equals the number of parts in all partitions of n with distinct odd parts, i.e.,

$$b_n = \sum_{\lambda \in \mathcal{Q}_o(n)} \ell(\lambda).$$

Let $n \geq 9$. We first create a length preserving bijection φ from $\{\lambda \in \mathcal{Q}_o(n) \mid 1 \notin \lambda\}$ to $\{\xi \in \mathcal{Q}_o(n-4) \mid \text{if } \ell(\xi) \geq 3, \text{ then } \xi_{\ell(\xi)-2} - \xi_{\ell(\xi)-1} > 2\}$.

Let $\lambda = (\lambda_1, \lambda_2, \ldots, \lambda_{\ell(\lambda)}) \in \mathcal{Q}_o(n)$, $1 \notin \lambda$ and define

$$\varphi(\lambda) = \begin{cases} (n-4), & \text{if } \ell(\lambda) = 1, \text{ i.e., } \lambda = (n), \\ (\lambda_1, \lambda_2, \ldots, \lambda_{\ell(\lambda)-1} - 2, \lambda_{\ell(\lambda)} - 2), & \text{if } \ell(\lambda) \geq 2. \end{cases}$$

Clearly, $\ell(\lambda) = \ell(\varphi(\lambda))$.

Next, we consider the bijection ψ from $\{\lambda \in \mathcal{Q}_o(n) \mid 1 \in \lambda\}$ to $\{\xi \in \mathcal{Q}_o(n-1) \mid 1 \notin \xi\}$ given by $\psi(\lambda) = \lambda \setminus (1)$. Clearly, $\ell(\lambda) = \ell(\psi(\lambda)) + 1$.

It remains to show that the number of parts equal to 1 in all partitions in $\mathcal{Q}_o(n)$ is less than or equal to the total number of parts in all partitions in $\mathcal{Y} := \{\xi \in \mathcal{Q}_o(n-1) \mid 1 \in \xi\} \cup \{\xi \in \mathcal{Q}_o(n-4) \mid \ell(\xi) \geq 3 \text{ and } \xi_{\ell(\xi)-2} - \xi_{\ell(\xi)-1} = 2\}$. We create an injection Ξ from $\{\lambda \in \mathcal{Q}_o(n) \mid 1 \in \lambda\}$ to the set of partitions in $\mathcal{Y}$ with exactly one marked part.

Let $\lambda \in \mathcal{Q}_o(n)$ be such that $1 \in \lambda$. Since $n \geq 9$, we have $\ell(\lambda) \geq 2$. Let $a := \lambda_{\ell(\lambda)-1}$.

If $a \geq 9$, $1 \notin \mu(a-1)$, and we define $\Xi(\lambda) = \lambda \setminus (a) \cup \mu(a-1) \in \mathcal{Q}_o(n-1)$ with part 1 marked. Note that $\Xi(\lambda)$ has at least three parts, and if it has exactly three parts, then the difference between the first and second parts is 2 or 4. The marked part of $\Xi(\lambda)$ is the last part.

Next we consider the case when $a = 3, 5$, or 7. Since $n \geq 9$, we have $\ell(\lambda) \geq 3$.

If $\ell(\lambda) \geq 4$, define $\Xi(\lambda) = \lambda \setminus (\lambda_1, a) \cup (\lambda_1 + a - 1) \in \mathcal{Q}_o(n-1)$ with marked first, second, or third part according to $a = 3, 5$, or 7, respectively. Note that if $a = 7$, the difference between the first and second parts in $\Xi(\lambda)$ is at least 8. Thus, if $\ell(\lambda) = 4$ and $a = 7$, then $\Xi(\lambda)$ has exactly three parts and the marked part is 1, but the obtained marked partition is different from the marked partitions obtained in the case $a \geq 9$.

If $\ell(\lambda) = 3$, then n is odd. We define $\Xi(n-8, 7, 1) = (\overline{n-2}, 1)$, $\Xi(n-6, 5, 1) = (n-2, \overline{1})$, and

$$\Xi(n-6, 5, 1) = \begin{cases} (\mu(n-5), \overline{1}), & \text{if } n \geq 17, n \equiv 1 \pmod 4, \\ (\mu(n-7), \overline{3}), & \text{if } n \geq 17, n \equiv 3 \pmod 4, \\ (\overline{n-2}, 1), & \text{if } 9 \leq n \leq 15. \end{cases}$$

Note that $(n-8, 7, 1)$ occurs only when $n \geq 17$ and that $(\mu(n-5), \overline{1}), (\mu(n-7), \overline{3}) \in \mathcal{Q}_o(n-4)$.

Analytic Proof of Proposition 5.4

To prove Proposition 5.4, it suffices to show that

$$(q^4 + q - 1)F_2(q) \succeq_{S'} 0, \tag{34}$$

where $S' := \{0, 1, 2, 3, 4, 5, 6, 7, 8\}$. Indeed, we will prove the stronger result with $S' = \{1, 3, 4, 6, 8\}$. Toward (34), we establish Lemma 5.6, which is stated in terms of the polynomials

$$f_m(q) := \begin{cases} -q + q^2 - q^4 + 2q^5 - q^6, & m = 1, \\ -q^3, & m = 2, \\ -q^5 + q^7 - q^8 + q^9 + 2q^{10} + q^{13} - q^{15}, & m = 3, \\ -q^{2m-1} + q^{2m+1} - q^{2m+2} + q^{2m+3} + q^{2m+4}, & m \geq 4. \end{cases}$$

Lemma 5.6 *For each integer $m \geq 1$, we have*

$$(q^4+q-1)q^{2m-1}\prod_{\substack{\ell=1\\ \ell\neq m}}^{\infty}(1+q^{2\ell-1}) = f_m(q)+g_m(q), \tag{35}$$

where $g_m(q) \succeq 0$, *and*

$$g_m(q) = \begin{cases} O(q^7), & m=1, \\ O(q^5), & m=2, \\ O(q^{16}), & m=3, \\ O(q^{2m+6}), & m \geq 4. \end{cases}$$

Using Lemma 5.6, we give an analytic proof of Proposition 5.4. Following its proof, the remainder of this section is devoted to proving Lemma 5.6.

Analytic Proof of Proposition 5.4 By Lemma 5.6, we have

$$(q^4+q-1)F_2(q) = \sum_{m=1}^{\infty}(f_m(q)+g_m(q)). \tag{36}$$

We rewrite $\sum_{m=1}^{\infty} f_m(q)$ as

$$\begin{aligned}
&\sum_{j=1}^{3} f_j(q) + \sum_{m\geq 4} q^{2m+3} - \sum_{m\geq 4}(q^{2m-1}+q^{2m+2}) + \sum_{m\geq 4}(q^{2m+1}+q^{2m+4}) \\
&= \sum_{j=1}^{3} f_j(q) + \sum_{m\geq 4} q^{2m+3} - \sum_{m\geq 4}(q^{2m-1}+q^{2m+2}) + \sum_{m\geq 5}(q^{2m-1}+q^{2m+2}) \\
&= \sum_{j=1}^{3} f_j(q) + \sum_{m\geq 4} q^{2m+3} - q^7 - q^{10} \\
&= -q+q^2-q^3-q^4+q^5-q^6-q^8+q^9+q^{10}+q^{11}+2q^{13}+\sum_{m\geq 7} q^{2m+3}.
\end{aligned}$$

Since $\sum_{m=1}^{\infty} g_m(q) \succeq 0$, we obtain the non-negativity of coefficients stated (34).

□

Proof of Lemma 5.6 We divide the proof into cases, depending on m.

Throughout the proof, we make use of the following calculations. Let a, i, j be positive integers. We express a product $\prod_{k\geq i}(1+q^{2k-1})$ in terms of the smallest, respectively largest, exponent appearing in monomials as

$$1+\sum_{k\geq i} q^{2k-1}\prod_{\ell>k}(1+q^{2\ell-1}) = 1+q^{2i-1}+\sum_{k\geq i+1} q^{2k-1}\prod_{i\leq \ell<k}(1+q^{2\ell-1}). \tag{37}$$

Then,

$$\begin{aligned} & q^{a+2j}\sum_{k\geq i} q^{2k-1}\prod_{\ell>k}(1+q^{2\ell-1}) - q^{a}\sum_{k\geq i+j} q^{2k-1}\prod_{\ell>k}(1+q^{2\ell-1}) \qquad (38)\\ &= q^{a+2j}\sum_{k\geq i} q^{2k-1}\prod_{\ell>k}(1+q^{2\ell-1}) - q^{a+2j}\sum_{k\geq i+j} q^{2(k-j)-1}\prod_{\ell>k}(1+q^{2\ell-1})\\ &= q^{a+2j}\sum_{k\geq i} q^{2k-1}\prod_{\ell>k}(1+q^{2\ell-1}) - q^{a+2j}\sum_{k\geq i} q^{2k-1}\prod_{\ell>k+j}(1+q^{2\ell-1}) \succeq 0. \end{aligned}$$

Similarly,

$$\begin{aligned} & q^{a}\sum_{k\geq i} q^{2k-1}\prod_{\ell>k}(1+q^{2\ell-1}) - q^{a+2}\sum_{k\geq i} q^{2k-1}\prod_{\ell>k}(1+q^{2\ell-1}) \qquad (39)\\ &= q^{a+2i-1} + q^{a+2i+1}(1+q^{2i-1}) + q^{a}\sum_{k\geq i+2} q^{2k-1}\prod_{i\leq\ell<k}(1+q^{2\ell-1})\\ &\quad - q^{a+2+2i-1} - q^{a+2}\sum_{k\geq i+1} q^{2k-1}\prod_{i\leq\ell<k}(1+q^{2\ell-1}) \succeq 0. \end{aligned}$$

The non-negativity of coefficients follows from the fact that

$$q^{a+2}\sum_{k\geq i+1} q^{2k-1}\prod_{i\leq\ell<k}(1+q^{2\ell-1}) = q^{a}\sum_{k\geq i+2} q^{2k-1}\prod_{i\leq\ell<k-1}(1+q^{2\ell-1}).$$

We continue with the proof of Lemma 5.6.

Case $m \geq 4$. We rewrite the left-hand side of (35) as

$$P_m(q)Q_m(q),$$

where

$$\begin{aligned} P_m(q) &:= (q^4+q-1)q^{2m-1}(1+q)(1+q^3)(1+q^5)\\ &= -q^{2m-1} + q^{2m+1} - q^{2m+2} + q^{2m+3} + q^{2m+4} + 2q^{2m+6} + q^{2m+8}\\ &\quad + 2q^{2m+9} + q^{2m+11} + q^{2m+12},\\ Q_m(q) &:= \prod_{\substack{\ell=4\\ \ell\neq m}}^{\infty}(1+q^{2\ell-1}) = 1+\sum_{\substack{k\geq 4\\ k\neq m}} q^{2k-1}\prod_{\substack{\ell>k\\ \ell\neq m}}(1+q^{2\ell-1}). \end{aligned}$$

Then,

$$g_m(q) = -\,(q^{2m-1}+q^{2m+2})\sum_{\substack{k\geq 4\\ k\neq m}} q^{2k-1}\prod_{\substack{\ell>k\\ \ell\neq m}}(1+q^{2\ell-1}) \tag{40}$$

$$+\,(q^{2m+1}+q^{2m+3}+q^{2m+4})\sum_{\substack{k\geq 4\\ k\neq m}} q^{2k-1}\prod_{\substack{\ell>k\\ \ell\neq m}}(1+q^{2\ell-1}) \tag{41}$$

$$+\,(2q^{2m+6}+q^{2m+8}+2q^{2m+9}+q^{2m+11}+q^{2m+12})Q_m(q). \tag{42}$$

Thus, $g_m(q) = O(q^{2m+6})$. To show that $g_m(q) \succeq 0$, we show that all terms in (40) appear with positive sign in (41) or (42).

We first consider the case $m = 4$. Then (40) equals

$$-\,(q^{16}+q^{19})\prod_{\ell>5}(1+q^{2\ell-1}) - (q^7+q^{10})\sum_{k\geq 6} q^{2k-1}\prod_{\ell>k}(1+q^{2\ell-1})$$
$$=: -\,C_1 - C_2.$$

All terms in C_1 appear in (42). Using (38) with $j = 1$, $i = 5$, and $a = 7, 10$ respectively, terms in C_2 cancel with terms in (41). Thus, $g_4(q) \succeq 0$.

For $m > 4$, we rewrite (40) separating the terms according to $k = 4$, $k \neq 4, m+1$, and $k = m + 1$. When $k \neq 4, m + 1$, we factor out q^2 and shift the index of summation. Thus, (40) equals

$$-\,(q^{2m+6}+q^{2m+9})\prod_{\substack{\ell>4\\ \ell\neq m}}(1+q^{2\ell-1})$$
$$-\,(q^{2m+1}+q^{2m+4})\sum_{\substack{k\geq 4\\ k\neq m-1,m}} q^{2k-1}\prod_{\substack{\ell>k+1\\ \ell\neq m}}(1+q^{2\ell-1})$$
$$-\,(q^{4m}\prod_{\ell>m+1}(1+q^{2\ell-1}) + q^{4m+3}\prod_{\ell>m+1}(1+q^{2\ell-1})).$$

Writing $q^{4m} = q^{2m+3}\cdot q^{2m-3}$ and $q^{4m+3} = q^{2m+6}\cdot q^{2m-3}$, we see that each term in (40) cancels with a corresponding positive term in (41) or (42) (and terms in (41) and (42) are used at most once in this cancellation). Hence, $g_m(q) \succeq 0$.

Case $m = 1$. Using (37), we rewrite the left-hand side of (35) as

$$
\begin{aligned}
(q^4+q-1)q\prod_{\ell\geq 2}(1+q^{2\ell-1})
&= q^5+q^2-q+(q^5+q^2-q)\sum_{k\geq 2}q^{2k-1}\prod_{\ell>k}(1+q^{2\ell-1})\\
&= -q+q^2-q^4+2q^5-q^6+g_1(q),
\end{aligned}
$$

where

$$
\begin{aligned}
g_1(q) =& q^5\sum_{k\geq 2}q^{2k-1}\prod_{\ell>k}(1+q^{2\ell-1})+q^5\sum_{k\geq 3}q^{2k-1}\prod_{\ell>k}(1+q^{2\ell-1})\\
&+q^2\sum_{k\geq 3}q^{2k-1}\prod_{\ell>k}(1+q^{2\ell-1})-q^4\sum_{k\geq 3}q^{2k-1}\prod_{\ell>k}(1+q^{2\ell-1})\\
&-q^6\sum_{k\geq 4}q^{2k-1}\prod_{\ell>k}(1+q^{2\ell-1})-q\sum_{k\geq 4}q^{2k-1}\prod_{\ell>k}(1+q^{2\ell-1})\\
&=: A_1+A_2+A_3-A_4-A_5-A_6.
\end{aligned}
$$

From this expression, it is clear that $g_1(q) = O(q^7)$. To show that $g_1(q) \succeq 0$, we first compute $A_1 - A_6$. This equals

$$
\begin{aligned}
&q^5\sum_{k\geq 2}q^{2k-1}\prod_{\ell>k}(1+q^{2\ell-1})-q^5\sum_{k\geq 2}q^{2k-1}\prod_{\ell>k+2}(1+q^{2\ell-1})\\
&=\sum_{k\geq 2}(q^{4k+5}+q^{4k+7}+q^{6k+8})\prod_{\ell>k+2}(1+q^{2\ell-1}) =: B_1+B_2+B_3.
\end{aligned}
$$

Next, separating terms by k even and odd, respectively, we rewrite A_5 as

$$
\sum_{j\geq 2}q^{4j+5}\prod_{\ell>2j}(1+q^{2\ell-1})+\sum_{j\geq 3}q^{4j+7}\prod_{\ell>2j+1}(1+q^{2\ell-1}).
$$

Since $k+2 \leq 2k$ if $k \geq 2$, we have $B_1+B_2-A_5 \succeq 0$. From (39) with $a = 2, i = 3$, it follows that $A_3 - A_4 \succeq 0$. Hence, $g_1(q) \succeq 0$.

Case $m = 2$. Using (37), we rewrite the left-hand side of (35) as

$$
(q^4+q-1)q^3(1+q)\prod_{\ell\geq 3}(1+q^{2\ell-1}) = -q^3+g_2(q),
$$

where

$$
g_2(q) = (q^5+q^7+q^8)\prod_{\ell\geq 3}(1+q^{2\ell-1})-q^3\sum_{k\geq 3}q^{2k-1}\prod_{\ell>k}(1+q^{2\ell-1}) = O(q^5).
$$

To show $g_2(q) \succeq 0$, we write

$$q^3 \sum_{k\geq 3} q^{2k-1} \prod_{\ell>k}(1+q^{2\ell-1}) = q^8 \prod_{\ell>3}(1+q^{2\ell-1}) + q^3 \sum_{k\geq 4} q^{2k-1} \prod_{\ell>k}(1+q^{2\ell-1}).$$

Using (37) and (38) with $a = 3$, $j = 1$, $i = 3$, it follows that $g_2(q) \succeq 0$.

Case $m = 3$. Using (37), we rewrite the left-hand side of (35) as

$$\begin{aligned}(q^4+q-1)q^5(1+q)(1+q^3)\prod_{\ell\geq 4}(1+q^{2\ell-1})\\ = -q^5+q^7-q^8+q^9+2q^{10}+q^{13}-q^{15}+g_3(q),\end{aligned}$$

where

$$\begin{aligned}g_3(q) =& q^{12}+q^{15}\\ &+(-q^5+q^7-q^8+q^9+2q^{10}+q^{12}+q^{13})\sum_{k\geq 4} q^{2k-1}\prod_{\ell>k}(1+q^{2\ell-1})\\ =&(q^9+2q^{10}+q^{12}+q^{13})\sum_{k\geq 4} q^{2k-1}\prod_{\ell>k}(1+q^{2\ell-1})\\ &+(q^7+q^{14})\sum_{k\geq 5} q^{2k-1}\prod_{\ell>k}(1+q^{2\ell-1})\\ &-(q^8+q^{15}+q^{12})\sum_{k\geq 5} q^{2k-1}\prod_{\ell>k}(1+q^{2\ell-1})\\ &-(q^5+q^{14})\sum_{k\geq 6} q^{2k-1}\prod_{\ell>k}(1+q^{2\ell-1}) = O(q^{16}).\end{aligned}$$

Using (38) with $a = 8$, $j = 1$, $i = 4$ and also with $a = 5$, $j = 1$, $i = 5$, as well as (39) with $a = 13$, $i = 5$, we obtain $g_3(q) \succeq 0$. □

Acknowledgments The authors thank the Banff International Research Station (BIRS) and the Women in Numbers 5 (WIN5) Program. The third author is partially supported by National Science Foundation Grants DMS-1901791 and DMS-2200728. The fifth author is partially supported by a FRQNT scholarship by Fonds de Recherche du Québec, and an ISM scholarship by Institut des Sciences Mathématiques.

References

1. G.E. Andrews, *The Theory of Partitions*. Reprint of the 1976 Original. Cambridge Mathematical Library (Cambridge University Press, Cambridge, 1998), 255 pp.

2. G.E. Andrews, Euler's partition identity and two problems of George Beck. Math. Student **86**(1–2), 115–119 (2017)
3. G.E. Andrews, Integer partitions with even parts below odd parts and the mock theta functions. Ann. Comb. **22**(3), 433–445 (2018)
4. G.E. Andrews, C. Ballantine, Almost partition identities. Proc. Natl. Acad. Sci. USA **116**(12), 5428–5436 (2019)
5. G.E. Andrews, A. Dixit, A.J. Yee, Partitions associated with the Ramanujan/Watson mock theta functions $\omega(q)$, $\nu(q)$, and $\phi(q)$. Res. Number Theory **1**(19), 25pp. (2015)
6. G.E. Andrews, A.J. Yee, Some identities associated with mock theta functions $\omega(q)$ and $\nu(q)$. Ramanujan J. **48**, 613–622 (2019)
7. T.M. Apostol, *Introduction to Analytic Number Theory*. Undergraduate Texts in Mathematics (Springer, Berlin, 1991)
8. C. Ballantine, R. Bielak, Combinatorial proofs of two Euler-type identities due to Andrews. Ann. Comb. **23**(3–4), 511–525 (2019)
9. C. Ballantine, H. Burson, A. Folsom, C.-Y. Hsu, I. Negrini, B. Wen, On a partition identity of Lehmer. Discrete Math. **345**(10), 112979 (2022)
10. C. Ballantine, A. Welch, Beck-type companion identities for Franklin's identity via a modular refinement. Discrete Math. **344**(8), 112480 (2021)
11. A. Berkovich, F.G. Garvan, Some observations on Dyson's new symmetries of partitions. J. Combin. Theory, Ser. A **100**, 61–93 (2002)
12. B.C. Berndt, A. Dixit, R. Gupta, Generalizations of the Andrews-Yee identities associated with the mock theta functions $\omega(q)$ *and* $\nu(q)$. J. Algebraic Combin. **55**(4), 1031–1062 (2022)
13. B.C. Berndt, R.A. Rankin, *Ramanujan. Letters and Commentary*. History of Mathematics, 9 (American Mathematical Society, Providence; London Mathematical Society, London, 1995), 347 pp.
14. B.C. Berndt, A.J. Yee, Combinatorial proofs of identities in Ramanujan's lost notebook associated with the rogers-fine identity and false theta functions. Ann. Comb. **7**, 409–423 (2003)
15. K. Bringmann, A. Folsom, K. Ono, L. Rolen, *Harmonic Maass Forms and Mock Modular Forms: Theory and Applications*. American Mathematical Society Colloquium Publications, 64 (American Mathematical Society, Providence, 2017), 391 pp.
16. W.D. Duke, Almost a century of answering the question: What is a mock theta function? Notices Am. Math. Soc. **61**(11), 1314–1320 (2014)
17. N.J. Fine, *Basic Hypergeometric Series and Applications*. Mathematical Surveys and Monographs, vol. 27 (American Mathematical Society, Providence, 1988)
18. A. Folsom, Perspectives on mock modular forms. J. Number Theory **176**, 500–540 (2017)
19. S. Garthwaite, The coefficients of the $\omega(q)$ mock theta function. Int. J. Number Theory **4**(6), 1027–1042 (2008)
20. G. Gasper, M. Rahman, *Basic Hypergeometric Series*, 2nd edn. Encyclopedia of Mathematics and its Applications, 96 (Cambridge University Press, Cambridge, 2004)
21. M. Griffin, L. Rolen, K. Ono, Ramanujan's mock theta functions. Proc. Nat. Acad. Sci., **110**(15), 5765–5768 (2013)
22. R. Li, A.Y.Z. Wang, Partitions associated with two fifth-order mock theta functions and Beck-type identities. Int. J. Number Theory **16**(4), 841–855 (2020)
23. F.Z.K. Li, J.Y.X. Yang, Combinatorial proofs for identities related to generalizations of the mock theta functions $\omega(q)$ *and* $\nu(q)$. Ramanujan J. **50**(3), 527–550 (2019)
24. H. Rademacher, *Topics in Analytic Number Theory,* ed. by E. Grosswald, J. Lehner, M. Newman. Die Grundlehren der mathematischen Wissenschaften, Band 169 (Springer-Verlag, New York-Heidelberg, 1973), 320 pp.
25. R.C. Rhoades, On Ramanujan's definition of mock theta function. Proc. Nat. Acad. Sci. **110**(19), 7592–7594 (2013)
26. J.Y.X. Yang, Combinatorial proofs and generalizations of conjectures related to Euler's partition theorem. Eur. J. Combin. **76**, 62–72 (2019)

27. D. Zagier, Ramanujan's mock theta functions and their applications, [d'après Zwegers and Bringmann-Ono], Séminaire Bourbaki 60ème année, 2006–2007, Exposés 982–996. Astérisque **326**, 22pp. (2009)
28. S. Zwegers, *Mock Theta Functions,* Ph.D. Thesis, Utrecht University (2002)

Transcendental Lattices of Certain Singular $K3$ Surfaces

M. J. Bertin and O. Lecacheux

1 Introduction

Most families of $K3$ surfaces have been studied in the past regarding their Shioda–Inose structure initiated by Shioda and Inose [28]. More precisely, these authors constructed when the Picard number, i.e., the rank of the Picard group, has the maximum possible value 20, an involution ι on the $K3$ surface X such that the quotient X/ι is isomorphic to a Kummer surface. Later on, this structure has been generalized to other Picard ranks by Morrison [13], Peters [18], and Dolgachev [8]. Morrison proved that $K3$ surfaces, with Picard numbers 19 and 20 and also for Picard numbers 18 and 17 with some additional conditions, possess a Shioda–Inose structure. Dolgachev developed the notion of M_n-polarized $K3$ surfaces for lattices $M_n \simeq E_8(-1)^2 \oplus U \oplus \langle 2n \rangle$. The M_n-polarization is the existence of a primitive embedding of the lattice M_n into the unimodular $K3$ lattice $U^3 \oplus E_8(-1)^2$. Here U denotes the hyperbolic lattice and E_8 the unique positive definite even unimodular lattice of rank 8. Notice that M_n-polarized $K3$ surfaces have Picard number 19 and transcendental lattice $U \oplus \langle 2n \rangle$. Narumiya and Shiga [14] were interested in a family of M_2-polarized $K3$ surfaces arising from the dual of the simplest 3-dimensional reflexive polytope whose corresponding mirror mapping was studied by Lian and Yau [12] and then by Verrill and Yui [34]. Ling Long [11] studied M_n-polarized $K3$ surfaces in regard to their Picard–Fuchs differential equations. Clingher and Doran [7] extended Morrison's result on Shioda–Inose structure for $K3$ surfaces with Picard number 18 and M polarization by $M \simeq E_8(-1)^2 \oplus U$. Recently Bertin and Lecacheux [4], considering a Picard number 18 family $S_{a,b}$ of $K3$ surfaces with

M. J. Bertin (✉) · O. Lecacheux
Institut de Mathématiques de Jussieu-Paris Rive Gauche, Sorbonne Université, Paris, France
e-mail: marie-jose.bertin@imj-prg.fr; odile.lecacheux@imj-prg.fr

A. Bucur et al. (eds.), *Research Directions in Number Theory*, Association for Women in Mathematics Series 33, https://doi.org/10.1007/978-3-031-51677-1_5

an elliptic fibration, a two-torsion section defining an involution ι, and two singular fibers of type I_4^*, recovered Narumiya and Shiga's family if $b = 1$.

In this chapter, we investigate the links between M_6, M_{12}, and M_3-polarized $K3$ surfaces, more precisely between their singular members of Picard number 20. Denote (Y_k), (Z_k), and (X_k) the generic member of pencils of $K3$ surfaces with defining polynomials, respectively, P_k, Q_k, and Q'_k. In a series of papers, Bertin tried to generalize to singular $K3$ surfaces some formulae valid for a certain type of two-variable polynomials defining elliptic curves, that is, their Mahler measures are expressed in terms of a L-series of a modular form that is also the L-series of the elliptic curve itself. For some values of k, the $K3$ surfaces Y_k, Z_k, and X_k are singular, and the L-series of their transcendental lattices are given by the L-series of a modular form suspected to be the same as in the expression of the Mahler measure.

The family of Laurent polynomials (P_k) defining (Y_k) with generic Picard number 19 is the well-known Apéry–Fermi pencil defined by

$$P_k = x + \frac{1}{x} + y + \frac{1}{y} + z + \frac{1}{z} - k, \qquad k \in \mathbb{C}.$$

It was studied first by Peters and Stienstra [19], then used by Bertin [1, 2] and others [21] to find explicit Mahler measures of some members, and finally studied by Bertin and Lecacheux [4]. In particular, Peters and Stienstra [19] gave their generic transcendental lattice $U \oplus \langle 12 \rangle$.

The family of polynomials (Q_k) defining (Z_k) is defined by

$$Q_k = (x + y + z + 1)(xy + xz + yz + xyz) - (k + 4)xyz, \qquad k \in \mathbb{C}.$$

It has been studied concerning the Mahler measure of some members of the family, namely those defining singular $K3$ surfaces, by Bertin [1, 2], Samart [21], but the link between the Mahler measure and the L-series of the $K3$ surface was only considered by Bertin [2] and her co-authors [3].

The family of polynomials (Q'_k) defining (X_k) with generic Picard number 19 is Verrill's family [32, 33] with generic transcendental lattice $U \oplus \langle 6 \rangle$ and defined by

$$Q'_k = (x + xy + xyz + 1)(1 + z + zy + zxy) - (k + 4)xyz, \qquad k \in \mathbb{C}.$$

The Mahler measure of Q_k is in fact the same as the Mahler measure of Q'_k since the change variables $x = X$, $y = XY$, $z = XYZ$ transform Q_k into Q'_k and thus get the Mahler measure unchanged. Hence, we can deduce that the generic Picard number of Z_k is 19 and that Z_k and X_k are singular $K3$ surfaces for the same values of k. These values and those corresponding to singular $K3$ surfaces Y_k have been computed long ago by Boyd [5].

It also leads to our first question: Is there a link between the two families (Z_k) and (X_k)? That is the object of Sect. 3 where we prove the following theorem.

Theorem 1.1

(1) The transcendental lattice of the generic member Z_k is $U \oplus \langle 24 \rangle$.
(2) There is a genus 1 fibration of Z_k whose Jacobian surface J_k is a $K3$ surface of the Verrill's family with generic transcendental lattice $U \oplus \langle 6 \rangle$.

We are also interested in another question: Among the singular members of these families, which ones belong to at least two families?

This question was suggested by an observation concerning the Mahler measures of their corresponding polynomials. In return, the answer to the question may give an expression of the Mahler measure in terms of the L-series of the $K3$ surface.

So all along Sects. 4–6, we shall prove the following theorem that completes some partial results already known. We recall that the transcendental lattice of a singular $K3$ surface has dimension 2. For simplification, we adopt Shimada and Zhang's notation [26], that is, $[a \quad b \quad c]$ means the transcendental lattice $\begin{pmatrix} a & b \\ b & c \end{pmatrix}$.

Theorem 1.2 *The transcendental lattices of the singular members of the previous families are given in the following table. In case we could not compute the transcendental lattice, we give the conjectural discriminant of the $K3$ surface obtained from results on Mahler's measure of the defining polynomial of the surface.*

Y_0	[4 2 4]	Z_{-36}	$?12a^2$	X_{-36}	[2 0 6]
Y_2	[2 0 4]	Z_{-12}	[10 2 10]	X_{-12}	[4 0 6]
Y_3	[2 1 8]	Z_{-6}	[8 4 8]	X_{-6}	[6 0 8]
Y_6	[2 0 12]	Z_{-3}	[4 1 4]	X_{-3}	[6 0 10]
Y_{10}	[6 0 12]	Z_0	[2 0 6]	X_0	[2 0 6]
Y_{18}	[10 0 12]	Z_4	[2 0 16]	X_4	[2 0 4]
Y_{102}	[12 0 26]	Z_{12}	[2 0 24]	X_{12}	[2 1 2]
Y_{198}	[12 0 34]	Z_{60}	$?15a^2$	X_{60}	[4 1 4]
$Y_{k^2=20}$	[2 0 10]				
$Y_{k^2=54}$	[4 0 12]				
$Y_{k^2=-12}$	[6 0 6]				
$Y_{k^2=-45}$	[8 2 8]				

The formulation $?12a^2$ (resp., $?15a^2$), $a \in \mathbb{Z}$, means we suspect only the discriminant of the $K3$ surface with such a value.

The method of computation of the transcendental lattices of singular $K3$ surfaces Y_k and X_k is similar. It uses the fact that the family is modular, meaning both that k is a modular function and their Picard–Fuchs equations have modular solutions. Since we ignore such a property for family (Z_k), we have to deduce the transcendental lattices from the Néron–Severi lattices thus require equations to compute the transcendental lattices of some members. Finally, in the last section, we deduce which $K3$ surfaces belong to several families. Computations were performed using partly the computer algebra system PARI [17], partly Sage [20], and mostly the computer algebra system MAPLE and the Maple Library "Elliptic Surface Calculator" written by Kuwata [10]. The first author is particularly indebted to David Boyd for computing k corresponding to singular $K3$ surfaces and allowing her to perform the first computations of Mahler measures.

Although motivated by quite different aims, Sarti's paper [22] must be cited here. In particular, we notice in her work the two singular $K3$ surfaces with transcendental lattices [4 1 4] and [6 0 10] that are, respectively, the surfaces Z_{-3} (or X_{60}) and X_{-3} in our work.

We thank Alessandra Sarti for catching our attention on these results.

Finally, we are very grateful to the anonymous referees for their remarks and suggestions that allowed us to improve the presentation of our paper.

2 Background

For definitions and main properties concerning elliptic K3 surfaces, see Schütt and Shioda's paper [23] and book [24].

2.1 *Computation of the Néron–Severi Lattice*

Recall the following Shioda's result [27]: Let $(S, \Phi, \mathbb{P}^1)$ be an elliptic surface with a section Φ, without exceptional curves of first kind.

Denote by $NS(S)$ the group of algebraic equivalence classes of divisors of S.

Let u be the generic point of $\mathbb{P}^1$ and $\Phi^{-1}(u) = E$ the elliptic curve defined over $K = \mathbb{C}(u)$ with a K-rational point $o = o(u)$. Then, $E(K)$ is an abelian group of finite type provided that $j(E)$ is transcendental over $\mathbb{C}$. Let r be the rank of $E(K)$ and $s_1, \ldots, s_r$ be generators of $E(K)$ modulo torsion. Besides, the torsion group $E(K)_{tors}$ is generated by at most two elements t_1 of order e_1 and t_2 of order e_2 such that $1 \leq e_2, e_2|e_1$, and $\mid E(K)_{tors} \mid = e_1e_2$.

The group $E(K)$ of K-rational points of E is canonically identified with the group of sections of S over $\mathbb{P}^1(\mathbb{C})$.

For $s \in E(K)$, we denote by (s) the curve image in S of the section corresponding to s.

Let us define

$$D_\alpha := (s_\alpha) - (o) \quad 1 \leq \alpha \leq r$$

$$D'_\beta := (t_\beta) - (o) \quad \beta = 1, 2.$$

Consider now the singular fibers of S over $\mathbb{P}^1$. We set

$$\Sigma := \{v \in \mathbb{P}^1 / C_v = \Phi^{-1}(v) \ \text{ be a singular fiber}\},$$

and for each $v \in \Sigma$, $\Theta_{v,i}, 0 \leq i \leq m_v - 1$, the m_v irreducible components of C_v.

Let $\Theta_{v,0}$ be the unique component of C_v passing through $o(v)$.

One gets

$$C_v = \Theta_{v,0} + \sum_{i \geq 1} \mu_{v,i} \Theta_{v,i}, \qquad \mu_{v,i} \geq 1.$$

Let A_v be the matrix of order $m_v - 1$ whose entry of index (i, j) is $(\Theta_{v,i}\Theta_{v,j})$, $i, j \geq 1$, where (DD') is the intersection number of the divisors D et D' along S. Finally, f will denote a non-singular fiber, i.e., $f = C_{u_0}$ for $u_0 \notin \Sigma$.

Theorem 2.1 (See Shioda [27] p. 23) *The Néron–Severi group $NS(S)$ of the elliptic surface S is generated by the following divisors:*

$$f, \Theta_{v,i} \quad (1 \leq i \leq m_v - 1, \quad v \in \Sigma)$$

$$(o), D_\alpha \quad 1 \leq \alpha \leq r, \quad D'_\beta \ \ \beta = 1, 2.$$

The only relations between these divisors are at most two relations

$$e_\beta D'_\beta \approx e_\beta (D'_\beta(o)) f + \sum_{v \in \Sigma} (\Theta_{v,1}, \ldots, \Theta_{v,m_v-1}) e_\beta A_v^{-1} \begin{pmatrix} (D'_\beta \Theta_{v,1}) \\ \cdot \\ \cdot \\ \cdot \\ (D'_\beta \Theta_{v,m_v-1}) \end{pmatrix},$$

where $\approx$ stands for the algebraic equivalence.

2.2 Computation of the Transcendental Lattice

We recall some results from Shimada [25].

Let L be an even lattice. The abelian group $G_L := L^*/L$ is called the discriminant group of L. Moreover, the symmetric bilinear form on L extends to a $\mathbb{Q}$-valued symmetric bilinear form on L^*. Hence, we get on G_L the following quadratic norm q_L:

$$\begin{aligned} q_L : G_L &\to \mathbb{Q}/2\mathbb{Z} \\ \bar{x} &\mapsto x^2 \qquad (\text{mod. } 2\mathbb{Z}). \end{aligned}$$

The pair (G_L, q_L) is by definition the discriminant form of the lattice L.

When the lattice L is given by its Gram matrix, we can compute its discriminant form as explained in Shimada's lemma [25].

Lemma 2.1 *Let A be the Gram matrix of L and $\tilde{U}, \tilde{V} \in Gl_n(\mathbb{Z})$ such that*

$$\tilde{U}A\tilde{V} = D = \begin{pmatrix} d_1 & & 0 \\ & \ddots & \\ 0 & & d_n \end{pmatrix}$$

with $1 = d_1 = \ldots = d_k < d_{k+1} \leq \ldots \leq d_n$. Then

$$G_L \simeq \oplus_{i>k}\mathbb{Z}/(d_i).$$

Moreover, the ith row vector of $\tilde{V}^{-1}$, regarded as an element of L^* with respect to the dual basis $e_1^*, \ldots, e_n^*$, generates the cyclic group $\mathbb{Z}/(d_i)$.

Let X be a $K3$ surface, and recall its $K3$ lattice $H^2(X, \mathbb{Z}) = U^3 \oplus E_8(-1)^2$ is unimodular. Here U denotes the hyperbolic lattice and E_8 the unique positive definite even unimodular lattice of rank 8. Recall also the definition of the Néron–Severi lattice $NS(X)$ and of the transcendental lattice $T(X)$ that is its orthogonal complement in $H^2(X, \mathbb{Z})$.

$$NS(X) := H^2(X) \cap H^{1,1}(X)$$

$$T(X) := NS(X)^{\perp}.$$

The transcendental lattice $T(X)$ is a birational invariant of algebraic surfaces satisfied by Nikulin ([16, Corollary 1.6.2]). since $H^2(X, \mathbb{Z})$ is unimodular, the following relation between the discriminant forms of $T(X)$ and $NS(X)$

$$(G_{T(X)}, q_{T(X)}) \equiv (G_{NS(X)}, -q_{NS(X)}).$$

3 Proof of Theorem 1.1

Notation 3.1 The singular fibers of type $I_n, I_n^*, IV^*, \ldots$ at $t = t_1, .., t_m$ or at roots of a polynomial $p(t)$ of degree m are denoted $mI_n(t_1, .., t_m)$ or $mI_n(p(t))$.

Theorem 3.1

(1) The transcendental lattice of the generic member Z_k is $U \oplus \langle 24 \rangle$.
(2) There is a genus 1 fibration of Z_k whose Jacobian surface J_k is a $K3$ surface of the Verrill's family with generic transcendental lattice $U \oplus \langle 6 \rangle$.

Proof

(1) We consider the family of $K3$ surfaces (Z_k) defined by the polynomials

$$Q_k = (x + y + z + 1)(xy + xz + yz + xyz) - (k + 4)xyz.$$

□

Lemma 3.1 *There is a generic rank 0 elliptic fibration of the generic member Z_k given by the Weierstrass equation*

$$(F_k) \qquad y^2 = x^3+((k^2-24)t^2+2(k-2)(k+4)^2t+k(k+4)^3)x^2-16t^4(t+k+3)x$$

with 2-torsion section and singular fibers III^*, I_8, I_3, I_2, $2I_1$.

Proof Taking for elliptic parameter $t = x + y + z$, with the change variables

$$x = \frac{(Y - X^2 + Yt)}{Y(k + 3 - t)}, \qquad y = \frac{Y + (t + 1)X}{X(k + 3 - t)},$$

it follows the Weierstrass equation

$$Y^2 + (t^2 - kt + 3)XY - (t + 1)^2(kt - (t - 1)^2)Y = X^3.$$

This defines a generic rank 1 elliptic fibration with 6-torsion section and singular fibers $I_6(-1)$, $2I_3((t - 1)^2 - kt)$, $2I_2(k + 3, 0)$, $2I_1((t + 3)^2 - kt)$, $I_6(\infty)$.

Now the new elliptic parameter $m = \frac{Y+(k+4)X}{(t+1)^2}$ leads to a rank 0 elliptic fibration with a 2-torsion section. First by elimination of Y, then with the new variable W defined by $X = (t+1)^2W$, it follows an equation of type $a(W)t^2+b(W)t+c(W) = 0$. Completing the square, we deduce the equation

$$\begin{aligned} Q^2 &= (W + 1)(W(k + 4) - m)(W^2(k(k + 4)^2 - 4m) \\ &\quad + W(k((k + 4)^2 - m(k + 8)) - 16m) - m(k^2 + 4k - 4m)). \end{aligned}$$

This is the equation of a quartic with a point of coordinates $W = -1$, $Q = 0$. So, with the following change $W = -1 + \frac{1}{x}$, $Q = \frac{y}{x^2}$, we get

$$\begin{aligned} &y^2 + ((m + k + 4)x - k - 4)(4m(m + k + 3)x^2 - (k(k + 4)^2 \\ &\quad + m(k^2 + 8k + 8))x + k(k + 4)^2 - 4m) = 0. \end{aligned}$$

Then putting $u = x(m + k + 4) - (k + 4)$, it follows

$$\begin{aligned} &(k + 4 + m)^2y^2 + 4m(m + k + 3)u^3 - ((k^2 - 24)m^2 + 2m(k - 2)(k + 4)^2 \\ &\quad + k(k + 4)^3)u^2 - 4m^3u = 0. \end{aligned}$$

Finally with

$$u = -\frac{\xi}{4m(m+k+3)}, \qquad \eta = 4m(m+k+3)(m+k+4)y,$$

we obtain

$$\eta^2 = \xi^3 + ((k^2-24)m^2 + 2(k-2)(k+4)^2 m + k(k+4)^3)\xi^2 - 16m^4(m+k+3)\xi.$$

This is precisely the equation (F_k).

□

Starting with the equation (F_k) and using Shioda's formula [27] (Corollary 1.7), the discriminant of Z_k is equal to

$$\Delta = \frac{2 \times 8 \times 3 \times 2}{4} = 24.$$

Our aim is to compute the transcendental lattice of the Picard number 19 surface Z_k.

Denote θ_i $1 \leq i \leq 7$, η_i, $1 \leq i \leq 7$, α_i, $i = 1, 2$, β_1 the generators corresponding, respectively, to the singular fibers I_8, III^*, I_3, I_2 and s_2 the 2-torsion section.

The relation is the following:

$$s_2 \approx \frac{1}{2}\theta_1 + \theta_2 + \frac{3}{2}\theta_3 + 2\theta_4 + \frac{3}{2}\theta_5 + \theta_6 + \frac{1}{2}\theta_7 + \frac{3}{2}\eta_1 + 2\eta_2 + \frac{5}{2}\eta_3 + 3\eta_4 + 2\eta_5 + \eta_6 + \frac{3}{2}\eta_7 + \frac{1}{2}\beta_1,$$

and since s_2 cuts θ_4, cuts η_1, does not cut α_i, $i = 1, 2$, but cuts β_1, taking the divisors in the following order (0), f, s_2, θ_i, $i = 2, .., 7$, η_i, $i = 1, .., 7$, α_i, $i = 1, 2$, β_1, we get the following Gram matrix:

$$\begin{pmatrix}
-2 & 1 & 0 & 0 & 0 & 0 & 0 & 0 & 0 & 0 & 0 & 0 & 0 & 0 & 0 & 0 & 0 & 0 & 0 \\
1 & 0 & 1 & 0 & 0 & 0 & 0 & 0 & 0 & 0 & 0 & 0 & 0 & 0 & 0 & 0 & 0 & 0 & 0 \\
0 & 1 & -2 & 0 & 0 & 1 & 0 & 0 & 0 & 1 & 0 & 0 & 0 & 0 & 0 & 0 & 0 & 0 & 1 \\
0 & 0 & 0 & -2 & 1 & 0 & 0 & 0 & 0 & 0 & 0 & 0 & 0 & 0 & 0 & 0 & 0 & 0 & 0 \\
0 & 0 & 0 & 1 & -2 & 1 & 0 & 0 & 0 & 0 & 0 & 0 & 0 & 0 & 0 & 0 & 0 & 0 & 0 \\
0 & 0 & 1 & 0 & 1 & -2 & 1 & 0 & 0 & 0 & 0 & 0 & 0 & 0 & 0 & 0 & 0 & 0 & 0 \\
0 & 0 & 0 & 0 & 0 & 1 & -2 & 1 & 0 & 0 & 0 & 0 & 0 & 0 & 0 & 0 & 0 & 0 & 0 \\
0 & 0 & 0 & 0 & 0 & 0 & 1 & -2 & 1 & 0 & 0 & 0 & 0 & 0 & 0 & 0 & 0 & 0 & 0 \\
0 & 0 & 0 & 0 & 0 & 0 & 0 & 1 & -2 & 0 & 0 & 0 & 0 & 0 & 0 & 1 & 0 & 0 & 0 \\
0 & 0 & 1 & 0 & 0 & 0 & 0 & 0 & 0 & -2 & 1 & 0 & 0 & 0 & 0 & 0 & 0 & 0 & 0 \\
0 & 0 & 0 & 0 & 0 & 0 & 0 & 0 & 0 & 1 & -2 & 1 & 0 & 0 & 0 & 0 & 0 & 0 & 0 \\
0 & 0 & 0 & 0 & 0 & 0 & 0 & 0 & 0 & 0 & 1 & -2 & 1 & 0 & 0 & 0 & 0 & 0 & 0 \\
0 & 0 & 0 & 0 & 0 & 0 & 0 & 0 & 0 & 0 & 0 & 1 & -2 & 1 & 0 & 1 & 0 & 0 & 0 \\
0 & 0 & 0 & 0 & 0 & 0 & 0 & 0 & 0 & 0 & 0 & 0 & 1 & -2 & 1 & 0 & 0 & 0 & 0 \\
0 & 0 & 0 & 0 & 0 & 0 & 0 & 0 & 0 & 0 & 0 & 0 & 0 & 1 & -2 & 0 & 0 & 0 & 0 \\
0 & 0 & 0 & 0 & 0 & 0 & 0 & 0 & 0 & 0 & 0 & 0 & 1 & 0 & 0 & -2 & 0 & 0 & 0 \\
0 & 0 & 0 & 0 & 0 & 0 & 0 & 0 & 0 & 0 & 0 & 0 & 0 & 0 & 0 & 0 & -2 & 1 & 0 \\
0 & 0 & 0 & 0 & 0 & 0 & 0 & 0 & 0 & 0 & 0 & 0 & 0 & 0 & 0 & 0 & 1 & -2 & 0 \\
0 & 0 & 1 & 0 & 0 & 0 & 0 & 0 & 0 & 0 & 0 & 0 & 0 & 0 & 0 & 0 & 0 & 0 & -2
\end{pmatrix}.$$

Its determinant is 24.

According to Lemma 2.1 and taking for A the Gram matrix of the Néron–Severi lattice, we find $D = I_d \oplus \langle 24 \rangle$ and the last row of $\tilde{V}^{-1}$ is

$$B = (-3, 0, 0, 0, 0, -3, 3, 0, 0, 0, 3, -6, 3, 0, 0, 0, -1, 1, -3).$$

Thus we can compute the norm of B

$$BA^{-1}B^t = -\frac{745}{24} = -\frac{25}{24} \text{ mod. } 2.$$

We deduce

$$(G_{NS(X)}, q_{NS(X)}) = (\mathbb{Z}/24\mathbb{Z})(-\frac{25}{24})$$

$$(G_{T(X)}, q_{T(X)}) = (\mathbb{Z}/24\mathbb{Z})(\frac{25}{24}).$$

With the same method if $T' = U \oplus \langle 24 \rangle$we find

$$(G_{T'}, q_{T'}) = (\mathbb{Z}/24\mathbb{Z})(\frac{1}{24}),$$

which means that a generator α of the cyclic group $\mathbb{Z}/24\mathbb{Z}$ has norm $\frac{1}{24}$. Replacing the generator α by 5α, we find

$$(G_{T'}, q_{T'}) = (\mathbb{Z}/24\mathbb{Z})(\frac{25}{24}).$$

Thus $T(X) = T'$.

(2) We start with the Weierstrass equation of Z_k

$$(F_k) \qquad y^2 = x^3 + ((k^2-24)t^2 + 2(k-2)(k+4)^2 t + k(k+4)^3)x^2 - 16t^4(t+k+3)x.$$

With the new parameter $m = \frac{-1}{4(k+4)^2}\frac{x}{(t+k+3)}$ and with $y = 4W(t+k+3)(k+4)$, we get

$$W^2 = -4t^4 m + (k^2-24)(k+4)^2 m^2 t^2 + (4m + 2(k-2))(k+4)^4 m^2 t$$
$$+ (4(k+3)m + k(k+4))(k+4)^4 m^2.$$

By classical trick given in Boyd [6] on page 57, its Jacobian J_k has Weierstrass equation changing m in t

$$y^2 = x^3 + t^2(k^2-24)x^2 + 16t^3(4(k+3)t - k(4+k))x$$
$$+ 64t^5((4+k)^2t^2 - (3k^2+24k+40)t + (4+k)(3k+4)).$$

Finally, the change x into $x + 8t^2$ leads to the equation

$$(J_k) \qquad y^2 = x^3 + t^2k^2x^2 + 16t^3k(k+4)(t-1)x + 64t^5(k+4)^2(t-1)^2,$$

with singular fibers $III^*(0)$, $I_3(1)$, $2I_1(432\,(k+4)^2\,t^2 + \big(k^6 - 72(4+k)(k^3+6k+24)\big)\,t + 64k^3(4+k))$, $II^*(\infty)$, hence, of type $A_2E_7E_8$. This equation of J_k is the equation of a rank zero elliptic fibration of discriminant 6 by Shioda's formula [27].

Now all the elliptic fibrations of the $K3$ surface with discriminant 6 and transcendental lattice

$$\begin{pmatrix} 0 & 0 & 1 \\ 0 & 6 & 0 \\ 1 & 0 & 0 \end{pmatrix}$$

are given by the primitive embeddings of $E_6 \oplus A_1$ into the Niemeier lattices (Sterk [29]). Indeed taking the primitive embedding of $E_6 \oplus A_1$ into the Niemeier lattice E_8^3 defined by the primitive embeddings $E_6 \hookrightarrow E_8$ and $A_1 \hookrightarrow E_8$, we get the above elliptic fibration since $(E_6)^{\perp}_{E_8} = A_2$ and $(A_1)^{\perp}_{E_8} = E_7$ [15]. It corresponds to the second fibration among the ten Jacobian elliptic fibrations obtained by Sterk [29].

□

4 Transcendental Lattices of Singular $K3$-Surfaces of Verrill's Family (X_k)

The singular $K3$ surfaces of Verrill's family (X_k) are given by the imaginary quadratic τ [31] such that

$$t = \frac{\eta(3\tau)^4\eta(12\tau)^8\eta(2\tau)^{12}}{\eta(\tau)^4\eta(4\tau)^8\eta(6\tau)^{12}}, \qquad k = -(t + \frac{1}{t}) - 2,$$

where η denotes the Dedekind eta function. They have been computed by Boyd [5] and are given in the following table.

Let us recall how to compute the transcendental lattices of the singular $K3$ surfaces of Verrill's family [33].

If Y_λ denotes a $K3$ surface, up to a scalar, there exists a unique 2-form $\omega_\lambda \in H^{2,0}(Y_\lambda, \mathbb{C})$. If $\{\gamma_{1,\lambda}, \cdots, \gamma_{22,\lambda}\}$ is a basis of $H_2(Y_\lambda, \mathbb{Z})$, then $\{\gamma^*_{1,\lambda}, \cdots, \gamma^*_{22,\lambda}\}$ is the corresponding dual basis in $H^2(Y_\lambda, \mathbb{C})$, dual for the intersection product. That means, if $\gamma \in H_2(Y_\lambda, \mathbb{Z})$, then $\gamma^* \in H^2(Y_\lambda, \mathbb{Z})$ is defined for every $\alpha \in H_2(Y_\lambda, \mathbb{C}))$ by

$$\int_\alpha \gamma^* = (\gamma \cdot \alpha),$$

k	τ	Equation of τ
-36	$\frac{3+\sqrt{-3}}{3}$	$3\tau^2 - 6\tau + 4 = 0$
-12	$\frac{\sqrt{-6}}{6}$	$6\tau^2 + 1 = 0$
-6	$\frac{\sqrt{-3}}{6}$	$12\tau^2 + 1 = 0$
-3	$\frac{3+\sqrt{-15}}{24}$	$24\tau^2 - 6\tau + 1 = 0$
0	$\frac{3+\sqrt{-3}}{12}$	$12\tau^2 - 6\tau + 1 = 0$
4	$\frac{-2+\sqrt{-2}}{6}$	$6\tau^2 + 4\tau + 1 = 0$
12	$\frac{3+\sqrt{-3}}{6}$	$3\tau^2 - 3\tau + 1 = 0$
60	$\frac{3+\sqrt{-15}}{6}$	$3\tau^2 - 3\tau + 2 = 0$

where $(\gamma \cdot \alpha)$ is the intersection number of the two cycles α and γ. The intersection product of homological cycles satisfies the Poincaré duality for the exterior product, i.e., $\forall \alpha, \gamma \in H_2(Y_\lambda, \mathbb{Z})$, it follows

$$(\gamma \cdot \alpha) = \int_{Y_\lambda} \alpha^* \wedge \gamma^*.$$

Hence, the dual basis intersection matrix is also the intersection matrix of the transcendental cycles lattice $T(Y_\lambda) = \mathrm{Pic}(Y_\lambda)^\perp$. For Verrill's family, the Gram matrix of its transcendental lattice in the basis $\{\gamma_1, \gamma_2, \gamma_3\}$ is given by

$$T = \begin{pmatrix} 0 & 0 & 1 \\ 0 & 6 & 0 \\ 1 & 0 & 0 \end{pmatrix}.$$

Since the integral of ω_λ along any algebraic cycle is zero, it follows from the identification of $H_2(Y_\lambda, \mathbb{Z}) \otimes \mathbb{C}$ with $H^2(Y_\lambda, \mathbb{C})$ that $\omega_\lambda \in T(Y_\lambda) \otimes \mathbb{C}$.

Hence, $\omega_\lambda \in H^{2,0}(Y_\lambda, \mathbb{C})$ can be written as

$$\omega_{\lambda(\tau)} = a(\tau)\gamma_1^* + b(\tau)\gamma_2^* + c(\tau)\gamma_3^*$$

with

$$6b(\tau)^2 + 2a(\tau)c(\tau) = 0,$$

since $\omega_\tau \wedge \omega_\tau \in H^{(4,0)} = \{0\}$.

It follows that

$$\omega(\tau) = G(\tau)\gamma_1^* + \tau G(\tau)\gamma_2^* - 3\tau^2 G(\tau)\gamma_3^*$$

belongs to $T(X_{\lambda(\tau)})$, where $\{G(\tau), \tau G(\tau), \tau^2 G(\tau)\}$ satisfies the Picard–Fuchs differential equation of periods of Verrill's family.

Thus the fact that the $K3$ surface X_λ has Picard number 20 is equivalent to the fact that there exists a vector $p\gamma_1 + q\gamma_2 + r\gamma_3 \in T_\lambda$ becoming algebraic, hence satisfying $\int_{p\gamma_1+q\gamma_2+r\gamma_3} \omega(\tau) = 0$, i.e.,

$$-3p\tau^2 + 6q\tau + r = 0.$$

This leads to the determination of the Gram matrix of the transcendental lattices of the singular $K3$ surfaces of Verrill's family.

4.1 Surface X_{-36}

The corresponding τ satisfies $3\tau^2 - 6\tau + 4 = 0$. Hence, $p = -1, q = -1, r = 4$, and the vector $-\gamma_1 - \gamma_2 + 4\gamma_3$ becomes algebraic. It follows that the vectors $\langle \gamma_1 + \gamma_2 - 2\gamma_3, \gamma_2 - 6\gamma_3 \rangle$ generate the transcendental lattice and

$$T(X_{-36}) = [2 \quad 0 \quad 6].$$

4.2 Surface X_{-12}

The corresponding τ satisfies $6\tau^2 + 1 = 0$. Hence, $p = -2, q = 0, r = 1$ and the vector $-2\gamma_1 + \gamma_3$ becomes algebraic. It follows that the vectors $\langle 2\gamma_1 + \gamma_3, \gamma_2 \rangle$ generate the transcendental lattice and

$$T(X_{-12}) = [4 \quad 0 \quad 6].$$

4.3 Surface X_{-6}

The corresponding τ satisfies $12\tau^2 + 1 = 0$. Hence, $p = -4, q = 0, r = 1$ and the vector $-4\gamma_1 + \gamma_3$ becomes algebraic. It follows that the vectors $\langle 4\gamma_1 + \gamma_3, \gamma_2 \rangle$ generate the transcendental lattice and

$$T(X_{-6}) = [6 \quad 0 \quad 8].$$

4.4 Surface X_{-3}

The corresponding τ satisfies $24\tau^2 + 6\tau + 1 = 0$.

Hence, $p = -8, q = -1, r = 1$ and the vector $-8\gamma_1 - \gamma_2 + \gamma_3$ becomes algebraic. It follows that the vectors $\langle \gamma_2 + 6\gamma_1, -\gamma_2 + 2\gamma_1 + \gamma_3 \rangle$ generate the transcendental lattice and

$$T(X_{-3}) = [6 \quad 0 \quad 10].$$

4.5 Surface X_0

The corresponding τ satisfies $12\tau^2 - 6\tau + 1 = 0$.

Hence, $p = -4, q = -1, r = 1$ and the vector $-4\gamma_1 - \gamma_2 + \gamma_3$ becomes algebraic. It follows that the vectors $\langle -2\gamma_1 - \gamma_2 + \gamma_3, \gamma_2 + 6\gamma_1 \rangle$ generate the transcendental lattice and

$$T(X_0) = [2 \quad 0 \quad 6].$$

4.6 Surface X_4

The corresponding τ satisfies $6\tau^2 + 4\tau + 1 = 0$.

Hence, $p = -6$, $q = 2$, $r = 3$ and the vector $-6\gamma_1 + 2\gamma_2 + 3\gamma_3$ becomes algebraic. It follows that the vectors $\langle \gamma_1 + 2\gamma_3, \gamma_2 - 2\gamma_1 + \gamma_3 \rangle$ generate the transcendental lattice and

$$T(X_4) = [2 \quad 0 \quad 4].$$

4.7 Surface X_{12}

The corresponding τ satisfies $3\tau^2 - 3\tau + 1 = 0$.

Hence, $p = -2$, $q = -1$, $r = 2$ and the vector $-2\gamma_1 - \gamma_2 + 2\gamma_3$ becomes algebraic. It follows that the vectors $\langle \gamma_1 + \gamma_3, \gamma_2 + 2\gamma_1 - \gamma_3 \rangle$ generate the transcendental lattice and

$$T(X_{12}) = [2 \quad 1 \quad 2].$$

4.8 Surface X_{60}

The corresponding τ satisfies $3\tau^2 - 3\tau + 2 = 0$.

Hence, $p = -2$, $q = -1$, $r = 4$ and the vector $-2\gamma_1 - \gamma_2 + 4\gamma_3$ becomes algebraic. It follows that the vectors $\langle \gamma_1 + 2\gamma_3, \gamma_2 + \gamma_1 - \gamma_3 \rangle$ generate the transcendental lattice and

$$T(X_{60}) = [4 \quad 1 \quad 4].$$

5 Transcendental Lattices of Singular $K3$ Surfaces of the Apéry–Fermi's Family

The singular $K3$ surfaces of the Apéry–Fermi's family (Y_k) are given by the imaginary quadratic τ [19] such that

$$t = \left(\frac{\eta(\tau)\eta(6\tau)}{\eta(2\tau\eta(3\tau)}\right)^6, \qquad k = t + \frac{1}{t}.$$

They have been computed by Boyd [5] and are given in the following table.

k	τ	Equation of τ
0	$\frac{-3+\sqrt{-3}}{6}$	$3\tau^2 + 3\tau + 1 = 0$
2	$\frac{-2+\sqrt{-2}}{6}$	$6\tau^2 + 4\tau + 1 = 0$
3	$\frac{-3+\sqrt{-15}}{12}$	$6\tau^2 + 3\tau + 1 = 0$
6	$\frac{\sqrt{-6}}{6}$	$6\tau^2 + 1 = 0$
10	$\frac{\sqrt{-2}}{2}$	$2\tau^2 + 1 = 0$
18	$\frac{\sqrt{-30}}{6}$	$6\tau^2 + 5 = 0$
102	$\frac{\sqrt{-6\times 13}}{6}$	$6\tau^2 + 13 = 0$
198	$\frac{\sqrt{-17\times 6}}{6}$	$6\tau^2 + 17 = 0$
$2\sqrt{5}$	$\frac{-1+\sqrt{-5}}{6}$	$6\tau^2 + 2\tau + 1 = 0$
$3\sqrt{6}$	$\frac{\sqrt{-3}}{3}$	$3\tau^2 + 1 = 0$
$2\sqrt{-3}$	$\frac{-1+\sqrt{-1}}{2}$	$2\tau^2 + 2\tau + 1 = 0$
$3\sqrt{-5}$	$\frac{-3+\sqrt{-15}}{6}$	$3\tau^2 + 3\tau + 2 = 0$

For Apéry–Fermi's family, the Gram matrix of its transcendental lattice in the basis $\{\gamma_1, \gamma_2, \gamma_3\}$ is given by

$$T = \begin{pmatrix} 0 & 0 & 1 \\ 0 & 12 & 0 \\ 1 & 0 & 0 \end{pmatrix}.$$

As previously, the fact that Y_λ has Picard number 20 is equivalent to the existence of a vector $p\gamma_1 + q\gamma_2 + r\gamma_3 \in T_\lambda$ becoming algebraic, hence satisfying

$\int_{p\gamma_1+q\gamma_2+r\gamma_3} \omega(\tau) = 0$, that is,

$$-6p\tau^2 + 12q\tau + r = 0.$$

In various papers [2, 3], it was proved

$$T(Y_0) = [4 \quad 2 \quad 4] \qquad T(Y_2) = [2 \quad 0 \quad 4] \qquad T(Y_6) = [2 \quad 0 \quad 12]$$

$$T(Y_3) = [2 \quad 1 \quad 8] \qquad T(Y_{10}) = [6 \quad 0 \quad 12] \qquad T(Y_{18}) = [10 \quad 0 \quad 12]$$
$$T(Y_{102}) = [12 \quad 0 \quad 26] \qquad T(Y_{198}) = [12 \quad 0 \quad 34].$$

We achieve with the remaining surfaces of the list.

5.1 Surface $Y_{2\sqrt{5}}$

The corresponding τ satisfies $6\tau^2 + 2\tau + 1 = 0$. Hence, $p = -6$, $q = 1$, $r = 6$ and the vector $-6\gamma_1 + \gamma_2 + 6\gamma_3$ becomes algebraic. It follows that the vectors $\langle \gamma_1 + \gamma_3, \gamma_2 - \gamma_1 + \gamma_3 \rangle$ generate the transcendental lattice and

$$T(Y_{2\sqrt{5}}) = [2 \quad 0 \quad 10].$$

5.2 Surface $Y_{3\sqrt{6}}$

The corresponding τ satisfies $3\tau^2 + 1 = 0$. Thus the vector $-\gamma_1 + 2\gamma_3$ becomes algebraic. It follows that the vectors $\langle \gamma_1 + 2\gamma_3, \gamma_2 \rangle$ generate the transcendental lattice and

$$T(Y_{3\sqrt{6}}) = [4 \quad 0 \quad 12].$$

5.3 Surface $Y_{2\sqrt{-3}}$

The corresponding τ satisfies $2\tau^2 + 2\tau + 1 = 0$. Thus the vector $-2\gamma_1 + \gamma_2 + 6\gamma_3$ becomes algebraic. It follows that the vectors $\langle \gamma_1 + 3\gamma_3, \gamma_2 - \gamma_1 + 3\gamma_3 \rangle$ generate the transcendental lattice and

$$T(Y_{2\sqrt{-3}}) = [6 \quad 0 \quad 6].$$

5.4 Surface $Y_{3\sqrt{-5}}$

The corresponding τ satisfies $3\tau^2 + 3\tau + 2 = 0$. Thus the vector $-2\gamma_1 + \gamma_2 + 8\gamma_3$ becomes algebraic. It follows that the vectors $\langle \gamma_1 + 4\gamma_3, \gamma_2 - \gamma_1 + 2\gamma_3 \rangle$ generate the transcendental lattice and

$$T(Y_{3\sqrt{-5}}) = [8 \quad 2 \quad 8].$$

6 Transcendental Lattices of Singular $K3$ Surfaces of the (Q_k) Family

Let us recall a Weierstrass equation for the $K3$ surface Z_k, see Lemma 3.1.

$$(F_k) \qquad y^2 = x^3 + ((k^2-24)t^2 + 2(k-2)(k+4)^2 t + k(k+4)^2)x^2 - 16t^4(t+k+3)x.$$

6.1 $T(Z_{-3})$

A Weierstrass equation is given by

$$(F_{-3}) \qquad y^2 = x^3 - x^2(15t^2 + 10t + 3) - 16t^5 x$$

with singular fibers $I_{10}(0)$,$I_3(-1)$, $2I_1(64t^2 + 33t + 9)$, $III^*(\infty)$. The rank is 0, and by Shimada and Zhang [26], we get

$$T(Z_{-3}) = [4 \quad 1 \quad 4].$$

6.2 $T(Z_0)$

A Weierstrass equation is given by

$$(F_0) \qquad y^2 = x^3 - x^2(24t^2 + 64t) - 16t^4(t + 3)x$$

with singular fibers $I_4^*(0)$, $I_3(-4)$, $I_2(-3)$, $III^*(\infty)$. The rank is 0, and by Shimada and Zhang [26],

$$T(Z_0) = [2 \quad 0 \quad 6].$$

6.3 $T(Z_{12})$

A Weierstrass equation is given by

$$(F_{12}) \qquad y^2 = x^3 + x^2(120t^2 + 5120t + 49152) - 16t^4(t+15)x$$

with singular fibers $I_8(0)$, $I_3(-16)$, $2I_2(-15,-96)$, $III^*(\infty)$. The rank is 0, and by Shimada and Zhang [26], it follows

$$T(Z_{12}) = [2 \quad 0 \quad 24].$$

In the following specializations of Z_k, we order the singular fibers as in the generic case Sect. 3.

6.4 $T(Z_{-6})$

A Weierstrass equation is given by

$$(F_{-6}) \qquad y^2 = x^3 + x^2(3t^2 - 16t + 12) - xt^4(t-3)$$

with singular fibers $I_8(0)$, $I_3(2)$, $I_2(3)$, $III^*(\infty)$. The rank is 1, and the infinite section given by the point P generates the Mordell–Weil group. We obtain

$$P = [16(t-3), 4i(t-3)(t^2-24)]$$

and find this infinite section P does not cut the zero section and cuts only η_1 and β_1. We deduce the Gram matrix of the Néron–Severi lattice

$$NS = \begin{pmatrix}
-2 & 1 & 0 & 0 & 0 & 0 & 0 & 0 & 0 & 0 & 0 & 0 & 0 & 0 & 0 & 0 & 0 & 0 & 0 & 0 \\
1 & 0 & 1 & 0 & 0 & 0 & 0 & 0 & 0 & 0 & 0 & 0 & 0 & 0 & 0 & 0 & 0 & 0 & 0 & 1 \\
0 & 1 & -2 & 0 & 0 & 1 & 0 & 0 & 0 & 1 & 0 & 0 & 0 & 0 & 0 & 0 & 0 & 0 & 1 & 0 \\
0 & 0 & 0 & -2 & 1 & 0 & 0 & 0 & 0 & 0 & 0 & 0 & 0 & 0 & 0 & 0 & 0 & 0 & 0 & 0 \\
0 & 0 & 0 & 1 & -2 & 1 & 0 & 0 & 0 & 0 & 0 & 0 & 0 & 0 & 0 & 0 & 0 & 0 & 0 & 0 \\
0 & 0 & 1 & 0 & 1 & -2 & 1 & 0 & 0 & 0 & 0 & 0 & 0 & 0 & 0 & 0 & 0 & 0 & 0 & 0 \\
0 & 0 & 0 & 0 & 0 & 1 & -2 & 1 & 0 & 0 & 0 & 0 & 0 & 0 & 0 & 0 & 0 & 0 & 0 & 0 \\
0 & 0 & 0 & 0 & 0 & 0 & 1 & -2 & 1 & 0 & 0 & 0 & 0 & 0 & 0 & 0 & 0 & 0 & 0 & 0 \\
0 & 0 & 0 & 0 & 0 & 0 & 0 & 1 & -2 & 0 & 0 & 0 & 0 & 0 & 0 & 1 & 0 & 0 & 0 & 0 \\
0 & 0 & 1 & 0 & 0 & 0 & 0 & 0 & 0 & -2 & 1 & 0 & 0 & 0 & 0 & 0 & 0 & 0 & 0 & 1 \\
0 & 0 & 0 & 0 & 0 & 0 & 0 & 0 & 0 & 1 & -2 & 1 & 0 & 0 & 0 & 0 & 0 & 0 & 0 & 0 \\
0 & 0 & 0 & 0 & 0 & 0 & 0 & 0 & 0 & 0 & 1 & -2 & 1 & 0 & 0 & 0 & 0 & 0 & 0 & 0 \\
0 & 0 & 0 & 0 & 0 & 0 & 0 & 0 & 0 & 0 & 0 & 1 & -2 & 1 & 0 & 1 & 0 & 0 & 0 & 0 \\
0 & 0 & 0 & 0 & 0 & 0 & 0 & 0 & 0 & 0 & 0 & 0 & 1 & -2 & 1 & 0 & 0 & 0 & 0 & 0 \\
0 & 0 & 0 & 0 & 0 & 0 & 0 & 0 & 0 & 0 & 0 & 0 & 0 & 1 & -2 & 0 & 0 & 0 & 0 & 0 \\
0 & 0 & 0 & 0 & 0 & 0 & 0 & 0 & 0 & 0 & 0 & 0 & 1 & 0 & 0 & -2 & 0 & 0 & 0 & 0 \\
0 & 0 & 0 & 0 & 0 & 0 & 0 & 0 & 0 & 0 & 0 & 0 & 0 & 0 & 0 & 0 & -2 & 1 & 0 & 1 \\
0 & 0 & 0 & 0 & 0 & 0 & 0 & 0 & 0 & 0 & 0 & 0 & 0 & 0 & 0 & 0 & 1 & -2 & 0 & 0 \\
0 & 0 & 1 & 0 & 0 & 0 & 0 & 0 & 0 & 0 & 0 & 0 & 0 & 0 & 0 & 0 & 0 & 0 & -2 & 0 \\
0 & 1 & 0 & 0 & 0 & 0 & 0 & 0 & 0 & 1 & 0 & 0 & 0 & 0 & 0 & 0 & 1 & 0 & 0 & -2
\end{pmatrix}$$

and its determinant -48. The elementary divisors are 4 and 12. Applying Lemma 2.1, we find

$$G_{NS} = \mathbb{Z}/(4) \oplus \mathbb{Z}/(12).$$

Moreover, the vector

$$B_4 = [-3, 0, -1, 0, 0, 0, 0, 0, 1, 0, 0, -3, 3, -3, 3, 0, -3, 3, -2, 3]$$

with quadratic norm $-\frac{1}{2}$ generates the cyclic group $\mathbb{Z}/(4)$ and the vector

$$B_{12} = [1, 0, 0, 0, 0, 0, 0, 0, 0, 0, 0, 1, -1, 1, -1, 0, 1, -1, 1, -1]$$

with quadratic norm $-\frac{7}{6}$ generates the cyclic group $\mathbb{Z}/(12)$. They also satisfy the relation $B_4.B_{12} = \frac{1}{4}$.

We are going to prove that $T(Z_{-6}) = [8 \quad 4 \quad 8]$.

Let

$$A = \begin{pmatrix} 8 & 4 \\ 4 & 8 \end{pmatrix}.$$

Applying Lemma 2.1, we get

$$D = \begin{pmatrix} 4 & 0 \\ 0 & 12 \end{pmatrix} \quad \tilde{U} = \begin{pmatrix} 0 & 1 \\ 1 & -2 \end{pmatrix} \quad \tilde{V} = \begin{pmatrix} 1 & 2 \\ 0 & -1 \end{pmatrix}.$$

We deduce $G_A = \mathbb{Z}/(4) \oplus \mathbb{Z}/(12)$ and get a vector b with quadratic norm $1/2$ generating $\mathbb{Z}/(4)$, a vector c with quadratic norm $1/6$ generating $\mathbb{Z}/(12)$ such that $b.c = -1/4$. It follows that the vector $2b+c$ is another generator of the cyclic group $\mathbb{Z}/(12)$ with quadratic norm $7/6$ and $b.(2b + c) \equiv -1/4(\text{mod. } \mathbb{Z})$, and hence,

$$T(Z_{-6}) = [8 \quad 4 \quad 8].$$

6.5 $T(Z_4)$

A Weierstrass equation is given by

$$(F_4) \qquad y^2 = x^3 + x^2(-8t^2 + 256t + 2048) - 16xt^4(t + 7)$$

with singular fibers $I_8(0)$, $I_3(-8)$, $I_2(-7)$, $III^*(\infty)$. The rank is 1, and the infinite section given by the point $P = [-256(t + 7), 64(t + 7)(t^2 - 64)]$ generates the Mordell–Weil group. We obtain that this infinite section does not cut the zero section and cuts only η_1, β_1, and α_1.

We deduce the Gram matrix of the Néron–Severi lattice

$$NS = \begin{pmatrix}
-2 & 1 & 0 & 0 & 0 & 0 & 0 & 0 & 0 & 0 & 0 & 0 & 0 & 0 & 0 & 0 & 0 & 0 & 0 & 0 \\
1 & 0 & 1 & 0 & 0 & 0 & 0 & 0 & 0 & 0 & 0 & 0 & 0 & 0 & 0 & 0 & 0 & 0 & 0 & 1 \\
0 & 1 & -2 & 0 & 0 & 1 & 0 & 0 & 0 & 1 & 0 & 0 & 0 & 0 & 0 & 0 & 0 & 0 & 1 & 0 \\
0 & 0 & 0 & -2 & 1 & 0 & 0 & 0 & 0 & 0 & 0 & 0 & 0 & 0 & 0 & 0 & 0 & 0 & 0 & 0 \\
0 & 0 & 0 & 1 & -2 & 1 & 0 & 0 & 0 & 0 & 0 & 0 & 0 & 0 & 0 & 0 & 0 & 0 & 0 & 0 \\
0 & 0 & 1 & 0 & 1 & -2 & 1 & 0 & 0 & 0 & 0 & 0 & 0 & 0 & 0 & 0 & 0 & 0 & 0 & 0 \\
0 & 0 & 0 & 0 & 0 & 1 & -2 & 1 & 0 & 0 & 0 & 0 & 0 & 0 & 0 & 0 & 0 & 0 & 0 & 0 \\
0 & 0 & 0 & 0 & 0 & 0 & 1 & -2 & 1 & 0 & 0 & 0 & 0 & 0 & 0 & 0 & 0 & 0 & 0 & 0 \\
0 & 0 & 0 & 0 & 0 & 0 & 0 & 1 & -2 & 0 & 0 & 0 & 0 & 0 & 0 & 1 & 0 & 0 & 0 & 0 \\
0 & 0 & 1 & 0 & 0 & 0 & 0 & 0 & 0 & -2 & 1 & 0 & 0 & 0 & 0 & 0 & 0 & 0 & 0 & 1 \\
0 & 0 & 0 & 0 & 0 & 0 & 0 & 0 & 0 & 1 & -2 & 1 & 0 & 0 & 0 & 0 & 0 & 0 & 0 & 0 \\
0 & 0 & 0 & 0 & 0 & 0 & 0 & 0 & 0 & 0 & 1 & -2 & 1 & 0 & 0 & 0 & 0 & 0 & 0 & 0 \\
0 & 0 & 0 & 0 & 0 & 0 & 0 & 0 & 0 & 0 & 0 & 1 & -2 & 1 & 0 & 1 & 0 & 0 & 0 & 0 \\
0 & 0 & 0 & 0 & 0 & 0 & 0 & 0 & 0 & 0 & 0 & 0 & 1 & -2 & 1 & 0 & 0 & 0 & 0 & 0 \\
0 & 0 & 0 & 0 & 0 & 0 & 0 & 0 & 0 & 0 & 0 & 0 & 0 & 1 & -2 & 0 & 0 & 0 & 0 & 0 \\
0 & 0 & 0 & 0 & 0 & 0 & 0 & 0 & 0 & 0 & 0 & 0 & 1 & 0 & 0 & -2 & 0 & 0 & 0 & 0 \\
0 & 0 & 0 & 0 & 0 & 0 & 0 & 0 & 0 & 0 & 0 & 0 & 0 & 0 & 0 & 0 & -2 & 1 & 0 & 1 \\
0 & 0 & 0 & 0 & 0 & 0 & 0 & 0 & 0 & 0 & 0 & 0 & 0 & 0 & 0 & 0 & 1 & -2 & 0 & 0 \\
0 & 0 & 1 & 0 & 0 & 0 & 0 & 0 & 0 & 0 & 0 & 0 & 0 & 0 & 0 & 0 & 0 & 0 & -2 & 1 \\
0 & 1 & 0 & 0 & 0 & 0 & 0 & 0 & 0 & 1 & 0 & 0 & 0 & 0 & 0 & 0 & 1 & 0 & 1 & -2
\end{pmatrix}$$

and its determinant -32. The elementary divisors are 2 and 16. Applying Lemma 2.1, we find

$$G_{NS} = \mathbb{Z}/(2) \oplus \mathbb{Z}/(16).$$

Moreover, the vector

$$B_2 = [0, 0, -1, 0, 0, 0, 1, -1, 1, 0, 0, 0, 0, 0, 0, 0, -1, 1, 0, 1]$$

with quadratic norm $-\frac{1}{2}$ generates the cyclic group $\mathbb{Z}/(2)$ and the vector

$$B_{16} = [-1, 0, -1, 0, 0, 0, -1, 1, 0, 0, 0, 0, 0, 0, 0, 0, 0, 1, 1, -2]$$

with quadratic norm $-\frac{1}{16}$ generates the cyclic group $\mathbb{Z}/(16)$. They also satisfy the relation $B_2.B_{16} = \frac{1}{2}$. Now the vector $B_2 - B_{16}$ is another generator of the cyclic group $\mathbb{Z}/(16)$ since its quadratic norm is equal to $-\frac{25}{16}$, and it follows $B_2.(B_2 - B_{16}) = -1 \equiv 0(\text{mod.}\mathbb{Z})$.

Hence,

$$T(Z_4) = [2 \quad 0 \quad 16].$$

6.6 $T(Z_{-12})$

Its Weierstrass equation is given by

$$(F_{-12}) \quad y^2 = x^3 + (30t^2 - 448t + 1536)x^2 - t^4(t-9)x$$

with 2-torsion and singular fibers $I_8(0)$, $I_3(8)$, $I_2(9)$ $III^*(\infty)$. Its rank is 1 and

$$P = [-\frac{t^4(t+72)^2}{1728(t-24)^2}, \frac{\sqrt{-3}t^4(t+72)}{124416(t-24)^3}(t^4+144t^3-20736t^2+138240t+995328)]$$

of infinite order generates the Mordell–Weil group. Moreover, this infinite section (P) cuts the 2-torsion section s_2, the zero section, and θ_4, and we deduce the Gram matrix of the Néron–Severi lattice of Z_{-12}.

$$NS = \begin{pmatrix}
-2 & 1 & 0 & 0 & 0 & 0 & 0 & 0 & 0 & 0 & 0 & 0 & 0 & 0 & 0 & 0 & 0 & 0 & 0 & 1 \\
1 & 0 & 1 & 0 & 0 & 0 & 0 & 0 & 0 & 0 & 0 & 0 & 0 & 0 & 0 & 0 & 0 & 0 & 0 & 1 \\
0 & 1 & -2 & 0 & 0 & 1 & 0 & 0 & 0 & 1 & 0 & 0 & 0 & 0 & 0 & 0 & 0 & 0 & 1 & 1 \\
0 & 0 & 0 & -2 & 1 & 0 & 0 & 0 & 0 & 0 & 0 & 0 & 0 & 0 & 0 & 0 & 0 & 0 & 0 & 0 \\
0 & 0 & 0 & 1 & -2 & 1 & 0 & 0 & 0 & 0 & 0 & 0 & 0 & 0 & 0 & 0 & 0 & 0 & 0 & 0 \\
0 & 0 & 1 & 0 & 1 & -2 & 1 & 0 & 0 & 0 & 0 & 0 & 0 & 0 & 0 & 0 & 0 & 0 & 0 & 1 \\
0 & 0 & 0 & 0 & 0 & 1 & -2 & 1 & 0 & 0 & 0 & 0 & 0 & 0 & 0 & 0 & 0 & 0 & 0 & 0 \\
0 & 0 & 0 & 0 & 0 & 0 & 1 & -2 & 1 & 0 & 0 & 0 & 0 & 0 & 0 & 0 & 0 & 0 & 0 & 0 \\
0 & 0 & 0 & 0 & 0 & 0 & 0 & 1 & -2 & 0 & 0 & 0 & 0 & 0 & 0 & 0 & 0 & 0 & 0 & 0 \\
0 & 0 & 1 & 0 & 0 & 0 & 0 & 0 & 0 & -2 & 1 & 0 & 0 & 0 & 0 & 0 & 0 & 0 & 0 & 0 \\
0 & 0 & 0 & 0 & 0 & 0 & 0 & 0 & 0 & 1 & -2 & 1 & 0 & 0 & 0 & 0 & 0 & 0 & 0 & 0 \\
0 & 0 & 0 & 0 & 0 & 0 & 0 & 0 & 0 & 0 & 1 & -2 & 1 & 0 & 0 & 0 & 0 & 0 & 0 & 0 \\
0 & 0 & 0 & 0 & 0 & 0 & 0 & 0 & 0 & 0 & 0 & 1 & -2 & 1 & 0 & 1 & 0 & 0 & 0 & 0 \\
0 & 0 & 0 & 0 & 0 & 0 & 0 & 0 & 0 & 0 & 0 & 0 & 1 & -2 & 1 & 0 & 0 & 0 & 0 & 0 \\
0 & 0 & 0 & 0 & 0 & 0 & 0 & 0 & 0 & 0 & 0 & 0 & 0 & 1 & -2 & 0 & 0 & 0 & 0 & 0 \\
0 & 0 & 0 & 0 & 0 & 0 & 0 & 0 & 0 & 0 & 0 & 0 & 1 & 0 & 0 & -2 & 0 & 0 & 0 & 0 \\
0 & 0 & 0 & 0 & 0 & 0 & 0 & 0 & 0 & 0 & 0 & 0 & 0 & 0 & 0 & 0 & -2 & 1 & 0 & 0 \\
0 & 0 & 0 & 0 & 0 & 0 & 0 & 0 & 0 & 0 & 0 & 0 & 0 & 0 & 0 & 0 & 1 & -2 & 0 & 0 \\
0 & 0 & 1 & 0 & 0 & 0 & 0 & 0 & 0 & 0 & 0 & 0 & 0 & 0 & 0 & 0 & 0 & 0 & -2 & 0 \\
1 & 1 & 1 & 0 & 0 & 1 & 0 & 0 & 0 & 0 & 0 & 0 & 0 & 0 & 0 & 0 & 0 & 0 & 0 & -2
\end{pmatrix}.$$

The determinant is -96. Applying Lemma 2.1, we find

$$G_L \simeq \mathbb{Z}/(2) \oplus \mathbb{Z}/(48).$$

Moreover, the vector

$$B_2 = [2, -9, 11, 0, 0, -7, -2, 0, 0, -11, 11, -9, -2, 2, 9, 0, 3, -3, 0, -2]$$

with quadratic norm $-\frac{1}{2}$ generates the cyclic group $\mathbb{Z}/(2)$ and the vector

$$B_{48} = [1, -3, 4, 0, 0, -2, -1, 0, 0, -4, 4, -3, -1, 1, 3, 0, 1, -1, 0, -1]$$

with quadratic norm $-\frac{29}{48}$ generates the cyclic group $\mathbb{Z}/(48)$. They also satisfy the relation

$$B_2.B_{48} = -171/2 \equiv -1/2(\text{mod}.\mathbb{Z}).$$

Let

$$A = \begin{pmatrix} 10 & 2 \\ 2 & 10 \end{pmatrix}$$

with Smith form $[2 \quad 0 \quad 48]$. Applying Lemma 2.1, we find $G_A \simeq \mathbb{Z}/(2) \oplus \mathbb{Z}/(48)$, a vector u with quadratic norm $1/2$ generating the cyclic group $\mathbb{Z}/(2)$, a vector v

with quadratic norm $5/48$ generating the cyclic group $\mathbb{Z}/(48)$ and satisfying $u.v = -1/2$. Since the vector $5v$ with quadratic norm $29/48$ is another generator of the cyclic group $\mathbb{Z}/(48)$ satisfying $u.5v = -5/2 \equiv 1/2(\text{mod.}\mathbb{Z})$, we deduce

$$T(Z_{-12}) = [10 \quad 2 \quad 10].$$

7 Jacobian Varieties of the Singular $K3$ Surfaces Z_k

Denote J_k the Jacobian variety of Z_k. Sometimes it is easy to get the transcendental lattice of J_k from a Weierstrass equation. For the remaining cases, we use the following Keum's results [9, Lemma 2.1 and Corollary 2.2]. More precisely, if S (resp., $J(S)$) denotes the $K3$ surface (resp., its Jacobian surface), we use the formula

$$\det(Pic(S)) = l^2 \det(Pic(J(S))$$

and its corollary: If $\det(Pic(S))$ is square-free, then every elliptic fibration on S is a Jacobian fibration, i.e., has a section.

By specialization of the Weierstrass equation of J_k, we get

$$(J_{12}) \qquad y^2 = x^3 + 144t^2x^2 + 3072t^3(t-1)x + 16384t^5(t-1)^2$$

with singular fibers $III^*(0)$, $I_3(1)$, $I_2(-1)$, $II^*(\infty)$. The rank is 0, and by Shimada and Zhang [26], it follows

$$T(J_{12}) = [2 \quad 0 \quad 6].$$

Also

$$(J_{-6}) \qquad y^2 = x^3 + 36t^2x^2 + 192t^3(t-1)x + 256t^5(t-1)^2$$

with singular fibers $III^*(0)$, $I_3(1)$, $I_2(-4)$, $II^*(\infty)$. The rank is 0, and by Shimada and Zhang [26], it follows

$$T(J_{-6}) = [2 \quad 0 \quad 6].$$

Moreover,

$$(J_4) \qquad y^2 = x^3 + 16t^2x^2 + 512t^3(t-1)x + 4096t^5(t-1)^2$$

with singular fibers $III^*(0)$, $I_4(1)$, $I_1(\frac{32}{27})$, $II^*(\infty)$. The rank is 0, and by Shimada and Zhang [26], it follows

$$T(J_4) = [2 \quad 0 \quad 4].$$

We also get

$$(J_0) \qquad y^2 = x^3 + 1024t^5(t-1)^2$$

with singular fibers $II^*(0)$, $IV(1)$, $II^*(\infty)$. The rank is 0, and by Shimada and Zhang [26] or by Utsumi [30], it follows

$$T(J_0) = [2 \quad 1 \quad 2].$$

Since $T(Z_{-3}) = [4 \quad 1 \quad 4]$, it follows $\det(Pic(Z_{-3})) = 15$ is square-free and every elliptic fibration on Z_{-3} has a section by Keum's Corollary 2.2 [9]. Hence,

$$Z_{-3} = J_{-3}.$$

Indeed the genus 1 fibration of Z_{-3} leading to J_{-3}

$$W^2 = -4mt^4 - 15m^2t^2 + m^2(4m-10)t - 3m^2$$

has the section $(t = 0, y = m\sqrt{-3})$.

Since $\det(Pic(Z_{-12})) = -24 \times 4$, according to the above formula, either $\det(Pic(J_{-12})) = -24 \times 4$ (impossible according to all the possible discriminants of singular $K3$ surfaces of Verrill's family), or $\det(Pic(J_{-12})) = -24$ and

$$T(J_{-12}) = [4 \quad 0 \quad 6].$$

8 Final Remarks

We can now complete the table.

Y_0	[4 2 4]	Z_{-36}	$?12a^2$	X_{-36}	[2 0 6]	J_{-36}	?[6 0 8]
Y_2	[2 0 4]	Z_{-12}	[10 2 10]	X_{-12}	[4 0 6]	J_{-12}	[4 0 6]
Y_3	[2 1 8]	Z_{-6}	[8 4 8]	X_{-6}	[6 0 8]	J_{-6}	[2 0 6]
Y_6	[2 0 12]	Z_{-3}	[4 1 4]	X_{-3}	[6 0 10]	J_{-3}	[4 1 4]
Y_{10}	[6 0 12]	Z_0	[2 0 6]	X_0	[2 0 6]	J_0	[2 1 2]
Y_{18}	[10 0 12]	Z_4	[2 0 16]	X_4	[2 0 4]	J_4	[2 0 4]
Y_{102}	[12 0 26]	Z_{12}	[2 0 24]	X_{12}	[2 1 2]	J_{12}	[2 0 6]
Y_{198}	[12 0 34]	Z_{60}	$?15a^2$	X_{60}	[4 1 4]	J_{60}	?[6 0 10]
$Y_{k^2=20}$	[2 0 10]						
$Y_{k^2=54}$	[4 0 12]						
$Y_{k^2=-12}$	[6 0 6]						
$Y_{k^2=-45}$	[8 2 8]						

Let us recall Morrison's result [13], Corollary 4.4 i).

Corollary 8.1 *Let Y be an algebraic $K3$ surface.*

If $\rho(Y) = 19$ or 20, then Y is a Kummer surface if and only if there is an even lattice T' with $T_Y \simeq T'(2)$.

Recall also that a $K3$ surface S has a Shioda–Inose structure (and this is the case if $\rho(Y) = 19$ or 20) if there is a symplectic involution ι on S with rational quotient map $S \xrightarrow{p} Y$ such that Y is a Kummer surface and p^* induces a Hodge isometry $T_S(2) \simeq T_Y$. In that case the Kummer surface is called the Kummer surface associated to S, for short the Kummer surface of S:

- From the previous table, we observe that $T_{Y_{k^2=-45}} = [8 \quad 2 \quad 8]$; thus $Y_{k^2=-45}$ is a Kummer surface. Since $T_{Z_{-3}} = [4 \quad 1 \quad 4]$, it follows that $Y_{k^2=-45}$ is the Kummer surface of Z_{-3}.
- Similarly, since $T_{Y_{k^2=54}} = [4 \quad 0 \quad 12]$ and $T_{X_0} = T_{X_{-36}} = [2 \quad 0 \quad 6]$, the $K3$ surface $Y_{k^2=54}$ is the Kummer surface of X_0 or of X_{-36}.
- The $K3$ surface Z_{-6} is the Kummer surface of Y_0, which is in turn the Kummer surface of J_0. Also Y_0 is the Kummer surface of X_{12}.
- The Kummer surface K_k associated to Y_k has discriminant $12 \times 2^3 = 24 \times 4$, and we found in [4] that its Jacobian $K3$ surface has transcendental lattice $\langle -2 \rangle \oplus \langle 6 \rangle \oplus \langle 2 \rangle$ of discriminant 24.
- The $K3$ surface Z_k with discriminant $24 = 6 \times 4$ has for Jacobian variety X_k with discriminant 6.

  These two results are illustrations of the previous Keum's result [9].
- Using Keum's corollary [9], since the discriminant of X_k is 6 without square factors, we can deduce that every elliptic fibration on X_k has a section.
- The $K3$ surface Z_0 is equal to X_{-36} and Y_2 is equal to X_4.
- We end with a conjecture

$$J_{60} \stackrel{?}{=} X_{-3} \qquad J_{-36} \stackrel{?}{=} X_{-6}.$$

  Since the singular $K3$ surfaces J_k are among the singular $K3$ surfaces X_k, assuming that Boyd's list in Sect. 4 is complete, J_{60} is either X_{-3} or X_{-6}. If $J_{60} = X_{-6}$, according to Keum's lemma, it would follow

$$15a^2 = \det(T_{Z_{60}}) = \Box \det(T_{J_{60}}) = \Box 48.$$

  And $\sqrt{5}$ should be a rational number, a contradiction.

Some of these remarks are useful to compute the L-series of the $K3$ surface.

References

1. M.-J. Bertin, Mesure de Mahler d'hypersurfaces K3. J. Number Theory **128**(11), 2890–2913 (2008). ArXivmath.NT/0501153
2. M.-J. Bertin, Mahler's measure and L-series of K3 hypersurfaces. AMS/IP Stud. Adv. Math. **38**, 3–18 (2006)
3. M.-J. Bertin, A. Feaver, J. Fuselier, M. Lalin, M. Manes, Mahler measure of some singular K3-surfaces, in *Women in Numbers 2: Research Directions in Number Theory*. Contemporary Mathematics, vol. 606 (American Mathematical Society, Providence, 2013), pp. 149–169. ArXiv 1208.6240, 1–17
4. M.-J. Bertin, O. Lecacheux, Apéry-Fermi pencil of K3-surfaces and 2-isogenies. J. Math. Soc. Jpn. **72**(2), 599–637 (2020)
5. D. Boyd, *Personal Communication*
6. D. Boyd, Mahler's measure and special values of L-functions. Experimental Math. **7**(1), 37–82 (1998)
7. A. Clingher, C. Doran, Modular invariants for lattice polarized $K3$ surfaces. Michigan Math. J. **55**, 355–393 (2007)
8. I. Dolgachev, Mirror symmetry for lattice polarized K3 surfaces. J. Math. Sci. **81**(3), 2599–2630 (1996)
9. J. Keum, A note on elliptic K3 surfaces. Trans. Am. Math. Soc. **352**(5), 2077–2086 (2000)
10. M. Kuwata, *Maple Library Elliptic Surface Calculator*. http://c-faculty.chuo-u.ac.jp/~kuwata/ESC.php.
11. Ling L., On Shioda-Inose structures of one-parameter families of K3 surfaces. J. Number Theory **109**, 299–318 (2004)
12. B. Lian, S.-T. Yau, Arithmetic properties of mirror map and quantum coupling. Commun. Math. Phys. **176**, 163–191 (1996)
13. D. Morrison, On K3 surfaces with large Picard number. Invent. Math. **75**, 105–121 (1984)
14. N. Narumiya, H. Shiga, The mirror map for a family of K3 surfaces induced from the simplest 3- dimensional reflexive polytope, in *Proceedings on Moonshine and related topics(Montreal, 1999)*. CRM Proceedings of the Lecture Notes, vol. 30 (American Mathematical Society, New York, 2001), pp. 139–161
15. K. Nishiyama, The Jacobian fibrations on some K3 surfaces and their Mordell-Weil groups. Jpn. J. Math. **22**(2), 293–347 (1996)
16. V. Nikulin, Integral symmetric bilinear forms and some of their applications. Izv. Math. **14**(1),103–167 (1980)
17. The PARI Group, Bordeaux GP/PARI version 2.7.3 (2015). http://pari.math.u-bordeaux.fr
18. C. Peters, Monodromy and Picard-Fuchs equations for families of K3 surfaces and elliptic curves. Ann. Sci. École Norm. Sup. **19**(4), 583–607 (1986)
19. C. Peters, J. Stienstra, A pencil of K3-surfaces related to Apéry's recurrence for $\zeta(3)$ and Fermi surfaces for potential zero, in *Arithmetic of Complex Manifolds, Proceedings of the Erlangen 1988*. Lecture Notes Mathematics, vol. 1399 (Springer, Barth, 1988)
20. SageMath—Open-Source Mathematical Software System. https://www.sagemath.org
21. D. Samart, Mahler measures as linear combinations of L-values of multiple modular forms. Can. J. Math. **67**(2), 424–449 (2015). ArXiv 2013
22. A. Sarti, Transcendental Lattices of some K3 surfaces. Math. Nachr. **281**(N○7), 1–17 (2008)
23. M. Schütt, T. Shioda, Elliptic surfaces, in *Algebraic Geometry in East Asia—Seoul 2008*. Advanced Studies in Pure Mathematics, vol. 60 (2010), pp. 51–160
24. M. Schütt, T. Shioda, Mordell-Weil Lattices, in *Ergebnisse der Mathematik und ihrer Grenzgebiete. 3. Folge (A Series of Modern Surveys in Mathematics)*, vol. 70 (Springer, Singapore, 2019)
25. I. Shimada, K3 surfaces and Lattice Theory, Talk Dec. 2, 2014, Hanoï Vietnam Institute for Advanced Study in Mathematics (Slides on Shimada's web page)

26. I. Shimada, D.Q. Zhang, Classification of extremal elliptic K3 surfaces and fundamental groups of open K3 surfaces. Nagoya Math. J. **161**, 23–54 (2001)
27. T. Shioda, On elliptic modular surfaces. J. Math. Soc. Jpn. **24**, 20–59 (1972)
28. T. Shioda, H. Inose, On singular K3 surfaces, in W.L. Baily Jr., T. Shioda, ed. by *Complex Analysis and Algebraic Geometry* (Iwanami Shoten, Tokyo, 1977), pp. 119–136
29. H. Sterk, Lattices and K3 surfaces of degree 6, in *Linear Algebra and Its Applications*, vol. 226–228 (1995), pp. 297–309
30. K. Utsumi, Jacobian fibrations on the singular K3 surface of discriminant 3. arXiv:1405.3577v1 [math.AG] (2014), pp. 1–13
31. H. Verrill, The L-series of Certain Rigid Calabi-Yau Threefolds. J. Number Theory **81**, 310–334 (2000)
32. H. Verrill, Root Lattices and pencils of varieties. J. Math. Kyoto Univ. (JMKYAZ) **36**(2), 423–446 (1996)
33. H. Verrill, *Root Lattices and Pencils of Varieties*, PhD (Cambridge University, Cambridge, 1994)
34. H. Verrill, N. Yui, Thompson series and the mirror map of pencils of K3 surfaces. CRM Proc. Lecture Notes **24**, 399–432 (2000)

Power-Saving Error Terms for the Number of D_4-Quartic Extensions over a Number Field Ordered by Discriminant

Alina Bucur, Alexandra Florea, Allechar Serrano López, and Ila Varma

2020 Mathematics Subject Classification 11R45, 11R16

1 Introduction

Obtaining asymptotics on the count of number fields ordered by discriminant has been a guiding force in the subfield of number theory that has been coined (originally by Mazur) as "arithmetic statistics."

The simplest case of this question arises from the study of quadratic fields: How many quadratic fields are there of bounded discriminant? In this lowest-degree case, it is fortunate that quadratic fields are uniquely determined by their discriminant, and so answering the above question boils down to understanding the possible discriminants for quadratic fields. Algebraic methods lead to the conclusion that quadratic (or fundamental) discriminants are exactly the integers of the form D, where $D \equiv 1 \bmod 4$ and D is square-free, or $4D$, where $D \equiv 2, 3 \bmod 4$ and D is square-free. Analytic methods can then be used to compute that the number of such integers with absolute value bounded by X is

A. Bucur
Department of Mathematics, University of California, San Diego, CA, USA
e-mail: alina@math.ucsd.edu; https://mathweb.ucsd.edu/~alina/

A. Florea
Department of Mathematics, University of California, Irvine, CA, USA
e-mail: floreaa@uci.edu; https://sites.google.com/view/alexandraflorea/home

A. Serrano López (✉)
Department of Mathematical Sciences, Montana State University, Bozeman, MT, USA
e-mail: allechar.serranolopez@montana.edu

I. Varma
Department of Mathematics, University of Toronto, Toronto, ON, USA
e-mail: ila@math.toronto.edu; http://www.math.toronto.edu/~ila/

A. Bucur et al. (eds.), *Research Directions in Number Theory*, Association for Women in Mathematics Series 33, https://doi.org/10.1007/978-3-031-51677-1_6

$$\frac{6}{\pi^2}X + O(\sqrt{X}).$$

Elementary arguments such as those found in [44, Theorem 2.2] studying the Mobius sum $\sum_{n\leq x, n\equiv a \bmod 4} \mu^2(n)$ can be used to prove the above asymptotic.[1] Improving the error to $o(\sqrt{X})$ can be accomplished by using more modern techniques such as contour integration and the fact that the Riemann zeta function $\zeta(s)$ does not vanish on $\Re(s) = 1$.

In the twentieth century, two main paths of generalizing this question on quadratic discriminants were studied: changing the degree of the number fields in question and changing the base field from $\mathbb{Q}$ to a general number field (or function field). Altogether, we can pose the general question that is studied to understand number field asymptotics.

Question Let F be a base number field and fix an algebraic closure $\overline{F}$ of F. Fix a degree n and a Galois group G equipped with an embedding into S_n. What is the number of (F-isomorphism classes of) degree n field extensions L of F such that the normal closure $\tilde{L}$ of L over F has Galois group isomorphic G and the absolute norm of the relative discriminant $\mathrm{Disc}(L/F)$ is bounded by X?

Here, we say two extensions L and L' of a number field F are *F-isomorphic* if there exists a field isomorphism between L and L' that restricts to the identity on F.

Let $N_{F,n}(G, X)$ denote the number of F-isomorphism classes of degree n extensions L over F such that $\mathrm{Gal}(\tilde{L}/F) \cong G$ and $\mathrm{Nm}_{F/\mathbb{Q}}(\mathrm{Disc}(L/F)) \leq X$. We also require that the Galois group $\mathrm{Gal}(\tilde{L}/F)$ of the Galois closure $\tilde{L}$ over F, viewed as permuting the set of embeddings of L into $\tilde{L}$, is isomorphic to the fixed choice of image of G in S_n.

Our focus is then on determining the constants of the main term of the asymptotic formula for $N_{F,n}(G, X)$; that is, finding $a_n(G)$, $b_{F,n}(G)$ and $c_{F,n}(G)$ such that

$$\lim_{X\to\infty} \frac{N_{F,n}(G, X)}{X^{a_n(G)} \log(X)^{b_{F,n}(G)}} = c_{F,n}(G). \tag{1}$$

The study of the non-negative rational constants $a_n(G) = a_{F,n}(G)$ and $b_{F,n}(G)$ is the focus of Malle's Conjecture [39, 40]. We first discuss the development of results and techniques for number field asymptotics predating the statement of Malle's Conjecture. After introducing our main result, we discuss the evolution of the study of number field asymptotics after Malle's statement of their conjecture.

---

[1] This result goes back many centuries and was at least known to Gauss, who in [28] provides evidence that that class number of imaginary quadratic rings grows like $|D|^{1/2}$. In particular, Gauss claimed (without proof) that the sum of class numbers of quadratic fields of discriminant up to T is $\sum_{|D|<T} H(D) \sim \frac{\pi}{18} T^{3/2}$, but Lagrange may have indexed these rings by their discriminant earlier. One can also think of this count as essentially reducing to counting square-free integers up to X, which might have even been done by Euler.

1.1 Number Field Asymptotics Predating Malle's Conjecture

Much of the work predating the statement of Malle's conjecture focused on the case when $F = \mathbb{Q}$. Past the $n = 2$ case described above, Cohn [19] determined in 1954 using class field theory and generating functions that Eq. (1) holds when $F = \mathbb{Q}$, $n = 3$, and $G = C_3$ with

$$a_{\mathbb{Q},3}(C_3) = 1/2, \quad b_{\mathbb{Q},3}(C_3) = 0, \quad c_{\mathbb{Q},3}(C_3) = \frac{11\sqrt{3}}{36\pi} \cdot \prod_{\substack{p \equiv 1 \bmod 6 \\ p \text{ prime}}} \frac{(p+2)(p-1)}{p(p+1)}.$$

In 1971, Davenport–Heilbronn [21] complemented this work by determining that the asymptotic in (1) holds for $F = \mathbb{Q}$, $n = 3$, and $G = S_3$ with

$$a_{\mathbb{Q},3}(S_3) = 1, \quad b_{\mathbb{Q},3}(S_3) = 0, \quad c_{\mathbb{Q},3}(S_3) = \frac{1}{3\zeta(3)}.$$

Their results utilized an algebraic parametrization of cubic rings in terms of $\mathrm{GL}_2(\mathbb{Z})$-orbits of integer binary cubic forms, geometry-of-numbers techniques to relate the count of certain orbits to a volume of a body in the fundamental domain for $\mathrm{GL}_2(\mathbb{Z})$ on real-coefficient binary cubic forms, and finally sieving methods to restrict to maximal rings and thus obtain an asymptotic count for S_3-cubic fields.

In 1979, Baily [6] proved the asymptotic (1) in the abelian degree 4 cases ($F = \mathbb{Q}$, $n = 4$, and $G = C_4$ or V_4) with

$$\begin{aligned} a_{\mathbb{Q},4}(C_4) = 1/2,\ b_{\mathbb{Q},4}(C_4) = 0,\ c_{\mathbb{Q},4}(C_4) \\ = \frac{3}{\pi^2}\left(-1 + \frac{25}{24}\prod_{\substack{p \equiv 1 \bmod 4 \\ p \text{ prime}}}\left(1 + \frac{2}{p^{3/2} + p^{1/2}}\right)\right), \end{aligned}$$

and

$$a_{\mathbb{Q},4}(V_4) = 1/2,\ b_{\mathbb{Q},4}(V_4) = 2,\ c_{\mathbb{Q},4}(V_4) = \frac{1}{48\zeta(2)^3}\prod_{p \text{ prime}}\left(1 + \frac{3p+1}{p^3 + 3p^2}\right)^{-1}.$$

Baily used class field theory and generating functions in a similar manner as the cyclic cubic case for the above results.

In addition, Baily proved that $cX < N_{\mathbb{Q},4}(D_4, X) < c'X$ for two constants c and c', where D_4 denotes the symmetries of the square. The upper bound is obtained using bounds on the number of quadratic ray class characters of quadratic fields that correspond under class field theory to D_4-quartic fields and summing these bounds over all quadratic fields with bounded discriminant. The lower bound was obtained by studying quadratic ray class characters of certain imaginary quadratic fields with even class number.

In the 1970s and early 1980s, a variety of authors (in many different languages!) worked on asymptotics for various families of p-groups, such as the case $G = C_{p^k}$ for $p \geq 3$, $k \in \mathbb{Z}_{\geq 2}$, $F = \mathbb{Q}$, and $n = p^k$, as well as $G = (C_p)^k$ and direct products of p-groups with p'-groups, where $p, p' \geq 3$ are distinct primes (see [47, 56, 62] building on work of [57] from the 1920s. For more detailed lists, see Pg. 7 of [38]).

In the late 1980s, a lot of different avenues saw progress, mostly surrounding work of Mäki and Wright. In 1985, Mäki [38] determined that for any abelian group G of order $n = |G|$,

$$a_{\mathbb{Q},|G|}(G) = \frac{1}{n}\left(1 - \frac{1}{p}\right), \quad b_{\mathbb{Q},|G|}(G) = \begin{cases} \frac{n_p}{p-1} - 1 & \text{if } p \geq 3 \\ n_p - 1 & \text{if } p = 2, \end{cases}$$

where p is the smallest prime dividing n and n_p denotes the number of elements of G of order p. In 1988, Datskovsky–Wright [20] determined for any number field F,

$$a_{F,3}(S_3) = 1, \; b_{F,3}(S_3) = 0, \; c_{F,3}(S_3) = \left(\frac{2}{3}\right)^{r_1(F)-1} \left(\frac{1}{6}\right)^{r_2(F)} \frac{\zeta_F(1)}{\zeta_F(3)},$$

where $\zeta_F(1)$ denotes the residue at $s = 1$ of $\zeta_F(s)$, the Dedekind zeta function associated to F, $r_1(F)$ denotes the number of real embeddings of F into $\overline{F}$, and $r_2(F)$ denotes the number of pairs of complex embeddings of F into $\overline{F}$. They also obtained analogous results for any global field of characteristic not equal to 2 or 3; see [20, Theorem I.1]. As a byproduct, they determined

$$a_{F,2}(S_2) = 1, \; b_{F,2}(S_2) = 0, \; c_{F,2}(S_2) = 2^{-r_2(F)-1} \frac{\zeta_F(1)}{\zeta_F(2)}.$$

The authors accomplished this work building on the work of Davenport–Heilbronn [21] and through the use of the adelic Shintani zeta functions [53, 54].

Building on work of Mäki, Wright [61] proved that for any finite abelian group G of cardinality n and any number field F, there exists a constant $c_{F,n}(G) > 0$ such that the asymptotic (1) holds with the exponents

$$a_n(G) = \frac{1}{n}\left(1 - \frac{1}{p}\right), \quad b_{F,n}(G) = \frac{n_p}{[F(\zeta_p) : F]} - 1,$$

where p is the smallest prime dividing n, and n_p denotes the number of elements of G of order p. Wright developed general techniques in the vein of [19] and [6], i.e., incorporating class field theory with the use of generating functions that happen to be related to Dirichlet series. Standard Tauberian theorems (most notably, Perron's formula) then allow for the extraction of number field asymptotics.

In the early 2000s, Cohen–Diaz y Diaz–Olivier [18] determined that, for $F = \mathbb{Q}$, $n = 4$, and $G = D_4$,

$$a_{\mathbb{Q},4}(D_4) = 1 \quad b_{\mathbb{Q},4}(D_4) = 0 \quad c_{\mathbb{Q},4}(D_4) = \sum_{[K:\mathbb{Q}]=2} \frac{2^{-r_2(K)-1}}{\mathrm{Disc}(K)^2} \frac{\zeta_K(1)}{\zeta_K(2)}. \tag{2}$$

They utilized Kummer theory in conjunction with the strategy of Baily [6] to relate the count of D_4-quartic extensions to the count of quadratic extensions of quadratic extensions. The authors proved [18, Corollary 6.1] by defining a Dirichlet series associated to counting quadratic extensions of quadratic extensions and related it to the generating function for D_4-quartic fields. They also remark that it is immediate to extend this result to the case of relative D_4-extensions over a general base field.

1.2 Main Results

In this chapter, we give a proof of the asymptotic (1) in the case of quartic dihedral extensions of a number field F, with $a_{F,4}(D_4) = 1$, $b_{F,4}(D_4) = 0$, and determine the constant $c_{F,4}(D_4)$. Further, we obtain an explicit formula for a power-saving error term in terms of the degree of F. More precisely, we prove the following result.

Theorem 1 *Let F be a number field with $[F : \mathbb{Q}] = d$. We have that the asymptotic number of (F-isomorphism classes of) quartic D_4-extensions of F whose norm of the relative F-discriminant is bounded by X is, as X grows,*

$$N_{F,4}(D_4, X) = \left(\sum_{[K:F]=2} \frac{1}{2^{r_2(K)+1}} \cdot \frac{1}{\mathrm{Nm}_{F/\mathbb{Q}}(\mathrm{Disc}(K/F))^2} \cdot \frac{\zeta_K(1)}{\zeta_K(2)} \right) \cdot X + O\left(X^{\frac{3}{4}+\epsilon}\right). \tag{3}$$

If we assume the Lindelöf hypothesis, then the error term above can be improved to $O\left(X^{\frac{1}{4}+\epsilon}\right)$.

Recall that for $\Re(s) > 1$ the Dedekind zeta function of any number field E is given by

$$\zeta_E(s) = \prod_{\mathfrak{p} \subseteq \mathcal{O}_E} \left(1 - \mathrm{Nm}_{E/\mathbb{Q}}(\mathfrak{p})^{-s}\right)^{-1},$$

where $\mathfrak{p}$ runs over prime ideals of $\mathcal{O}_E$. It has meromorphic continuation to the full complex plane and is analytic everywhere except for a simple pole at $s = 1$. By abuse of notation, we denote by $\zeta_E(1)$ the residue at $s = 1$.

The main term in Theorem 1 is proven by Klüners [32] without a power-saving error term. We show that the proof argument utilized by Cohen–Diaz y Diaz–Olivier [18] to prove that $a_{\mathbb{Q},4}(D_4) = 1$ and $b_{\mathbb{Q},4}(D_4) = 0$ hold true for general F and that the formula (3) is the same as the one stated by them in Remark (4) after [18, Corollary 6.1] with $\alpha = \frac{1}{4}$. We point out that our power-saving error term does not depend on the degree of F, as the aforementioned authors suggest!

1.3 Malle's Conjecture, Counterexamples, Refinements, and Proven Cases

In 2004, Malle [39, 40] made a prediction for the constants $a_{F,n}(G) = a_n(G)$ and $b_{F,n}(G)$ for all transitive groups $G \hookrightarrow S_n$. The prediction was that

$$a_n(G) = \frac{1}{\min_{g \in G \setminus 1} \mathrm{Ind}(g)},$$

where the index of an element g is defined as $\mathrm{Ind}(g) := n - \#$ orbits of $\{1, 2, \ldots n\}$ when acted upon by g via the embedding into S_n, and

$$b_{F,n}(G) = b_{F,n}^{\mathrm{Malle}}(G) := \#\{\mathrm{Gal}(\overline{F}/F)$$

$$\text{- orbits of } G\text{-conjugacy classes of minimal index } \mathrm{Ind}(C)\} - 1,$$

where the notation is defined as follows: Note that if g and g' are in the same conjugacy class C of G, then $\mathrm{Ind}(g) = \mathrm{Ind}(g')$, and furthermore, the absolute Galois group $\mathrm{Gal}(\overline{F}/F)$ acts on the set of conjugacy classes $\{C\}$ of G via the action of the cyclotomic character, i.e., $\mu : \mathrm{Gal}(\overline{\mathbb{Q}}/\mathbb{Q}) \to \hat{\mathbb{Z}}^\times$ so that $\sigma \in \mathrm{Gal}(\overline{\mathbb{Q}}/\mathbb{Q})$ acts on roots of unity as $\sigma(\zeta_n) = \zeta_n^{\mu(\sigma)}$. For $g \in G$, we can define $\sigma(g) = g^{\mu(\sigma)}$, or in fact on conjugacy classes C of G: $\sigma(C) = C^{\mu(\sigma)}$. We call the orbits under this action on the G-conjugacy classes $\{C\}$ by $\mathrm{Gal}(\overline{F}/F)$-orbits.

Note that in the trivial case when $n = 1$ and G is the trivial group, we have $a_{1,F}(1) = b_{1,F}(1) = 0$ and $c_{1,F}(1) = 1$ for any field F.

It is also worth pointing out that when the image of G in S_n contains a simple transposition (ab), then $\mathrm{Ind}((ab)) = 1$ since (ab) has $n - 1$ orbits in S_n, and hence, $a_n(G)$ is predicted to be 1. The number of G-conjugacy classes of minimal index 1 will be 1, since all simple transpositions are conjugate to one another and no other permutation type can have $n - 1$ orbits, and hence, $b_{F,n}^{\mathrm{Malle}}(G) = 0$ in this case.

In 2005, two different directions of progress arose from the development of averaging techniques in geometry-of-numbers counting due to Bhargava and a noteworthy counterexample and subsequent work to Malle's Conjecture due to Klüners.

Bhargava [9] determined for S_4-quartic fields over $\mathbb{Q}$ that $a_{\mathbb{Q},4}(S_4) = 1$ and $b_{\mathbb{Q},4}(S_4) = 0$ and obtained an Euler product formula for

$$c_{\mathbb{Q},4}(S_4) = \frac{5}{24} \prod_p \left(1 + \frac{1}{p^2} - \frac{1}{p^3} - \frac{1}{p^4}\right).$$

Bhargava's methods built on the work of Davenport–Heilbronn [21], Delone–Faddeev [22], and Sato–Kimura [49] using the same proof strategy of a parametrization (this time, in terms of pairs of ternary quadratic forms) allowing for geometry-of-numbers methods followed by sieving techniques. However, Bhargava produced a general machine of averaging techniques for cutting off cusps when computing a volume in the fundamental domain and relating it to the number of integral orbits or lattice points that enumerate quartic rings. This allowed for generalizations to the $n = 5$ case [10], in which he determined in 2010 that $a_{\mathbb{Q},5}(S_5) = 1$ and $b_{\mathbb{Q},5}(S_5) = 0$ and obtained an Euler product formula for

$$c_{\mathbb{Q},5}(S_5) = \frac{13}{120} \prod_p \left(1 + \frac{1}{p^2} - \frac{1}{p^4} - \frac{1}{p^5}\right).$$

In addition, Bhargava–Wood [13] and, independently, Belabas–Fouvry [7] verified for S_3-sextic fields that $a_{\mathbb{Q},6}(S_3) = 1/3$ and $b_{\mathbb{Q},6}(S_3) = 0$ and furthermore determined that

$$c_{\mathbb{Q},6}(S_3) = \frac{1}{3}\left(1 + \frac{1}{3} + \frac{1}{3^{5/3}} + \frac{2}{3^{7/3}}\right)\left(1 - \frac{1}{3}\right) \prod_{p \neq 3} \left(1 + \frac{1}{p} + \frac{1}{p^{4/3}}\right)\left(1 - \frac{1}{p}\right).$$

While studying these questions, Bhargava [8] formulated a conjecture for the constant in the case when $G = S_n$. Namely, he conjectured that

$$c_{F,n}(S_n) = \frac{1}{2} \cdot \zeta_F(1) \left(\frac{\#S_n[2]}{n!}\right)^{r_1(F)} \left(\frac{1}{n!}\right)^{r_2(F)} \prod_{\mathfrak{p}} \left(\sum_{k=0}^{n} \frac{q(k, n-k) - q(k-1, n-k+1)}{\mathrm{Nm}_{F/\mathbb{Q}}(\mathfrak{p})}\right)$$

for all n, where $q(k, n)$ denotes the number of partitions of k into at most n parts (see conjecture A in the Appendix of [8]). A conjecture for $c_{F,n}(G)$ for general G has not been formulated as the direct analogy to Bhargava's heuristic is contradicted by the result of [18] in the case of $n = 4$ and $G = D_4$ (see [18] and [4, Section §3.3]). Bhargava's quartic and quintic results were generalized to all number fields F by Bhargava–Shankar–Wang [12], including formulas for the constants $c_{F,4}(S_4)$ and $c_{F,5}(S_5)$ that verified Bhargava's conjecture along with Malle's conjecture.

Also in 2005, Klüners [30] provided a counterexample to Malle's Conjecture by proving a lower bound for $N_{\mathbb{Q},6}(C_3 \wr C_2, X)$ that implies that $b_{\mathbb{Q},6}(C_3 \wr C_2) > 0$, contradicting the formula given by Malle, which would have predicted that $b_{\mathbb{Q},6}(C_3 \wr C_2) = 0$. The contradiction seems to arise from the fact that while there are not third roots of unity in the base field, third roots of unity can arise in an intermediate extension, and $C_3 \wr C_2$-extensions of $\mathbb{Q}$ that contain $\mathbb{Q}(\zeta_3)$ and have discriminant bounded by X are themselves at least $c \cdot X^{1/2} \log(X)$ for some positive constant c. Nevertheless, Klüners [32] did prove that Malle's conjecture holds for $G = C_2 \wr H$, where H is a transitive subgroup of $S_{n/2}$ such that over F, there is at least one extension of F with Galois group H and $N_{F,n/2}(H, X) = O_{F,H,\epsilon}(X^{1+\epsilon})$ for every $\epsilon > 0$. Building on [18], Klüners [32] computed

$$c_{F,n}(C_2 \wr H) = \sum_{\substack{[K:F]=n/2 \\ \mathrm{Gal}(\tilde{K}/F)=H}} \frac{1}{2^{r_2(K)}\,\mathrm{Nm}_{F/\mathbb{Q}}(\mathrm{Disc}(K)^2)} \cdot \frac{\zeta_K(1)}{\zeta_K(2)}.$$

In particular, when $H = C_2$, Klüners [32] proved the generalization of (2) in the D_4-quartic case from over $\mathbb{Q}$ to over a general base number field F without a power-saving error term. Furthermore, Klüners [33] verified Malle's Conjecture for generalized quaternion groups of order $4m$, determining that $a_{F,4m}(Q_{4m}) = 1/2m$ and $b_{F,4m}(Q_{4m}) = 0$.

In 2015, Türkelli [55] formulated function field heuristics for $N_{F,n}(G, X)$ (i.e., when F is taken to be a function field $\mathbb{F}_q(t)$ instead of a number field) and came up with a way to salvage Malle's conjecture that is consistent with Klüners's counterexample. Namely, consider first for a group G the set $\mathcal{N}(G)$ of normal subgroups N in G such that G/N is abelian and $a_n(N) = a_n(G)$ (and where $a_n(\{1\}) = 0$). Define

$$b_{F,n}(G, N) = \max_{\varphi \in \mathrm{Sur}(\mathrm{Gal}(F(\mu_\infty)/F), G/N))} b_{F,n}(G, N, \varphi),$$

where $b_{F,n}(G, N, \varphi)$ is the number of twisted F-orbits of N-conjugacy classes of minimal index, where the action of $\mathrm{Gal}(\overline{F}/F)$ is defined now with a twist: For a conjugacy class C of N, $\sigma(C) = \varphi(\mathrm{Res}(\sigma))^{-1} \cdot C^{\mu(\sigma)}$, where $\mathrm{Res} : \mathrm{Gal}(\overline{F}/F) \to \mathrm{Gal}(F(\mu_\infty)/F)$ and $\varphi(\mathrm{Res}(\sigma))^{-1}$ acts by conjugation on the N-conjugacy class $C^{\mu(\sigma)}$.[2] Then, we can define

$$b_{F,n}(G) = -1 + \max_{N \in \mathcal{N}(G)} b_{F,n}(G, N)$$

as the salvage to $b_{F,n}^{\mathrm{Malle}}(G)$.

[2] It is worth pointing out that in Malle's original formulation, he only considered the case of $N = G$, and so $b_{F,n}^{\mathrm{Malle}}(G) = b_{F,n}(G, G)$. This only coincides with $b_{F,n}(G)$ when $b_{F,n}(G, G)$ is the maximum over all $b_{F,n}(G, N)$ for $N \in \mathcal{N}(G)$.

Türkelli's heuristics [55] seemed to contradict earlier predictions made by Ellenberg–Venkatesh [24] for counting extensions of the function field $\mathbb{F}_q(t)$ that agree with Malle's original conjecture. However, Ellenberg–Venkatesh focused on extensions that do not contain any nontrivial constant intermediate extensions, while Türkelli [55] allowed for those to be counted in their heuristic, which has a stronger parallel to Malle's conjecture where extensions containing cyclotomic subfields are included in the asymptotics. (On the geometric side, Türkelli [55] includes connected, but not geometrically connected, covers of the projective line over $\mathbb{F}_q$, while Ellenberg–Venkatesh [24] require covers to be geometrically connected as well.)

More recently, Wang [59] has made substantial progress on techniques for proving Malle's conjecture in compositums of number fields. The first application of these methods was to prove Malle's Conjecture for certain direct product groups of the form $G = S_n \times A$, including the case when $n = 3$ and A is an odd abelian group, the case when $n = 4$ and A is an abelian group of order coprime to 6, and the case when $n = 5$ and A is an abelian group of order relatively prime to 30 (see [58]). Later generalizations came from joint work of Masri–Thorne–Tsai–Wang [42] to proving Malle's conjecture for $G = S_n \times A$, where A is any abelian group and $n = 3, 4, 5$.

Most recently, Fouvry–Koymans [27] proved Malle's conjecture for nonic Heisenberg extensions by estimating a certain character sum. They determined that there exists a constant $c_{\mathbb{Q},9}(\mathrm{Heis}_3) > 0$, which they determine explicitly (see Pgs. 17–18 of [27]), such that (1) holds with

$$a_{\mathbb{Q},9}(\mathrm{Heis}_3) = 1/4, \;\; b_{\mathbb{Q},9}(\mathrm{Heis}_3) = 0.$$

Additionally, Koymans–Pagano [36] proved that for any number field F and any finite nilpotent group G of cardinality n such that all elements of order p are central, where p is the smallest prime dividing n, there exists a constant $c_{F,n}(G) > 0$, and determined

$$a_n(G) = \frac{1}{n}\left(1 - \frac{1}{p}\right), \;\; b_{F,n}(G) = \frac{n_p}{[F(\zeta_p) : F]} - 1,$$

where n_p denotes the number of elements of G of order p.

At the time of writing this chapter, the authors have heard of forthcoming work of Alberts–Lemke Oliver–Wang–Wood [3] in which the authors develop inductive counting methods to prove Malle's conjecture for groups G for which all elements in G of sufficiently small index are contained in a normal subgroup that is either abelian or generated by conjugates of a single subgroup of G isomorphic to S_3.

Additionally, work-in-progress at the time of writing this chapter includes Shankar–Varma [51] for verifying Malle's conjecture in the Galois octic case by determining $a_{\mathbb{Q},8}(D_4) = 1/4$ and $b_{\mathbb{Q},8}(D_4) = 2$ and computing an Euler product for $c_{\mathbb{Q},8}(D_4)$.

We do not claim that this is a full history of the study of number field asymptotics; we tried to focus on the results that specifically proved (the strong form of) Malle's conjecture and study of the constants $c_{F,n}(G)$ in those cases. Much work has been developed in proving the weak form of Malle's conjecture [39], i.e., that there exist positive real constants $c^{(1)}_{F,n}(G)$ and $c^{(2)}_{F,n}(G, \epsilon)$ such that

$$c^{(1)}_{F,n}(G)X^{a(G)} \leq N_{F,n}(G, X) < c^{(2)}_{F,n}(G, \epsilon)X^{a(G)+\epsilon}$$

for large enough X and for all $\epsilon > 0$ (see [1, 2, 34]). Many analogous results have been obtained over function fields, the case of quartic D_4-extensions by Keliher [29], the case of cubic S_3-extensions studied by Wood [60], work of Ellenberg–Venkatesh–Westerland [26] on degree p D_p-extensions for odd primes p, the case of C_2-extensions (with added local conditions) studied by Kurlberg–Rudnick [37], the case of degree p C_p-extensions (with local conditions) for odd primes p studied by Bucur–David–Feigon–Lalín [15] and Bucur–David–Feigon–Kaplan–Lalín–Ozman–Wood [14].

In addition, much work has been done in proving and improving upper and lower bounds for $N_{F,n}(G, X)$ (see [5, 11, 23, 25, 31, 35, 41, 43, 46, 48, 50, 52]) as well as other important work on building methods and techniques for answering these questions.

1.4 Summary of Our Methods

Fix a number field F of degree d, and let $\overline{F}$ denote a fixed algebraic closure of F. In the sequel, all extensions of F will be assumed to be contained in $\overline{F}$.

Recall that $N_{F,n}(G, X)$ denotes the number of F-isomorphism classes of degree n extensions L over F such that the normal closure $\tilde{L}$ of L over F has Galois group $\mathrm{Gal}(\tilde{L}/F) \cong G$ and $\mathrm{Nm}_{F/\mathbb{Q}}(\mathrm{Disc}(L/F)) \leq X$, and we say two extensions L and L' of a number field F are *F-isomorphic* if there exists a field isomorphism between L and L' that restricts to the identity on F.

We are interested in counting the number of (F-isomorphism classes of) quartic extensions L whose normal closure over F has Galois group D_4, the symmetries of a square, when such extensions are ordered by the norms of their relative discriminants over F. We will call such an L a *D_4-quartic extension* over F, and let $\mathrm{Disc}(L/F)$ denote the relative discriminant. Denote by $\mathcal{F}_{F,4}(D_4)$ the set of F-isomorphism classes of D_4-quartic extensions over F.

To obtain the order of magnitude for D_4-quartic extensions over a general base number field F with discriminant norm bounded by X, we must understand the poles of the Dirichlet series $\Phi_{F,4}(D_4, s)$ associated to the count $N_{F,4}(D_4, X)$:

$$\Phi_{F,4}(D_4, s) = \sum_{L \in \mathcal{F}_{F,4}(D_4)} \frac{1}{\mathrm{Nm}_{F/\mathbb{Q}}(\mathrm{Disc}(L/F))^s},$$

where $\mathcal{F}_{F,4}(D_4)$ is defined as in the previous paragraph. Perron's formula implies that

$$N_{F,4}(D_4, X) = \frac{1}{2\pi i}\int_{\Re(s)=2} \Phi_{F,4}(D_4, s)\frac{X^s}{s}\,ds. \tag{4}$$

Determining the residue of the poles of the above integral after shifting the contours of integration gives the main term contribution for $N_{F,4}(D_4, X)$, and we additionally obtain (a power-saving) error term from bounding the contribution of the integrals on the other sides of the contour of integration.

To study the residue, we follow the same strategy as Cohen–Diaz y Diaz–Olivier utilized in [18, Corollary 1.4] to prove the case $F = \mathbb{Q}$. Specifically, we will relate the count $N_{F,4}(D_4, X)$ to the count of quadratic extensions of quadratic extensions of F and therefore relate the Dirichlet series $\Phi_{F,4}(D_4, s)$ to a Dirichlet series

$$\frac{1}{2}\Phi(s) = \frac{1}{2}\sum_{[K:F]=2}\frac{1}{\mathrm{Nm}(\mathrm{Disc}(K/F))^{2s}}\left(1+\sum_{[L:K]=2}\frac{1}{\mathrm{Nm}(\mathrm{Disc}(L/K))^{2s}}\right),$$

which has the same residue at the simple pole at $s = 1$ as $\Phi_{F,4}(D_4, s)$. We can then use Cohen–Diaz y Diaz–Olivier's results on counting quadratic extensions of a number field to give a summation formula for the constant that verifies [18, Remark 4].

2 Notation and Background

In this section, we introduce the notation and background information we will be using throughout the chapter.

For any number field E, let $\mathcal{O}_E$ denote its ring of integers, let $r_1(E)$ denote the number of real embeddings of E, and let $r_2(E)$ denote the number of pairs of complex conjugate embeddings.

2.1 L-Functions Associated to a Number Field

For $s \in \mathbb{C}$ such that $\Re(s) > 1$, the Dedekind zeta function of a number field E is given by the Euler product running over nonzero primes ideals of $\mathcal{O}_E$

$$\zeta_E(s) = \prod_{\mathfrak{p}\subseteq\mathcal{O}_E}\frac{1}{1-\mathrm{Nm}_{E/\mathbb{Q}}(\mathfrak{p})^{-s}}.$$

Hecke proved that $\zeta_E(s)$ has meromorphic continuation to the entire complex plane with only a simple pole at $s = 1$.

For a character χ of a ray class group $\mathcal{C}\ell_{\mathfrak{m}}(F)$ of conductor $\mathfrak{m}$ of E, let $L_E(s, \chi)$ denote the L-function associated to the character χ, namely for $\Re(s) > 1$,

$$L_E(s, \chi) = \sum_{\substack{\mathfrak{a} \subset \mathcal{O}_E \\ (\mathfrak{a}, \mathfrak{m})=1}} \frac{\chi(\mathfrak{a})}{\mathrm{Nm}_{E/\mathbb{Q}}(\mathfrak{a})^s}.$$

Lemma 2.1.1 *If $\chi = \chi_0$ is the trivial character, then $L_E(s, \chi_0)$ has a simple pole at $s = 1$. Otherwise, $L_E(s, \chi)$ is analytic everywhere.*

Proof We see that

$$L_E(s, \chi_0) = \zeta_E(s) \prod_{\mathfrak{p} | \mathfrak{m}} \left(1 - \mathrm{Nm}_{E/\mathbb{Q}}(\mathfrak{p})^{-s}\right), \tag{5}$$

and the conclusion follows from the properties of $\zeta_E(s)$ as the product over $\mathfrak{p} \mid \mathfrak{m}$ is finite. For nontrivial χ, the result follows from [45, Chapter VII, Theorem 8.5]. □

If there is a pole at $s = s_0$, and we write $L_E(s_0, \chi)$ or $\zeta_E(s_0)$, we are abusing notation and referring to the residue at the pole.

The Lindelöf hypothesis would give that

$$\zeta_E\left(\frac{1}{2} + it\right) \ll t^{\epsilon}, \tag{6}$$

and a similar bound holds for $s = \sigma + it$ and $\sigma > \frac{1}{2}$.

2.2 *Counting G-Extensions of a Number Field and Associated Dirichlet Series*

Fix a finite transitive subgroup G of S_n for some $n \geq 2$. Let $\mathcal{F}_{E,m}(G)$ denote the set of E-isomorphism classes of extensions E'/E of degree m (in a fixed algebraic closure of E) such that the Galois group of the Galois closure of E'/E is isomorphic to G.

Denote by $N_{E,m}(G, X)$ the number of elements of $\mathcal{F}_{E,m}(G)$ whose discriminant is bounded by X; namely

$$N_{E,m}(G, X) = \#\{E' \in \mathcal{F}_{E,m}(G);\ \mathrm{Nm}_{E/\mathbb{Q}}(\mathrm{Disc}(E'/E)) \leq X\}.$$

In order to obtain asymptotic formulas for $N_{E,m}(G, X)$, we need to understand the analytic properties of the function $\Phi_{E,m}(G, s)$ defined by

$$\Phi_{E,m}(G,s) = \sum_{E' \in \mathcal{F}_{E,m}(G)} \frac{1}{\mathrm{Nm}_{E/\mathbb{Q}}(\mathrm{Disc}(E'/E))^s}.$$

Using Perron's formula, we will obtain the main term in the asymptotic formula for $N_{E,m}(G, X)$ from the right-most pole of $\Phi_{E,m}(G, s)$, while the error term will be dictated by how much information we have about the meromorphic continuation of $\Phi_{E,m}(G, s)$.

3 Counting Quadratic Extensions of Quadratic Extensions of a Number Field

For the rest of the chapter, we fix a number field F and set $d = [F : \mathbb{Q}]$. We are interested in computing an asymptotic formula for $N_{F,4}(D_4, X)$, and our strategy is to study $\Phi_{F,4}(D_4, s)$ by relating it to the Dirichlet series associated to counting quadratic extensions of quadratic extensions of F of bounded discriminant norm. The equivalent of $\Phi_{F,m}(G, s)$ for quadratic extensions of quadratic extensions is the function

$$\tilde{\Phi}(s) = \sum_{[K:F]=2} \sum_{[L:K]=2} \frac{1}{\mathrm{Nm}_{F/\mathbb{Q}}(\mathrm{Disc}(L/F))^s}. \tag{7}$$

We will first study the relationship of $\Phi_{F,4}(D_4, s)$ and $\tilde{\Phi}(s)$, showing that they only differ by a holomorphic function. We will then study the poles and residues of $\tilde{\Phi}(s)$ by instead considering a slightly altered Dirichlet series $\Phi(s)$ that again only differs from $\tilde{\Phi}(s)$ (and therefore also from $\Phi_{F,4}(D_4, s)$) by a holomorphic function.

3.1 Quadratic Extensions of Quadratic Extensions

For a quadratic extension K of F, let $\mathcal{F}_2(K)$ denote the K-isomorphism classes of quadratic extensions L of K. There are three possibilities for the extension L/F in this case:

- L/F is a degree 4 non-Galois extension whose normal closure has Galois group D_4 over F. These are the extensions we want to count.
- L/F is a Galois extension with Galois group C_4, the cyclic group of order 4.
- L/F is a Galois extension with Galois group $V_4 \simeq \mathbb{Z}/2 \times \mathbb{Z}/2$, the Klein four group.

By Cohen et al. [18, Corollary 2.3(3)],

$$\sum_{K\in\mathcal{F}_2(F)} \sum_{\substack{L\in\mathcal{F}_2(K)\\ \mathrm{Nm}_{F/\mathbb{Q}}(\mathrm{Disc}(L/F))\le X}} 1 = 2N_{F,4}(D_4,X)+N_{F,4}(C_4,X)+3N_{F,4}(V_4,X). \tag{8}$$

To see this note that for a fixed quadratic extension K of F, conjugate D_4-extensions of F will be counted separately by the sum on the left, and every D_4-extension contains a single quadratic subextension (shared with its unique F-conjugate). The sum on the left counts C_4-extensions up to F-isomorphism just once since these extensions are Galois and contain a single quadratic subextension. Each V_4-extension L/F contains three quadratic subfields, so the sum on the left is counting each V_4-extension up to F-isomorphism three times.

Hence, we can express the function $\Phi_{F,4}(D_4,s)$ we are interested in as

$$\Phi_{F,4}(D_4,s) = \frac{1}{2}\left(\tilde{\Phi}(s) - \Phi_{F,4}(C_4,s) - 3\Phi_{F,4}(V_4,s)\right). \tag{9}$$

On the other hand, note that by the relative discriminant formula, if $K \in \mathcal{F}_2(F)$ and $L \in \mathcal{F}_2(K)$, then $\mathrm{Disc}(L/F) = \mathrm{Disc}(K/F)^2\,\mathrm{Nm}_{K/F}(\mathrm{Disc}(L/K))$. Thus, we can write

$$\begin{aligned}
\tilde{\Phi}(s) &= \sum_{[K:F]=2}\sum_{[L:K]=2} \frac{1}{\mathrm{Nm}_{F/\mathbb{Q}}(\mathrm{Disc}(K/F))^{2s}\,\mathrm{Nm}_{K/\mathbb{Q}}(\mathrm{Disc}(L/K))^{s}}\\
&= \sum_{[K:F]=2} \frac{1}{\mathrm{Nm}_{F/\mathbb{Q}}(\mathrm{Disc}(K/F))^{2s}} \sum_{[L:K]=2} \frac{1}{\mathrm{Nm}_{K/\mathbb{Q}}(\mathrm{Disc}(L/K))^{s}}\\
&= \sum_{[K:F]=2} \frac{\Phi_{K,2}(C_2,s)}{\mathrm{Nm}_{F/\mathbb{Q}}(\mathrm{Disc}(K/F))^{2s}}.
\end{aligned} \tag{10}$$

3.2 Counting Using Dirichlet Series

The previous section indicates that in order to study $\tilde{\Phi}(s)$, we will be compelled to deal with not just $\Phi_{F,4}(D_4,s)$, but also $\Phi_{F,4}(C_4,s)$ and $\Phi_{F,4}(V_4,s)$ that appears in (9), which were first considered by Wright [61]. By Wright [61, Theorem I.1], it is known that $\Phi_{F,4}(C_4,s)$ and $\Phi_{F,4}(V_4,s)$ are absolutely convergent for $\Re(s) > 1/2$. Using these functions, Wright proved that

$$N_{F,4}(C_4,X) = O\left(X^{1/2}\right)$$

and

$$N_{F,4}(V_4, X) = O\left(X^{1/2}\log^2(X)\right).$$

The functions $\Phi_{F,4}(C_4, s)$ and $\Phi_{F,4}(V_4, s)$ can also be rewritten in terms of L-functions as given below.

Proposition 3.2.1 ([17, Theorem 4.3]) *We have that*

$$\Phi_{F,4}(C_4, s) = \frac{2^{r_1(F)+r_2(F)-1}}{4^{2[F:\mathbb{Q}]s}\zeta_F(2s)} \sum_{\mathfrak{c}|4\mathcal{O}_F} \mathrm{Nm}(\mathfrak{c})^{2s} \sum_{\chi} L_F(2s, \chi) F_{\mathfrak{c},\chi}(s), \tag{11}$$

where the sum over χ is over quadratic characters of $\mathrm{Cl}_{\mathfrak{c}}(F)$, the sum over $\mathfrak{c}$ is over ideals dividing (4), and where $F_{\mathfrak{c},\chi}(s)$ is defined at the bottom of page 499 in [17] and converges for $\Re(s) > 1/3$. (Note that $F_{\mathfrak{c},\chi}(s)$ is given in terms of a sum over quadratic extensions k/F that are embeddable into a C_4-extension.)

Proposition 3.2.2 ([16, Theorem 5.1]) *We have that*

$$\Phi_{F,4}(V_4, s) = \frac{1}{6}\Psi(s) - \frac{1}{2}\Phi_{F,2}(C_2, 2s) - \frac{1}{6},$$

where

$$\begin{aligned}\Psi(s) = {} & \frac{2^{2r_1(F)+2r_2(F)}}{2^{6[F:\mathbb{Q}]s}} \\ & \times \sum_{\substack{\mathfrak{n}|2\mathcal{O}_F \\ \mathfrak{m}|2\mathcal{O}_F}} \mathrm{Nm}(\mathfrak{n}\mathfrak{m})^{2s-1} \sum_{\mathfrak{c}|\mathfrak{n}} \frac{\mu_F(\mathfrak{n}/\mathfrak{c})}{\mathrm{Nm}(\mathfrak{n}/\mathfrak{c})} \mathrm{Nm}((\mathfrak{m},\mathfrak{c}))^{2s} \\ & \times \prod_{\substack{\mathfrak{p}\ s.t. \\ 1\le v_{\mathfrak{p}}(\mathfrak{m})\le v_{\mathfrak{p}}(\mathfrak{c})}} \left(1-\mathrm{Nm}(\mathfrak{p})^{-4s}\right) \prod_{\mathfrak{p}|\mathfrak{m}/(\mathfrak{m},\mathfrak{c})} \left(1-\mathrm{Nm}(\mathfrak{p})^{-2s}\right) \\ & \times \sum_{\chi,\psi} \sum_{\mathfrak{q}|\mathfrak{m}/(\mathfrak{m},\mathfrak{c})} L_F(2s, \chi_{1,\mathfrak{q}}\chi_{2,\mathfrak{q}}) L_F(2s, \chi_{1,\mathfrak{q}}\psi) L_F(2s, \chi_{2,\mathfrak{q}}\psi) H_{\mathfrak{q}}(s, \chi, \psi),\end{aligned} \tag{12}$$

where $\mathfrak{n}$, $\mathfrak{m}$, and $\mathfrak{c}$ run over integral ideals, $\mathfrak{p}$ and $\mathfrak{q}$ run over prime ideals, χ and ψ run over all quadratic characters of $\mathrm{Cl}_{\mathfrak{m}}(F)$ and $\mathrm{Cl}_{\mathfrak{n}}(F)$, respectively, and $H_{\mathfrak{q}}(\chi, \psi, s)$ is given by an Euler product that converges for $\Re(s) > 1/4$.

Corollary 3.2.3 *The poles and the residues of $\Phi_{F,4}(D_4, s)$ are the same as the poles of $\frac{1}{2}\tilde{\Phi}(s)$.*

Proof This follows directly from Eq. (9) and work of Wright (see [61, Theorem I.1]) or Propositions 3.2.1 and 3.2.2. □

It will be more convenient to work with the function

$$\Phi(s) = \sum_{[K:F]=2} \frac{1}{\mathrm{Nm}(\mathrm{Disc}(K/F))^{2s}} \left(1 + \Phi_{K,2}(C_2, s)\right) \tag{13}$$

rather than the function $\tilde{\Phi}(s)$ defined in (7). From [18, Theorem 1.1], we have

$$1 + \Phi_{K,2}(C_2, s) = \frac{2^{-r_2(K)}}{\zeta_K(2s)} \sum_{\mathfrak{c}|2\mathcal{O}_K} \mathrm{Nm}(2/\mathfrak{c})^{1-2s} \sum_{\chi} L_K(s, \chi), \tag{14}$$

where $\mathfrak{c}$ runs over integral ideals and χ runs over all the quadratic characters of the ray class group $\mathrm{Cl}_{\mathfrak{c}^2}(K)$. Note that we would not expect a summation formula in terms of L-functions like (14) to exist if we did not add 1 to $\Phi_{K,2}(C_2, s)$, and so computing the residue of $\tilde{\Phi}(s)$ would be much more difficult to do directly as we would not be able to utilize the study of $L_K(s, \chi)$.

By Eq. (9), $\Phi(s)$ can be expressed as

$$\Phi(s) = 2\Phi_{F,4}(D_4, s) + \Phi_{F,2}(C_2, 2s) + \Phi_{F,4}(C_4, s) + 3\Phi_{F,4}(V_4, s). \tag{15}$$

From Eq. (14), we also have

$$\Phi_{F,2}(C_2, s) = -1 + \frac{2^{-r_2(F)}}{\zeta_F(2s)} \sum_{\mathfrak{c}|2\mathcal{O}_F} \frac{\mathrm{Nm}(2/\mathfrak{c})}{\mathrm{Nm}(2/\mathfrak{c})^{2s}} \sum_{\chi} L_F(s, \chi), \tag{16}$$

where $\mathfrak{c}$ runs over all integral ideals of F dividing 2 and χ runs over all quadratic characters of the ray class group $\mathcal{C}\ell_{\mathfrak{c}^2}(F)$ modulo $\mathfrak{c}^2$.

We now focus on $\Phi(s)$. It follows from (14) and Lemma 2.1.1 that $1 + \Phi_{K,2}(C_2, s)$ is meromorphic in the complex plane and has a simple pole at $s = 1$. If we denote by R the residue of $\Phi(s)$ at $s = 1$, by using the convexity bound

$$L_K(s, \chi) \ll \mathrm{Nm}(\mathrm{Disc}(K/F))^{\frac{1}{2}},$$

combined with Eqs. (13) and (14), it follows that $\Phi(s) - R$ is absolutely convergent for $\Re(s) \geq 3/4 + \epsilon$. From Eqs. (15) and (16), in conjunction with Propositions 3.2.1 and 3.2.2, we also get the following.

Proposition 3.2.4 *The function $\Phi_{F,4}(D_4, s)$ has meromorphic continuation to the whole complex plane and has a simple pole at $s = 1$. Furthermore, $\frac{1}{2}\Phi(s)$ and $\Phi_{F,4}(D_4, s)$ have the same residue at the simple pole at $s = 1$. If $\Phi(1)$ denotes the residue of $\Phi(s)$ at the pole $s = 1$, then $\Phi_{F,4}(D_4, s) - \frac{\Phi(1)}{2}$ is absolutely convergent for $\Re(s) \geq 3/4 + \epsilon$.*

4 Proof of the Main Theorem

We will now prove Theorem 1, originally stated in Remark 4 of [18, Corollary 6.1]. We denote $c_{F,4}(D_4)$ the constant in the main term, i.e.,

$$c_{F,4}(D_4) = \sum_{[K:F]=2} \frac{1}{2^{r_2(K)+1}\,\mathrm{Nm}(\mathrm{Disc}(K/F))^2} \frac{\zeta_K(1)}{\zeta_K(2)}.$$

Recall that $\zeta_K(1)$ denotes the value of the residue of the Dedekind zeta function of K at $s = 1$. With this notation, we need to prove that

$$N_{F,4}(D_4, X) = c_{F,4}(D_4)X + O(X^{\frac{3}{4}+\epsilon}),$$

where $d = [F : \mathbb{Q}]$.

By Perron's formula, we have that

$$N_{F,4}(D_4, X) = \frac{1}{2\pi i} \int_{\Re(s)=2} \Phi_{F,4}(D_4, s) \frac{X^s}{s}\, ds. \tag{17}$$

When we shift the contour to the left to the line $\Re(s) = 1 + \epsilon$, we do not encounter any poles. So we can rewrite

$$N_{F,4}(D_4, X) = \frac{1}{2\pi i} \int_{\Re(s)=1+\epsilon} \Phi_{F,4}(D_4, s) \frac{X^s}{s}\, ds.$$

Using standard contour integration, we get that

$$N_{F,4}(D_4, X) = \frac{1}{2\pi i} \int_{1+\epsilon-iT}^{1+\epsilon+iT} \Phi_{F,4}(D_4, s) \frac{X^s}{s}\, ds + O\left(\frac{X^{1+\epsilon}}{T}\right), \tag{18}$$

and we will choose T suitably. We shift the line of integration to $\Re(s) = 3/4+\epsilon$ and encounter the simple pole at $s = 1$. The residue of the pole at $s = 1$ will contribute the main term that will be of size X, and the error term will come from bounding the contribution of the integrals on the other three sides of the contour of integration.

Next we bound the integral on the two horizontal lines of the contour and on the vertical line from $3/4 + \epsilon - iT$ and $3/4 + \epsilon + iT$. The integral on the horizontal lines will be bounded by $X^{1+\epsilon}/T$, and for the integral on the vertical line, we have

$$\frac{1}{2\pi i} \int_{3/4+\epsilon-iT}^{3/4+\epsilon+iT} \Phi_{F,4}(D_4, s) \frac{X^s}{s}\, ds \ll X^{3/4+\epsilon}, \tag{19}$$

where we have used Proposition 3.2.4.

Note that using the Lindelöf bound (6) to bound $\Phi(s)$ yields an upper bound of size $X^{1/4+\epsilon}$ for the vertical integrals.

Finally, we evaluate the residue of the simple pole at $s = 1$, which will give us the constant $c_{F,4}(D_4)$, following [18]. Using (15) and Proposition 3.2.4, we see that, since the difference between $\Phi_{F,4}(D_4, s)$ and $\frac{1}{2}\Phi(s)$ is holomorphic for $\Re(s) > 1/2$, the two functions $\Phi_{F,4}(D_4, s)$ and $\frac{1}{2}\Phi(s)$ have the same residue at $s = 1$. From Eqs. (5), (13), (14), and (15), it follows that the value of the residue at $s = 1$ is equal to

$$\frac{1}{2}\sum_{[K:F]=2} \frac{2^{-r_2(K)}\zeta_K(1)}{\mathrm{Nm}(\mathrm{Disc}(K/F))^2\zeta_K(2)} \sum_{\mathfrak{c}|2\mathcal{O}_K} \mathrm{Nm}(2/\mathfrak{c})^{-1}\prod_{\mathfrak{p}|\mathfrak{c}}(1-\mathrm{Nm}(\mathfrak{p})^{-1}), \tag{20}$$

where $\mathfrak{c}$ runs over integral ideals of $\mathcal{O}_K$ and $\mathfrak{p}$ runs over prime ideals of $\mathcal{O}_K$. Combining (18), (19), (20), and Lemma 4.0.1 below and choosing $T = X^{1/4}$, the conclusion follows.

Note that if we assume the Lindelöf hypothesis, then using the observation following (19), it follows that we can choose $T = X^{3/4}$ and then the improved bound in the second part of Theorem 1 follows.

Lemma 4.0.1 *For a number field E,*

$$\sum_{\mathfrak{c}|2\mathcal{O}_E} \mathrm{Nm}(2/\mathfrak{c})^{-1}\prod_{\mathfrak{p}|\mathfrak{c}}(1-\mathrm{Nm}(\mathfrak{p})^{-1}) = 1,$$

where $\mathfrak{c}$ runs over integral ideals of $\mathcal{O}_E$ and $\mathfrak{p}$ runs overs prime ideals of $\mathcal{O}_E$.

Proof Following §3.3 of [18], the above formula can be proven by writing

$$\sum_{\mathfrak{c}|2}\mathrm{Nm}(2/\mathfrak{c})^{-1}\prod_{\mathfrak{p}|\mathfrak{c}}(1-\mathrm{Nm}(\mathfrak{p})^{-1}) = \frac{1}{\mathrm{Nm}(2)}\sum_{\mathfrak{c}|2}\mathrm{Nm}(\mathfrak{c})\prod_{\mathfrak{p}|\mathfrak{c}}(1-\mathrm{Nm}(\mathfrak{p})^{-1})$$

$$= \frac{1}{\mathrm{Nm}(2)}\prod_{\mathfrak{p}|2}\left(1+\sum_{j=1}^{v_{\mathfrak{p}}(2)}\mathrm{Nm}(\mathfrak{p}^j)(1-\mathrm{Nm}(\mathfrak{p})^{-1})\right),$$

where $v_{\mathfrak{p}}(2)$ is the valuation of 2 at $\mathfrak{p}$. The previous formula simplifies to

$$\sum_{\mathfrak{c}|2}\mathrm{Nm}(2/\mathfrak{c})^{-1}\prod_{\mathfrak{p}|\mathfrak{c}}(1-\mathrm{Nm}(\mathfrak{p})^{-1}) = \frac{1}{\mathrm{Nm}(2)}\prod_{\mathfrak{p}|2}\mathrm{Nm}(\mathfrak{p})^{v_{\mathfrak{p}}(2)} = 1.$$

□

Acknowledgments We would like to thank Brandon Alberts, Henri Cohen, Roger Heath-Brown, Erik Holmes, Jordan Ellenberg, Ofir Gorodetsky, Dick Gross, Kiran S. Kedlaya, Gunter Malle, Robert Lemke Oliver, Arul Shankar, Jiuya Wang, Melanie Matchett Wood, and the referee for helpful discussions. This work was initiated as part of the Women in Numbers (WIN) 5 workshop.

We thank the workshop organizers, Wei Ho and Renate Scheidler, for providing the impetus for this project, and we thank NSF for making WIN5 possible in the form of the grant DMS-2012061.

Alina Bucur was supported by Simons Foundation grant 524015 and NSF grant DMS-2002716. Alexandra Florea was supported by NSF grant DMS-2101769.

References

1. B. Alberts, Statistics of the first Galois cohomology group: a refinement of Malle's conjecture. Algebra Number Theory **15**(10), 2513–2569 (2021). ISSN: 1937-0652. https://doi.org/10.2140/ant.2021.15.2513
2. B. Alberts, The weak form of Malle's conjecture and solvable groups. Res. Number Theory **6**(1), Paper No. 10, 23 (2020). ISSN: 2522-0160. https://doi.org/10.1007/s40993-019-0185-7
3. B. Alberts, R.J. Lemke Oliver, J. Wang, M.M. Wood, Untitled Manuscript. Preprint (2022)
4. S.A. Altuğ, A. Shankar, I. Varma, K.H. Wilson, The number of D4-fields ordered by conductor. J. Eur. Math. Soc. (JEMS) **23**(8), 2733–2785 (2021). ISSN: 1435-9855. https://doi.org/10.4171/jems/1070
5. T.C. Anderson, A. Gafni, K. Hughes, R.J. Lemke Oliver, D. Lowry-Duda, F. Thorne, J. Wang, R. Zhang, *Improved Bounds on Number Fields of Small Degree* (2022). https://doi.org/10.48550/ARXIV.2204.01651. https://arxiv.org/abs/2204.01651
6. A.M. Baily, On the density of discriminants of quartic fields. J. Reine Angew. Math **315**, 190–210 (1980). ISSN: 0075-4102. https://doi.org/10.1515/crll.1980.315.190
7. K. Belabas, É. Fouvry, Discriminants cubiques et progressions arithmétiques. Int. J. Number Theory **6**(7), 1491–1529 (2010). ISSN: 1793-0421. https://doi.org/10.1142/S1793042110003605
8. M. Bhargava, Mass formulae for extensions of local fields, and conjectures on the density of number field discriminants. Int. Math. Res. Not. IMRN 17 **2007**, Art. ID rnm052, 20 (2007). ISSN: 1073-7928. https://doi.org/10.1093/imrn/rnm052
9. M. Bhargava, The density of discriminants of quartic rings and fields. Ann. Math. (2) **162**(2), 1031–1063 (2005). ISSN: 0003-486X. https://doi.org/10.4007/annals.2005.162.1031
10. M. Bhargava, The density of discriminants of quintic rings and fields. Ann. of Math. (2) **172**(3), 1559–1591 (2010). ISSN: 0003-486X. https://doi.org/10.4007/annals.2010.172.1559
11. M. Bhargava, A. Shankar, X. Wang, *An Improvement on Schmidt's Bound on the Number of Number Fields of Bounded Discriminant and Small Degree* (2022). https://doi.org/10.48550/ARXIV.2204.01331. https://arxiv.org/abs/2204.01331
12. M. Bhargava, A. Shankar, X. Wang, *Geometry-of-numbers Methods Over Global Fields I: Prehomogeneous Vector Spaces* (2015). arXiv: `1512.03035 [math.NT]`
13. M. Bhargava, M.M. Wood, The density of discriminants of S_3-sextic number fields. Proc. Am. Math. Soc. **136**(5), 1581–1587 (2008). ISSN: 0002-9939. https://doi.org/10.1090/S0002-9939-07-09171-X
14. A. Bucur, C. David, B. Feigon, N. Kaplan, M. Lalín, E. Ozman, M.M. Wood, The distribution of $\mathbb{F}_q$-points on cyclic 'ℓ-covers of genus g. Int. Math. Res. Not. IMRN **14**, 4297–4340 (2016). ISSN: 1073-7928. https://doi.org/10.1093/imrn/rnv279. https://doi-org.myaccess.library.utoronto.ca/10.1093/imrn/rnv279
15. A. Bucur, C. David, B. Feigon, M. Lalín, Statistics for ordinary Artin-Schreier covers and other p-rank strata. Trans. Am. Math. Soc. **368**(4), 2371–2413 (2016). ISSN: 0002-9947. https://doi.org/10.1090/tran/6410
16. H. Cohen, F.D. y Diaz, M. Olivier, Counting biquadratic extensions of a number field. Preprint (2022)
17. H. Cohen, F.D. y Diaz, M. Olivier, Counting cyclic quartic extensions of a number field. J. Théor. Nombres Bordeaux **17**(2), 475–510 (2005). ISSN: 1246-7405. URL: http://jtnb.cedram.org/item?id=JTNB_2005_17_2_475_0

18. H. Cohen, F.D. y Diaz, M. Olivier, Enumerating quartic dihedral extensions of $\mathbb{Q}$. Compos. Math. **133**(1), 65–93 (2002). ISSN: 0010-437X. https://doi.org/10.1023/A:1016310902973
19. H. Cohn, The density of abelian cubic fields. Proc. Am. Math. Soc. **5**, 476–477 (1954). ISSN: 0002-9939. https://doi.org/10.2307/2031963
20. B. Datskovsky, D.J. Wright, Density of discriminants of cubic extensions. J. Reine Angew. Math. **386**, 116–138 (1988). ISSN: 0075-4102. https://doi.org/10.1515/crll.1988.386.116
21. H. Davenport, H. Heilbronn, On the density of discriminants of cubic fields. II. Proc. R. Soc. London, Ser. A **322**(1551), 405–420 (1971). ISSN: 0962-8444. https://doi.org/10.1098/rspa.1971.0075
22. B.N. Delone, D.K. Faddeev, The theory of irrationalities of the third degree, in *Translations of Mathematical Monographs*, vol. 10 (American Mathematical Society, Providence, 1964), pp. xvi+509
23. E.P. Dummit, Counting G-extensions by discriminant. Math. Res. Lett. **25**(4), 1151–1172 (2018). ISSN: 1073-2780. https://doi.org/10.4310/MRL.2018.v25.n4.a5. https://doi-org.myaccess.library.utoronto.ca/10.4310/MRL.2018.v25.n4.a5
24. J.S. Ellenberg, A. Venkatesh, Counting extensions of function fields with bounded discriminant and specified Galois group, in *Geometric Methods in Algebra and Number Theory*, vol. 235 (Progress in Mathematics Birkhäuser Boston, Boston, 2005), pp. 151–168. https://doi.org/10.1007/0-8176-4417-2_7
25. J.S. Ellenberg, A. Venkatesh, The number of extensions of a number field with fixed degree and bounded discriminant. Ann. Math. (2) **163**(2), 723–741 (2006). ISSN: 0003-486X. https://doi.org/10.4007/annals.2006.163.723. https://doi-org.myaccess.library.utoronto.ca/10.4007/annals.2006.163.723
26. J.S. Ellenberg, A. Venkatesh, C. Westerland, Homological stability for Hurwitz spaces and the Cohen-Lenstra conjecture over function fields. Ann. Math. (2) **183**(3), 729–786 (2016). ISSN: 0003-486X. https://doi.org/10.4007/annals.2016.183.3.1
27. É. Fouvry, P. Koymans, *Malle's Conjecture for Nonic Heisenberg Extensions* (2021). https://doi.org/10.48550/ARXIV.2102.09465. https://arxiv.org/abs/2102.09465
28. C.F. Gauss. *Disquisitiones Arithmeticae*. Translated and with a preface by A.A. Clarke, Revised by W.C. Waterhouse, C. Greither, A.W. Grootendorst and with a preface by Waterhouse (Springer, New York, 1986), pp. xx+472. ISBN: 0-387-96254-9
29. D. Keliher, Enumerating D4 Quartics and a Galois Group Bias over Function Fields. preprint (2020). URL: https://arxiv.org/abs/2009.09274
30. J. Klüners, A counterexample to Malle's conjecture on the asymptotics of discriminants. C. R. Math. Acad. Sci. Paris **340**(6), 411–414 (2005). ISSN: 1631-073X. https://doi.org/10.1016/j.crma.2005.02.010
31. J. Klüners, Asymptotics of number fields and the Cohen-Lenstra heuristics. J. Théor. Nombres Bordeaux **18**(3), 607–615 (2006). ISSN: 1246-7405. URL: http://jtnb.cedram.org.myaccess.library.utoronto.ca/item?id=JTNB_2006__18_3_607_0
32. J. Klüners, The distribution of number fields with wreath products as Galois groups. Int. J. Number Theory **8**(3), 845–858 (2012). ISSN: 1793-0421. https://doi.org/10.1142/S1793042112500492
33. J. Klüners, *Über die Asymptotik von Zahlkörpern mit vorgegebener Galoisgruppe*. Habilitation thesis University Kassel (Shaker Verlag, Herzogenrath, 2005)
34. J. Klüners, G. Malle, Counting nilpotent Galois extensions. J. Reine Angew. Math. **572**, 1–26 (2004). ISSN: 0075-4102. https://doi.org/10.1515/crll.2004.050
35. J. Klüners, J. Wang, ℓ-torsion bounds for the class group of number fields with an 'ℓ-group as Galois group. Proc. Am. Math. Soc. **150**(7), 2793–2805 (2022). ISSN: 0002-9939. https://doi.org/10.1090/proc/15882
36. P. Koymans, C. Pagano, On Malle's conjecture for nilpotent groups, I (2021). https://doi.org/10.48550/ARXIV.2103.17223. https://arxiv.org/abs/2103.17223
37. P. Kurlberg, Z. Rudnick, The fluctuations in the number of points on a hyperelliptic curve over a finite field. J. Number Theory **129**(3), 580–587 (2009). ISSN: 0022-314X. https://doi.org/10.1016/j.jnt.2008.09.004

38. S. Mäki, On the density of abelian number fields. Ann. Acad. Sci. Fenn. Ser. A I Math. Dissertationes **54**, 104 (1985). ISSN: 0355-0087
39. G. Malle, On the distribution of Galois groups. J. Number Theory **92**(2), 315–329 (2002). ISSN: 0022-314X. https://doi.org/10.1006/jnth.2001.2713
40. G. Malle, On the distribution of Galois groups. II. Experiment. Math. **13**(2), 129–135 (2004). ISSN: 1058-6458. URL: http://projecteuclid.org/euclid.em/1090350928
41. G. Malle, B.H. Matzat, *Inverse Galois Theory*. Springer Monographs in Mathematics, 2nd edn. [MR1711577] (Springer, Berlin, 2018), pp. xvii+532. ISBN: 978-3-662-55419-7; 978-3-662-55420-3
42. R. Masri, F. Thorne, W.-L. Tsai, J. Wang, Malle's Conjecture for $G \times A$, with $G = S_3, S_4, S_5$ (2020). https://doi.org/10.48550/ARXIV.2004.04651. https://arxiv.org/abs/2004.04651
43. B.H. Matzat, A. Zeh-Marschke, Realisierung der Mathieugruppen M_{11} und M_{12} als Galoisgruppen über Q. J. Number Theory **23**(2), 195–202 (1986). ISSN: 0022-314X. https://doi.org/10.1016/0022-314X(86)90089-2. https://doi-org.myaccess.library.utoronto.ca/10.1016/0022-314X(86)90089-2
44. H.L. Montgomery, R.C. Vaughan. *Multiplicative Number Theory I: Classical Theory*. Cambridge Studies in Advanced Mathematics (Cambridge University, Cambridge, 2006). https://doi.org/10.1017/CBO9780511618314
45. J. Neukirch, Algebraic number theory. *Grundlehren der Mathematischen Wissenschaften*, vol. 322. Translated from the 1992 German original and with a note by Norbert Schappacher, With a foreword by G. Harder (Springer, Berlin, 1999), pp. xviii+571. ISBN: 3-540-65399-6. https://doi.org/10.1007/978-3-662-03983-0. https://doi-org.ezp-prod1.hul.harvard.edu/10.1007/978-3-662-03983-0
46. R.J.L. Oliver, F. Thorne, *Upper Bounds on Number Fields of Given Degree and Bounded Discriminant* (2020). https://doi.org/10.48550/ARXIV.2005.14110. https://arxiv.org/abs/2005.14110
47. B. Oriat, Étude arithmétique des corps cycliques de degré pr sur le corps des nombres rationnels. Enseign. Math. (2) **18**, 57–104 (1972). ISSN: 0013-8584
48. L.B. Pierce, C.L. Turnage-Butterbaugh, M.M. Wood, An effective Chebotarev density theorem for families of number fields, with an application to 'ℓ-torsion in class groups. Invent. Math. **219**(2), 701–778 (2020). ISSN: 0020-9910. https://doi.org/10.1007/s00222-019-00915-z.
49. M. Sato, T. Kimura, A classification of irreducible prehomogeneous vector spaces and their relative invariants. Nagoya Math. J. **65**, 1–155 (1977). ISSN: 0027-7630. URL: http://projecteuclid.org.myaccess.library.utoronto.ca/euclid.nmj/1118796150
50. W.M. Schmidt, Number fields of given degree and bounded discriminant, in *228. Columbia University Number Theory Seminar (New York, 1992)* (1995), pp. 4, 189–195
51. A. Shankar, I. Varma, Preprint (2022)
52. K.-y. Shih, On the construction of Galois extensions of function fields and number fields. Math. Ann. **207**, 99–120 (1974). ISSN: 0025-5831. https://doi.org/10.1007/BF01362150
53. T. Shintani, On Dirichlet series whose coefficients are class numbers of integral binary cubic forms. J. Math. Soc. Jpn. **24**, 132–188 (1972). ISSN: 0025-5645. https://doi.org/10.2969/jmsj/02410132
54. T. Shintani, On zeta-functions associated with the vector space of quadratic forms. J. Fac. Sci. Univ. Tokyo Sect. I A Math. **22**, 25–65 (1975)
55. S. Türkelli, Connected components of Hurwitz schemes and Malle's conjecture. J. Number Theory **155**, 163–201 (2015). ISSN: 0022-314X. https://doi.org/10.1016/j.jnt.2015.03.005
56. B.M. Urazbaev, The distribution of cyclic fields. Izv. Akad. Nauk Kazah. SSR Ser. Fiz.-Mat. **5**, 66–70, 93 (1977). ISSN: 0002-3191
57. J. Värmon, *Über Abelsche Körper, deren alle Gruppeninvarianten aus einer Primzahl bestehen, und über Abelsche Körper als Kreiskörper*. German. Lund, Hakan Ohlssons Boktryckeri (1925). 1925.
58. J. Wang, Malle's conjecture for $S_n \times A$ for $n = 3, 4, 5$. Compos. Math. **157**(1), 83–121 (2021). ISSN: 0010-437X. https://doi.org/10.1112/s0010437x20007587

59. J. Wang, *Malle's Conjecture for the Compositum of Number Fields*. Thesis (Ph.D.) The University of Wisconsin–Madison (ProQuest LLC, Ann Arbor, 2018), p. 84. ISBN: 979-8569-97420-7. URL: http://gateway.proquest.com/openurl?url_ver=Z39.88-2004&rft_val_fmt=info:ofi/fmt:kev:mtx:dissertation&res_dat=xri:pqm&rft_dat=xri:pqdiss:10843152
60. M.M. Wood, The distribution of the number of points on trigonal curves over $\mathbb{F}_q$. Int. Math. Res. Not. IMRN **23**, 5444–5456 (2012). ISSN: 1073-7928. https://doi.org/10.1093/imrn/rnr256
61. D.J. Wright, Distribution of discriminants of abelian extensions. Proc. Lond. Math. Soc. (3) **58**(1), 17–50 (1989). ISSN: 0024-6115. https://doi.org/10.1112/plms/s3-58.1.17
62. X.K. Zhang, On number fields of type $(l, l, \cdots, l)$. Sci. Sin. Ser. A **27**(10), 1018–1026 (1984). ISSN: 0253-5831.

Dynamical Mahler Measure: A Survey and Some Recent Results

Annie Carter, Matilde Lalín, Michelle Manes, and Alison Beth Miller

2020 Mathematics Subject Classification Primary: 11R06; Secondary: 11G50, 37P15, 37P30

1 Introduction

The inspiration for our investigation comes from the following result that relates the canonical height $\hat{h}_f$ (see Definition 2.6) of a point $\alpha \in \mathbb{P}^1(\overline{\mathbb{Q}})$ relative to some monic polynomial $f \in \mathbb{Z}[z]$ with an integral of the minimal polynomial of α relative to an invariant measure defined by f.

Theorem 1.1 ([30]) *Let $f \in \mathbb{Z}[z]$ be a monic polynomial, and let $\mathcal{J}_f$ denote the Julia set of f. Let K be a number field with $\alpha \in \mathbb{P}^1(K)$, and let $P \in \mathbb{Z}[x]$ be the minimal polynomial for α. Then:*

$$[\mathbb{Q}(\alpha) : \mathbb{Q}]\hat{h}_f(\alpha) = \int_{\mathcal{J}_f} \log|P(z)|d\mu_f(z). \tag{1}$$

A. Carter
Panadura, Sri Lanka
e-mail: carter21@alumni.iu.edu

M. Lalín (✉)
Département de mathématiques et de statistique, Université de Montréal, Montreal, QC, Canada
e-mail: matilde.lalin@umontreal.ca

M. Manes
American Institute of Mathematics, Caltech, Pasadena, CA, USA
e-mail: mmanes@aimath.org

A. B. Miller
Mathematical Reviews, Ann Arbor, MI, USA
e-mail: alimil@umich.edu

A. Bucur et al. (eds.), *Research Directions in Number Theory*, Association for Women in Mathematics Series 33, https://doi.org/10.1007/978-3-031-51677-1_7

Compare this with a standard formula relating the Mahler measure of the minimal polynomial P and the height of a root of that polynomial:

$$[\mathbb{Q}(\alpha):\mathbb{Q}]h(\alpha) = \underbrace{\frac{1}{2\pi i}\int_{\mathbb{T}^1} \log|P(z)|\frac{dz}{z}}_{\mathrm{m}(P)}. \tag{2}$$

The tantalizing similarities in these formulas—in fact, (2) is just (1) for the particular case $f(z) = z^d$—lead naturally to questions about extending classical results of Mahler measure to this new "dynamical Mahler measure" relative to a fixed polynomial f. In [9], the authors define a multivariate dynamical Mahler measure and prove several preliminary results with this flavor. This survey article presents background, motivation, examples, and strengthening of those results, both illustrating and expanding on the work begun in [9].

In Sect. 2, we provide background on Mahler measure, arithmetic dynamics, and equilibrium measures. In Sect. 3, we define the multivariate dynamical Mahler measure and give examples where it is possible to compute it exactly. Section 4 provides a summary of results from [9], drawing explicit connections between classical Mahler measure and the dynamical setting. In Sect. 5, we prove the existence of dynamical Mahler measure as defined in the previous section; these proofs also appear in [9] but are reiterated here (with a bit more detail) to provide a self-contained reference to the subject. Section 6 contains a survey of recent results on properties that are either known or conjectured to be equivalent to a multivariate polynomial having dynamical Mahler measure zero. Section 7 contains the proof of a new implication of this sort, and Sect. 8 contains strengthening of one of the results from [9]. In particular, in the proof of the two-variable Dynamical Kronecker's Lemma, we replace the (rather strong) assumption of Dynamical Lehmer's Conjecture with an assumption about the preperiodic points for the polynomial f. Finally, Sect. 9 investigates that condition on unicritical polynomials.

2 Basic Notions

In this section, we provide preliminary material on both Mahler measure and arithmetic dynamics. We refer the interested reader to [4] for a more comprehensive article describing the history and applications of Mahler measure in arithmetic geometry and to [3] for background and motivation from the arithmetic dynamics perspective.

2.1 *Mahler Measure*

The (logarithmic) Mahler measure of a nonzero polynomial $P \in \mathbb{C}[x]$, originally defined by Lehmer [24], is a height function given by

$$\mathrm{m}(P) = \mathrm{m}\left(a\prod_j (x-\alpha_j)\right) = \log|a| + \sum_j \log\max\{1, |\alpha_j|\}. \tag{3}$$

If $P \in \mathbb{Z}[x]$, the formula above makes it clear that $\mathrm{m}(P) \geq 0$. In such a case, it is natural to ask which polynomials $P \in \mathbb{Z}[x]$ satisfy $\mathrm{m}(P) = 0$. A result of Kronecker [22] gives the answer.

Lemma 2.1 (Kronecker's Lemma) *Let $P \in \mathbb{Z}[x]$. Then $\mathrm{m}(P) = 0$ if and only if P is monic and can be decomposed as a product of a monomial and cyclotomic polynomials.*

Lehmer [24] computed

$$\mathrm{m}(x^{10}+x^9-x^7-x^6-x^5-x^4-x^3+x+1) = \log(1.176280818\ldots) = 0.162357612\ldots$$

and asked the following:

Question 2.2 Is there a constant $C > 0$ such that for every polynomial $P \in \mathbb{Z}[x]$ with $\mathrm{m}(P) > 0$, then $\mathrm{m}(P) \geq C$?

Lehmer's Conjecture is the statement that the answer to the above question is yes. It remains open, and Lehmer's degree-10 polynomial remains the integer polynomial with the smallest known positive measure.

Jensen's formula [20] relates an average of a linear polynomial over the unit circle with the size of its root:

$$\frac{1}{2\pi i}\int_{\mathbb{T}^1} \log|x-\alpha|\frac{dx}{x} = \log\max\{1, |\alpha|\}. \tag{4}$$

Applying Jensen's formula to the definition of Mahler measure in (3), we find a formula that can be extended naturally to multivariate polynomials and rational functions. Following Mahler [26], we have:

Definition 2.3 The (logarithmic) Mahler measure of a nonzero rational function $P \in \mathbb{C}(x_1, \ldots, x_n)$ is defined by

$$\mathrm{m}(P) := \frac{1}{(2\pi i)^n}\int_{\mathbb{T}^n} \log|P(x_1, \ldots, x_n)|\frac{dx_1}{x_1}\cdots\frac{dx_n}{x_n},$$

where $\mathbb{T}^n = \{(x_1, \ldots, x_n) \in \mathbb{C}^n : |x_1| = \cdots = |x_n| = 1\}$.

The above integral converges, and for $P \in \mathbb{Z}[x_1, \ldots, x_n]$, we still have $\mathrm{m}(P) \geq 0$ (see Proposition 5.6). It is natural, then, to consider whether Kronecker's Lemma has an extension to multivariate polynomials. Recall that a polynomial in $\mathbb{Z}[x_1, \ldots, x_n]$ is said to be **primitive** if the coefficients have no non-trivial factor. We have the following result.

Theorem 2.4 ([12, Theorem 3.10]) *For any primitive polynomial P in $\mathbb{Z}[x_1, \ldots, x_n]$, we have $\mathrm{m}(P) = 0$ if and only if P is the product of monomial and cyclotomic polynomials evaluated on monomials.*

A connection between the single-variable case and the multivariate case is given by a result due to Boyd [6] and Lawton [23].

Theorem 2.5 ([23, Theorem 2]) *If $P \in \mathbb{C}(x_1, \ldots, x_n)^\times$, then*

$$\lim_{q(\mathbf{k})\to\infty} \mathrm{m}(P(x, x^{k_2}, \ldots, x^{k_n})) = \mathrm{m}(P(x_1, \ldots, x_n)), \tag{5}$$

where

$$q(\mathbf{k}) = \min\left\{ H(\mathbf{s}) : \mathbf{s} = (s_2, \ldots, s_n) \in \mathbb{Z}^{n-1}, \mathbf{s} \neq (0, \ldots, 0), \text{ and } \sum_{j=2}^{n} s_j k_j = 0 \right\}$$

and $H(\mathbf{s}) = \max\{|s_j| : 2 \le j \le n\}$.

Intuitively, the second equation says that the limit is taken while $k_2, \ldots, k_n$ go to infinity independently from each other.

Mahler measure often yields special values of interesting number-theoretic functions, such as the Riemann zeta function and L-functions associated to arithmetic–geometric objects such as elliptic curves. For more on these connections, see [4, 7].

2.2 *Arithmetic Dynamics*

A discrete dynamical system is a set X together with a self-map: $f : X \to X$, allowing for iteration. Here we focus on polynomials $f : \mathbb{C} \to \mathbb{C}$. For such an f and for $L \in \mathrm{Aut}(\mathbb{C})$ (so $L = az + b \in \mathbb{C}[x]$), we write

$$f^n = \underbrace{f \circ f \circ \cdots \circ f}_{n\text{-fold composition}}, \quad \text{and} \quad f^L := L^{-1} \circ f \circ L. \tag{6}$$

We will say that f^L and f are affine conjugate over K when $L \in K[x]$. This conjugation is a natural dynamical equivalence relation because it respects iteration: $(f^L)^n = (f^n)^L$.

A fundamental goal of dynamics is to study the behavior of points of X under iteration. For example, a point $\alpha \in X$ is said to be:

- **Periodic** if $f^n(\alpha) = \alpha$ for some $n > 0$
- **Preperiodic** if $f^n(\alpha) = f^m(\alpha)$ for some $n > m \ge 0$, and
- **Wandering** if it is not preperiodic

We write

$$\mathrm{PrePer}(f) = \{\alpha \in X : \alpha \text{ is preperiodic under } f\}.$$

As usual, we say that α is a **critical** point if $f'(\alpha) = 0$. Critical points play an important role in analyzing the dynamics of the function f.

Questions in arithmetic dynamics are often motivated by an analogy between arithmetic geometry and dynamical systems in which, for example, rational and integral points on varieties correspond to rational and integral points in orbits, and torsion points on abelian varieties correspond to preperiodic points. It should be no surprise, then, that heights are an essential tool in the study of arithmetic dynamics.

We recall that the classical (logarithmic) height of a rational number $\alpha = \frac{a}{b} \in \mathbb{Q}$, written in lowest terms, is $h(\alpha) = \log\max\{|a|, |b|\}$. This can be extended naturally to a height on algebraic numbers.

One way of making such an extension is to consider the naïve height. Let α be an algebraic number. We consider its minimal polynomial $P_\alpha(z) = \sum_{j=0}^{n} a_j z^n$ normalized such that it has integral coefficients and is primitive. Then

$$h_{\text{naïve}}(\alpha) := \log\max_{j} |a_j|.$$

Another possible extension of the classical height is given by the Weil height. For $\alpha \in K$, with K a number field, the (absolute logarithmic) Weil height is given by

$$h_{\text{Weil}}(\alpha) = \frac{1}{[K:\mathbb{Q}]} \sum_{v \in M_K} [K_v : \mathbb{Q}_v] \log\max\{||\alpha||_v, 1\}, \tag{7}$$

where M_K is an appropriately normalized set of inequivalent absolute values on K, so that the product formula is satisfied:

$$\prod_{v \in M_K} ||x||_v^{\frac{[K_v:\mathbb{Q}_v]}{[K:\mathbb{Q}]}} = 1.$$

More concretely, for $K = \mathbb{Q}$, we can take $|\cdot|_\infty$ to be the usual absolute value and $|\cdot|_p$ to be the p-adic absolute value, normalized so that $|p|_p = 1/p$. Then for $v \in M_K$ lying over a prime p,

$$||x||_v = |N_{K_v/\mathbb{Q}_p}(x)|_p^{\frac{1}{[K_v:\mathbb{Q}_p]}}.$$

The factor of $[K : \mathbb{Q}]$ in (7) ensures that $h_{\text{Weil}}(\alpha)$ is well-defined, with the same answer for any field K containing α.

While the naïve height is very natural to consider, the Weil height is the canonical height for the power map $z \mapsto z^d$. The two heights are commensurate in the sense that for $\alpha \in \mathbb{P}^1(\overline{\mathbb{Q}})$, there is a constant $C(d)$ depending only on the degree d of α such that

$$|h_{\text{naïve}}(\alpha) - h_{\text{Weil}}(\alpha)| \leq C(d). \tag{8}$$

If $f(x) \in K(x)$ is a rational function of degree d, then $h(f(\alpha))$ should be approximately $dh(\alpha)$. The dynamical canonical height makes this an equality. The definition is reminiscent of the Néron–Tate height on an abelian variety, and the proofs of the statements below follow exactly as in this more familiar case.

Definition 2.6 If $f \in \overline{\mathbb{Q}}(z)$ is a rational map of degree d (i.e., the maximum of the degrees of the numerator and denominator is d), and $\alpha \in \mathbb{P}^1(\overline{\mathbb{Q}})$, we then define

$$\hat{h}_f(\alpha) = \lim_{n\to\infty} \frac{h(f^n(\alpha))}{d^n},$$

where h may be taken as either the naïve height $h_{\text{naïve}}$ or the Weil height h_{Weil} by Eq. (8).

It is known that this limit exists, that $\hat{h}_f(f(\alpha)) = d\hat{h}_f(\alpha)$, and that $\hat{h}_f(\alpha) = 0$ if and only if α is a preperiodic point for f. See Section 3.4 of [35] for details.

Definition 2.7 Let $f \in \mathbb{C}[z]$. The **filled Julia set** of f is

$$\mathcal{K}_f = \{z \in \mathbb{C} \ : \ f^n(z) \not\to \infty \text{ as } n \to \infty\}.$$

The **Julia set** $\mathcal{J}_f$ of f is the boundary of the filled Julia set. That is, $\mathcal{J}_f = \partial\mathcal{K}_f$.

It follows from these definitions that for a polynomial $f \in \mathbb{C}[z]$, both $\mathcal{K}_f$ and $\mathcal{J}_f$ are compact. We denote by F_∞ the complement of $\mathcal{K}_f$ in $\mathbb{P}^1(\mathbb{C})$, which is also the attracting basin of ∞ for f, that is, the set of points in $\mathbb{P}^1(\mathbb{C})$ whose orbits go off to ∞.

For example, for $f(z) = z^d$, we see that $f^n(z) = z^{d^n}$. So for $d \geq 2$, we have three cases:

- If $|\alpha| > 1$, then $|\alpha^{d^n}| \to \infty$ as $n \to \infty$.
- If $|\alpha| < 1$, then $|\alpha^{d^n}| \to 0$ as $n \to \infty$.
- If $|\alpha| = 1$, then $|\alpha^{d^n}| = 1$ for all n.

So for pure power maps, we can understand the Julia sets completely: $\mathcal{K}_f$ is the unit disc, and $\mathcal{J}_f$ is the unit circle. In general, however, these sets are quite complex. (See Fig. 1.)

It is clear from Definition 2.7 that PrePer$(f) \subseteq \mathcal{K}_f$. When all of the critical points of a polynomial f have unbounded orbits, the Julia set $\mathcal{J}_f$ is totally disconnected, while $\mathcal{J}_f$ is connected if and only if all the critical orbits are bounded [13, 21]. A polynomial of degree 2 has a unique critical point, and we have only these two cases (see Fig. 1). This situation is known as the Fatou–Julia dichotomy. However, for higher-degree polynomials, the situation is complicated by having more than one critical point, and there are cases of polynomials f for which $\mathcal{J}_f$ is disconnected but not totally disconnected (see Fig. 2).

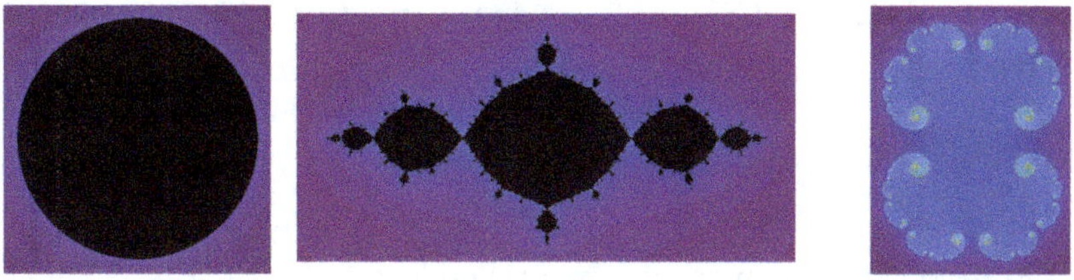

(A) (B) (C)

Fig. 1 The black area shows the filled Julia set $\mathcal{K}_f$, and its boundary is the Julia set $\mathcal{J}_f$. In the third case, the Julia set has empty interior, so $\mathcal{K}_f = \mathcal{J}_f$. (**a**) Filled Julia set for $f(z) = z^2$. (**b**) Filled Julia set for $f(z) = z^2 - 1$. (**c**) (Filled) Julia set for $f(z) = z^2 + 0.3$

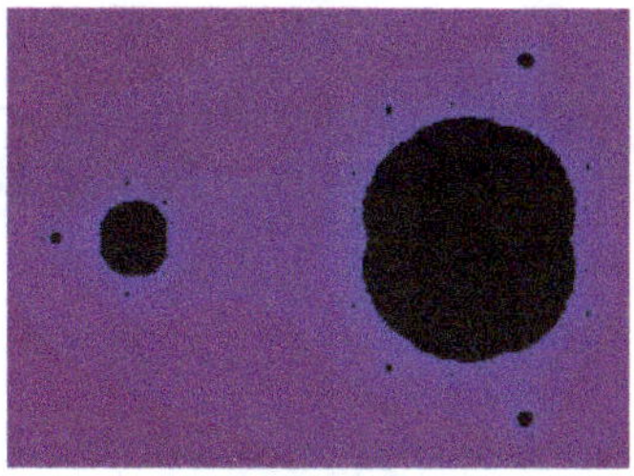

Fig. 2 The filled Julia set for $f(z) = z^3 - z + 1$. The Julia set $\mathcal{J}_f$ is disconnected but not totally disconnected

We saw in Eqs. (1) and (2) that the Julia set $\mathcal{J}_f$ for a polynomial f will play the role of the unit torus $\mathbb{T}^1$ when studying dynamical Mahler measure.

2.3 *Equilibrium Measures*

Definition 2.8 Given a compact subset $K \subseteq \mathbb{C}$, an **equilibrium measure** for K is a Borel probability measure μ on K that has maximal **energy**

$$I(\mu) := \int_K \int_K \log|z - w|\, d\mu(z)\, d\mu(w)$$

among all Borel probability measures on K.

Every compact set $K \subseteq \mathbb{C}$ has an equilibrium measure [31, Theorem 3.3.2], and if f denotes a polynomial of degree $d \geq 2$, then the equilibrium measure μ_f on its Julia set $\mathcal{J}_f$ is unique. (This is a consequence of a more general result stating that the equilibrium measure of any compact, *non-polar* set is unique [31, Theorem 3.7.6]; the Julia set $\mathcal{J}_f$ of any polynomial f is non-polar [31, Theorem 6.5.1], which is to say that there is a non-trivial finite Borel measure with compact support such that $I(\mu) > -\infty$.) A result due to Freire, Lopes, and Mañé and to Lyubich says that we can characterize the equilibrium measure μ_f as follows:

Theorem 2.9 ([14, 25]) *Let $w \in \mathcal{J}_f$, and for $n \geq 1$, define the Borel probability measures*

$$\mu_n := \frac{1}{d^n} \sum_{f^n(\zeta)=w} \delta_\zeta,$$

where δ_ζ denotes the unit mass at ζ, and the preimages ζ of w under f^n are taken with multiplicity. Then $\mu_n \xrightarrow{w^*} \mu_f$ (weak-convergence) as $n \to \infty$.*

3 Dynamical Mahler Measure: Definition and Examples

Inspired by Definition 2.3 and the parallels between Mahler measure and the dynamical setting in Eqs. (1) and (2), we define the following:

Definition 3.1 Let $f \in \mathbb{Z}[z]$ be a monic polynomial of degree $d \geq 2$, and let $P \in \mathbb{C}(x_1, \ldots, x_n)^\times$. The f**-dynamical Mahler measure** of P is the number

$$\mathrm{m}_f(P) := \int_{\mathcal{J}_f} \cdots \int_{\mathcal{J}_f} \log |P(z_1, \ldots, z_n)| \, d\mu_f(z_1) \cdots d\mu_f(z_n). \tag{9}$$

Note that as μ_f is a probability measure, the value of this integral is not affected by omitted variables, so in this sense the value of m_f is independent of n.

Remark 3.2 Note that one could generalize this notion even further by letting the polynomials vary. For $F = (f_1, f_2, \ldots, f_n)$ with each $f_i \in \mathbb{Z}[z]$ monic, define

$$\mathrm{m}_F(P) := \int_{\mathcal{J}_{f_n}} \cdots \int_{\mathcal{J}_{f_1}} \log |P(z_1, \ldots, z_n)| \, d\mu_{f_1}(z_1) \cdots d\mu_{f_n}(z_n).$$

It might be interesting to explore this higher-dimensional dynamical Mahler measure. However, results in this chapter, specifically a dynamical version of the Boyd–Lawton Theorem, only make sense in the split case where $F = (f, f, \ldots, f)$.

For a projective integral scheme X over a number field F and an ample line bundle L together with a semi-positive adelic metric in the sense of Zhang [38], Chambert-Loir and Thuillier [10] gave a formula expressing local heights (relative to L) of a Cartier divisor D on X in terms of integrals of Green functions for D against measures attached to L. This formula reduces to the classical Mahler measure when $X = \mathbb{P}^n$. As we discuss in Sect. 6.2, we believe that this construction is the same as the definition given in (9), but we do not prove this.

It is not clear a priori that the integral in (9) converges. It does, and we prove this in Proposition 5.6. But before discussing these details, we provide some examples where the dynamical Mahler measure can be explicitly computed. The following lemmas will prove useful throughout.

Lemma 3.3 *If $f \in \mathbb{Z}[z]$ and $P, Q \in \mathbb{C}(x_1, \ldots, x_n)^\times$, then*

$$\mathrm{m}_f(PQ) = \mathrm{m}_f(P) + \mathrm{m}_f(Q).$$

Proof This follows immediately from the corresponding fact about logarithms. □

Lemma 3.4 *If f and g are nonlinear polynomials that commute under composition, then $\mathrm{m}_f = \mathrm{m}_g$.*

Proof If f and g commute, then they have the same Julia set (see [1]), and the equilibrium measure is determined by this set. □

Proposition 3.5 *If $f(z) = z^d$ with $d \geq 2$, then $\mathrm{m}_f(P) = \mathrm{m}(P)$ for $P \in \mathbb{C}(x)^\times$.*

Proof In this case, we have seen that $\mathcal{J}_f$ is given by the circle $\mathbb{T}^1$, and Theorem 2.9 tells us that the equilibrium measure is the uniform measure on the circle:

$$\frac{\chi_{\mathbb{T}^1} dz}{2\pi i z},$$

where $\chi_{\mathbb{T}^1}$ is the characteristic function on the unit circle. Taking $z = e^{i\theta}$, we then have

$$\mathrm{m}_f(P) = \frac{1}{2\pi} \int_0^{2\pi} \log |P(e^{i\theta})| \, d\theta = \frac{1}{2\pi i} \int_{\mathbb{T}^1} \log |P(z)| \frac{dz}{z} = \mathrm{m}(P).$$

Proposition 3.6 *Define the dth Chebyshev polynomial to be the polynomial $T_d(z) \in \mathbb{Z}[z]$ that satisfies*

$$T_d(z + z^{-1}) = z^d + z^{-d}. \tag{10}$$

Then

$$\mathrm{m}_{T_d}(P) = \mathrm{m}(P \circ w)$$

for $P \in \mathbb{C}(x)^\times$, where $w(z) = z + z^{-1}$.

Proof Note that $T_d \circ w = w \circ f$, where $f(z) = z^d$ (and note the analogy with Proposition 3.9). The function w maps the Julia set $\mathcal{J}_f$ onto the Julia set $\mathcal{J}_{T_d}$, so that $\mathcal{J}_{T_d}$ is the segment $[-2, 2]$ (traversed twice as z proceeds around the unit circle). It follows from Theorem 2.9 that the equilibrium measure on $\mathcal{J}_{T_d}$ is the pushforward $w_* \mu_f$ of the equilibrium measure on $\mathcal{J}_f$, so

$$\mathrm{m}_f(P) = \int_{\mathcal{J}_{T_d}} \log |P(z)| \, d\mu_{T_d}(z)$$

$$
\begin{aligned}
&= \int_{w(\mathcal{J}_f)} \log |P(z)| \, dw_* \mu_f(z) \\
&= \int_{\mathcal{J}_f} \log |P(w(z))| \, d\mu_f(z) \\
&= \mathrm{m}_f(P \circ w) \\
&= \mathrm{m}(P \circ w)
\end{aligned}
$$

by Proposition 3.5. □

Remark 3.7 Incidentally, writing $u = z + z^{-1} = e^{i\theta} + e^{-i\theta} = 2\cos\theta$, we have $du = -2\sin\theta \, d\theta$, which is to say

$$
-\frac{du}{\sqrt{4-u^2}} = d\theta.
$$

Thus we can write

$$
\begin{aligned}
\mathrm{m}_{T_d}(P) &= \frac{1}{\pi} \int_0^{\pi} \log |P(e^{i\theta} + e^{-i\theta})| \, d\theta \\
&= \frac{1}{\pi} \int_{-2}^{2} \log |P(u)| \frac{du}{\sqrt{4-u^2}},
\end{aligned}
$$

from which we see that the equilibrium measure on the segment $[-2, 2]$ is

$$
\frac{\chi_{[-2,2]} dz}{\pi \sqrt{4-z^2}}.
$$

More generally, we have:

Proposition 3.8 *Let $\alpha, \beta \in \mathbb{C}$, and let $f(z) = \frac{\beta-\alpha}{4} T_d\left(\frac{4z-2(\alpha+\beta)}{\beta-\alpha}\right) + \frac{\alpha+\beta}{2}$, where T_d is a Chebyshev polynomial defined in Eq.* (10) *for $d \geq 2$. For $P \in \mathbb{C}[x]$, we have*

$$
\mathrm{m}_f(P) = \mathrm{m}\left(P \circ \left(\frac{\beta-\alpha}{4}(z+z^{-1}) + \frac{\alpha+\beta}{2}\right)\right). \tag{11}
$$

Proof A change of variables shows that the Julia sets of these polynomials are given by $\mathcal{J}_f = [\alpha, \beta]$, where this is to be understood as the line segment connecting α and β in the complex plane. The equilibrium measure is then given by

$$
\frac{\chi_{[\alpha,\beta]} dz}{\pi \sqrt{(z-\alpha)(\beta-z)}},
$$

where $\chi_{[\alpha,\beta]}$ is the characteristic function on the segment $[\alpha, \beta]$. This gives

$$\mathrm{m}_f(P) = \frac{1}{\pi}\int_\alpha^\beta \log|P(z)|\frac{dz}{\sqrt{(z-\alpha)(\beta-z)}}.$$

Setting $z = \frac{\beta-\alpha}{2}\cos(\pi\theta) + \frac{\alpha+\beta}{2}$ gives

$$\begin{aligned}\mathrm{m}_f(P) &= \int_0^1 \log\left|P\left(\frac{\beta-\alpha}{2}\cos(\pi\theta) + \frac{\alpha+\beta}{2}\right)\right| d\theta \\ &= \frac{1}{2}\int_{-1}^1 \log\left|P\left(\frac{\beta-\alpha}{2}\cos(\pi\theta) + \frac{\alpha+\beta}{2}\right)\right| d\theta.\end{aligned}$$

Substituting $w = e^{i\theta}$, we turn the domain of integration into the unit circle, and we conclude that m_f is given by (11). □

We provide the details of Proposition 3.8 because it is so difficult, in general, to calculate dynamical Mahler measure exactly. However, we note that the result can also be viewed as a consequence of Proposition 3.6 about Chebyshev polynomials and the following result on dynamical Mahler measure for conjugate maps.

Proposition 3.9 *Let $f \in \mathbb{C}[z]$ and $P \in \mathbb{C}[z]$, and let $L(z) = az + b \in \mathbb{C}[z]$ with $a \neq 0$. Then*

$$\mathrm{m}_{f^L}(P) = \mathrm{m}_f(P \circ L^{-1}).$$

Proof Note first that the Julia sets of f and f^L are related by

$$\mathcal{J}_{f^L} = L^{-1}(\mathcal{J}_f).$$

Next, fixing $w \in \mathcal{J}_f$, observe that by Theorem 2.9, the measure μ_{f^L} is the limit in the weak* topology of the sequence of measures

$$d^{-n} \sum_{(f^L)^n(\zeta)=w} \delta_\zeta = d^{-n} \sum_{f^n(L(\zeta))=L(w)} (L^{-1})_*\delta_{L(\zeta)},$$

where d denotes the degree of f and $(L^{-1})_*\delta_{L(\zeta)}$ denotes the measure defined by $(L^{-1})_*\delta_{L(\zeta)}(X) = \delta_{L(\zeta)}(L(X))$. But by Theorem 2.9, this sequence also has weak*-limit $(L^{-1})_*\mu_f$. So $\mu_{f^L} = (L^{-1})_*\mu_f$. Making the substitution $w = L(z)$, we then have

$$\begin{aligned}\mathrm{m}_{f^L}(P) &= \int_{\mathcal{J}_{f^L}} \log|P(z)|\, d\mu_{f^L}(z) \\ &= \int_{L^{-1}(\mathcal{J}_f)} \log|P(z)|\, d(L^{-1})_*\mu_f(z)\end{aligned}$$

$$= \int_{\mathcal{J}_f} \log |P(L^{-1}(w))| \, d\mu_f(w)$$

$$= \mathrm{m}_f(P \circ L^{-1}),$$

as desired. □

4 Dynamical Versions of Classical Results

In this section, we summarize results from [9], focusing on the connections between these results and classical Mahler measure as outlined in Sect. 2.1. We provide more detail and refine some of these results in the next sections.

Jensen's formula gave us an equivalent definition of Mahler measure that extended naturally to higher dimensions:

$$\mathrm{m}(P) = \log|a| + \sum_{|\alpha_i|>1} \log|\alpha_i| = \frac{1}{2\pi i} \int_{\mathbb{T}^1} \log|P(z)| \frac{dz}{z}.$$

Dynamical Jensen's formula [9, Lemma 3.1] plays a similar role, allowing us to prove that the integral in (9) converges and that when $P \in \mathbb{Z}[x_1, \ldots, x_n]$, we have $\mathrm{m}_f(P) \geq 0$. Here p_{μ_f} is the potential function (see Definition 5.1 and Proposition 5.2).

$$\mathrm{m}_f(P) = \log|a| + \sum_i p_{\mu_f}(\alpha_i) = \log|a| + \int_{\mathcal{K}_f} \log|z - w| \, d\mu_f(w).$$

Kronecker's Lemma tells us which integer polynomials have Mahler measure zero: Let $P \in \mathbb{Z}[x]$. If $\mathrm{m}(P) = 0$, then the roots of P are either zero or roots of unity. Conversely, if P is primitive and its roots either zero or roots of unity, then $\mathrm{m}(P) = 0$.

Dynamical Kronecker's Lemma [9, Lemmas 1.2 and 4.3] answers the same question for dynamical Mahler measure. Recalling the driving analogy of arithmetic dynamics, that preperiodic points are like torsion points in arithmetic geometry, the result feels natural.

Lemma 4.1 (Dynamical Kronecker's Lemma) *Let $f \in \mathbb{Z}[z]$ be monic of degree $d \geq 2$ and let $P \in \mathbb{Z}[x]$. Then we have $\mathrm{m}_f(P) = 0$ if and only if $P(x) = \pm \prod_i (x - \alpha_i)$ with each α_i a preperiodic point of f.*

The Boyd–Lawton Theorem relates single-variable and multivariate Mahler measure. For $P \in \mathbb{C}(x_1, \ldots, x_n)^{\times}$,

$$\lim_{k_2\to\infty}\cdots\lim_{k_n\to\infty} \mathrm{m}(P(x, x^{k_2}, \ldots, x^{k_n})) = \mathrm{m}(P(x_1, \ldots, x_n))$$

with $k_2, \ldots, k_n \to \infty$ independently from each other.

The Weak Dynamical Boyd–Lawton Theorem [9, Proposition 1.3] provides a partial analogue in the dynamical setting for polynomials in two variables.

Proposition 4.2 (Weak Dynamical Boyd–Lawton) *Let $f \in \mathbb{Z}[z]$ monic of degree $d \geq 2$ and let $P \in \mathbb{C}[x, y]$. Then*

$$\limsup_{n\to\infty} \mathrm{m}_f(P(x, f^n(x))) \leq \mathrm{m}_f(P(x, y)).$$

Lehmer's Conjecture proposes that there are not integer polynomials with arbitrarily small Mahler measure, or that the Mahler measure of $P \in \mathbb{Z}[x]$ with $\mathrm{m}(P) \neq 0$ is bounded away from zero.

Dynamical Lehmer's Conjecture [35, Conjecture 3.25] proposes the same for dynamical Mahler measure. More precisely:

Conjecture 4.3 (Dynamical Lehmer's Conjecture) There is some $\delta = \delta_f > 0$ such that any single-variable polynomial $P \in \mathbb{Z}[x]$ with $\mathrm{m}_f(P) > 0$ satisfies $\mathrm{m}_f(P) > \delta$.

Higher-dimensional Kronecker's Lemma can be stated quite simply and has a similar feel to the one-dimensional version:

Theorem 2.4 ([12, Theorem 3.10]) *For any primitive polynomial P in $\mathbb{Z}[x_1, \ldots, x_n]$, $\mathrm{m}(P) = 0$ if and only if P is the product of monomial and cyclotomic polynomials evaluated on monomials.*

The main result of [9] provides a partial two-variable Kronecker's Lemma for dynamical Mahler measure, but the statement and hypotheses are significantly more delicate than in the one-variable case.

Theorem 4.4 ([9, Theorem 1.5]) *Assume the Dynamical Lehmer's Conjecture.*

Let $f \in \mathbb{Z}[z]$ be a monic polynomial of degree $d \geq 2$ that is not conjugate to z^d or to $\pm T_d(z)$, where $T_d(z)$ is the dth Chebyshev polynomial. Then any polynomial $P \in \mathbb{Z}[x, y]$ that is irreducible in $\mathbb{Z}[x, y]$ (but not necessarily irreducible in $\mathbb{C}[x, y]$) with $\mathrm{m}_f(P) = 0$ and that contains both variables x and y divides a product of complex polynomials of the following form:

$$\tilde{f}^n(x) - L(\tilde{f}^m(y)),$$

where $m, n \geq 0$ are integers, $L \in \mathbb{C}[z]$ is a linear polynomial commuting with an iterate of f, and $\tilde{f} \in \mathbb{C}[z]$ is a nonlinear polynomial of minimal degree commuting with an iterate of f (with possibly different choices of L, $\tilde{f}$, n, and m for each factor).

As a partial converse, suppose there exists a product of complex polynomials F_j *such that:*

(1) Each F_j *has the form* $\tilde{f}^n(x) - L(\tilde{f}^m(y))$*, where* L *and* $\tilde{f}$ *are as above (with possibly different choices of* L, $\tilde{f}$, n, *and* m *for each* j*).*
(2) $\prod F_j \in \mathbb{Z}[x, y]$.
(3) P *divides* $\prod F_j$ *in* $\mathbb{Z}[x, y]$.

Then $\mathrm{m}_f(P) = 0$.

In Sect. 6, we discuss several other statements that are either known or conjectured to be equivalent to having dynamical Mahler measure 0, and in Sect. 7, we prove a new implication in this family of results. In Sect. 8, we prove a new version of Theorem 4.4 in which we replace the assumption of Dynamical Lehmer's Conjecture with the assumption that $\mathrm{PrePer}(f) \subseteq \mathcal{J}_f$. This is a strengthening of the result in some respects, since the hypothesis on the preperiodic points is much easier to check, when it holds, than Dynamical Lehmer's Conjecture. However, there are certainly polynomials $f \in \mathbb{Z}[z]$ for which that assumption does not hold. This is discussed in Sect. 9.

5 Convergence and Positivity

In this section, we give an introduction to potentials and then use them to prove the existence of the dynamical Mahler measure.

Definition 5.1 The **potential** of a finite Borel measure μ with compact support K is the function $p_\mu : \mathbb{C} \to [-\infty, \infty)$ given by

$$p_\mu(z) = \int_K \log|z - w| \, d\mu(w).$$

We can see the relationship between potentials and dynamical Mahler measure in the following result, which should be considered the dynamical analogue of Jensen's formula:

Proposition 5.2 *Suppose* $P(x)$ *factors over* $\mathbb{C}$ *as* $P(x) = a\prod_i(x - \alpha_i)$. *Then*

$$\mathrm{m}_f(P) = \log|a| + \sum_i p_{\mu_f}(\alpha_i).$$

Proof We have

$$\mathrm{m}_f(P) = \int_{\mathcal{J}_f} \log\left|a\prod_i(z - \alpha_i)\right| d\mu_f$$

$$= \log|a| + \sum_i \int_{\mathcal{J}_f} \log|z - \alpha_i|\, d\mu_f$$

$$= \log|a| + \sum_i p_{\mu_f}(\alpha_i).$$

□

Remark 5.3 If f is a monic polynomial, the potential p_{μ_f} of the equilibrium measure on its Julia set is equal to the *Green's function* $g_{F_\infty}(z, \infty)$ on F_∞, the complement of the filled Julia set $\mathcal{K}_f$ in the Riemann sphere. For non-monic f, we have $p_{\mu_f}(z) = g_{F_\infty}(z, \infty) + I(\mu_f)$. (See [31, Theorem 6.5.1] and the proof of [31, Theorem 4.4.2].)

Let us make a few more observations about potentials before returning to dynamical Mahler measure.

Proposition 5.4 *Let $f \in \mathbb{C}[z]$ be a nonlinear polynomial. Then the potential p_{μ_f} is Hölder continuous and thus continuous.*

Proof It is known that the function that is equal to the Green's function $g_{F_\infty}(z, \infty)$ on $\mathbb{C} \setminus \mathcal{K}_f$ and identically zero on $\mathcal{K}_f$ is Hölder continuous (see [8, Theorem VIII.3.2] and the comment thereafter). When f is monic, this function is exactly the potential p_{μ_f}, and in general differs from p_{μ_f} by a constant by Remark 5.3, so p_{μ_f} is also Hölder continuous. □

Proposition 5.5 *Let $f \in \mathbb{C}[z]$ be a nonlinear, monic polynomial. Then $p_{\mu_f}(z) \geq 0$ for all $z \in \mathbb{C}$, and $p_{\mu_f}(z) = 0$ if and only if $z \in \mathcal{K}_f$.*

Proof It follows from [31, Theorem 6.5.1] that $I(\mu_f) = 0$ if f is monic. The proof of Proposition 5.4 then shows that $p_{\mu_f}(z) = 0$ for $z \in \mathcal{K}_f$, while $p_{\mu_f}(z) > 0$ for $z \notin \mathcal{K}_f$ [31, Theorem 4.4.3]. □

We now return to the dynamical Mahler measure.

Proposition 5.6 *Let $f \in \mathbb{Z}[z]$ be a monic, nonlinear polynomial, and let P in $\mathbb{C}(x_1, \ldots, x_n)^\times$. Then the integral defining the f-dynamical Mahler measure of P converges, and if P is furthermore a nonzero integer polynomial, then* $\mathrm{m}_f(P) \geq 0$.

Remark 5.7 This result appears in [9, Proposition 3.2]. In the interest of providing a self-contained introduction to the key ideas in the subject, we present here an expanded and more detailed proof. The argument is based on the proof of [12, Lemma 3.7] for the classical Mahler measure.

Proof It suffices to consider the case of P a polynomial, since $\mathrm{m}_f(F/G) = \mathrm{m}_f(F) - \mathrm{m}_f(G)$ by Lemma 3.3.

We induct on the number of variables. When $n = 1$, we can factor P over $\mathbb{C}$ as $a \prod_i (x - \alpha_i)$. By Proposition 5.2, we have

$$\mathrm{m}_f(P) = \log|a| + \sum_i p_{\mu_f}(\alpha_i).$$

Since the potential p_{μ_f} is nonnegative on $\mathbb{C}$, we can immediately conclude that the integral defining $\mathrm{m}_f(P)$ converges and that it is nonnegative when P has integer coefficients.

Now assume the result holds for polynomials in $n - 1$ variables, and let P in $\mathbb{C}[x_1, \ldots, x_n]$. Write P as a polynomial in x_1 with coefficients in $\mathbb{C}[x_2, \ldots, x_n]$:

$$P(x_1, \ldots, x_n) = a_d(x_2, \ldots, x_n)x_1^d + \cdots + a_0(x_2, \ldots, x_n).$$

Factor this as

$$a_d(x_2, \ldots, x_n) \prod_{j=1}^{d} (x_1 - g_j(x_2, \ldots, x_n))$$

for some algebraic functions g_j. We then have

$$\begin{aligned}
\mathrm{m}_f(P) &= \mathrm{m}_f(a_d) + \int_{\mathcal{J}_f} \cdots \int_{\mathcal{J}_f} \log\left|\prod_{j=1}^{d} (z_1 - g_j(z_2, \ldots, z_n))\right| \\
&\quad \times d\mu_f(z_1) \cdots d\mu_f(z_n) \\
&= \mathrm{m}_f(a_d) + \int_{\mathcal{J}_f} \cdots \int_{\mathcal{J}_f} \sum_{j=1}^{d} p_{\mu_f}(g_j(z_2, \ldots, z_n))\, d\mu_f(z_2) \cdots d\mu_f(z_n).
\end{aligned} \tag{12}$$

By the induction hypothesis, $\mathrm{m}_f(a_d)$ exists and is nonnegative if P, and thus a_d, has integer coefficients. While the g_j may not be continuous, the multiset of values $\{g_j(z_2, \ldots, z_n)\}$ is, so it follows from Propositions 5.4 and 5.5 that the integrand

$$\sum_{j=1}^{d} p_{\mu_f}(g_j(z_2, \ldots, z_n))$$

is nonnegative and continuous away from any poles of the g_j. On the other hand, as $\mathcal{J}_f$ is compact, the polynomial $P(z_1, \ldots, z_n)$ is bounded above on $\mathcal{J}_f^n$, so the integral defining $\mathrm{m}_f(P)$ is also bounded. The same can then be said for the integral

$$\int_{\mathcal{J}_f} \cdots \int_{\mathcal{J}_f} \sum_{j=1}^{d} p_{\mu_f}(g_j(z_2, \ldots, z_n))\, d\mu_f(z_2) \cdots d\mu_f(z_n)$$

by the finiteness of $\mathrm{m}_f(a_d)$; it follows that this integral converges, despite the presence of any poles of the g_j, and therefore, the integral defining $\mathrm{m}_f(P)$ does also. □

6 Multivariable Analogues of Dynamical Kronecker's Lemma

Multivariate dynamical Mahler measure was defined in [9], but similar ideas have appeared in the literature in recent years. In this section, we present a summary of some results in arithmetic dynamics. These statements, which include the statement that a polynomial has dynamical Mahler measure zero, are all known or conjectured to be equivalent.

Assume as usual that $f \in \mathbb{Z}[x]$ is monic of degree d, and $P \in \mathbb{Z}[x_1, \ldots, x_n]$. As in the statement of Theorem 4.4, L always denotes a linear polynomial in $\mathbb{C}[z]$ commuting with an iterate of f, and $\tilde{f}$ always denotes a nonlinear polynomial in $\mathbb{C}[z]$ of minimal degree commuting with an iterate of f:

(a) $\mathrm{m}_f(P) = 0$.
(b) $h\left(\overline{\{P = 0\}} \subseteq X\right) = 0$, where $\overline{\{P = 0\}}$ is the Zariski closure of the hypersurface $\{P = 0\} \subset \mathbb{A}^n \subseteq X$, X equals either $\mathbb{P}^n$ or $(\mathbb{P}^1)^n$, and h is a dynamical height for subvarieties of X of the type introduced in [38] (note there is some ambiguity here: h will depend on the choice of X and also on the choice of a line bundle on X).
(c) The hypersurface $\{P = 0\} \subseteq \mathbb{A}^n(\mathbb{C})$ is preperiodic under the map $(x_1, \ldots, x_n) \mapsto (f(x_1), \ldots, f(x_n))$. (A subvariety V of a variety X is preperiodic for a map $\Phi : X \to X$ if $\Phi^m(V) = \Phi^n(V)$ for some $m \neq n$.)
(d) The hypersurface $\{P = 0\} \subseteq \mathbb{A}^n(\mathbb{C})$ contains a Zariski dense subset of points that are preperiodic for the map $(x_1, \ldots, x_n) \mapsto (f(x_1), \ldots, f(x_n))$ (equivalently, with all coordinates preperiodic for f).
(e1) P is primitive (gcd of coefficients $= 1$) and, inside the ring $\mathbb{C}[x_1, \ldots, x_n]$, $P(x)$ divides some polynomial in $\mathbb{C}[x_1, \ldots, x_n]$ that is a product of factors of the form $\tilde{f}^n(x_i) - L(\tilde{f}^m(x_j))$ (here i can equal j).
(e2) Inside the ring $\mathbb{Z}[x_1, \ldots, x_n]$, $P(x)$ divides some polynomial in $\mathbb{Z}[x_1, \ldots, x_n]$, which is a product of factors of the form $\tilde{f}^n(x_i) - L(\tilde{f}^m(x_j))$ (the factors do not need to be in $\mathbb{Z}[x]$, but the product does).

The known relationships among these statements are summarized in Fig. 3.

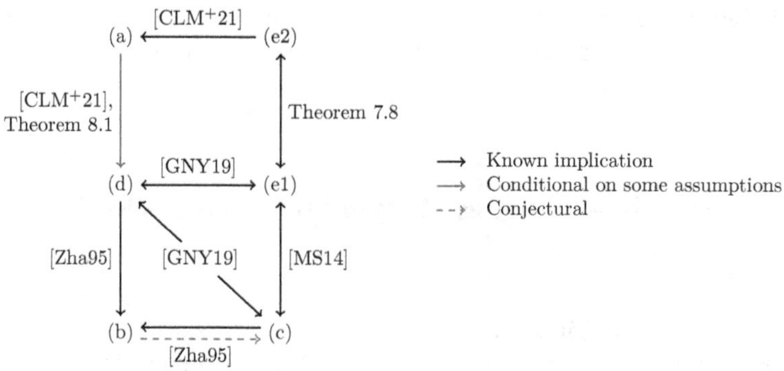

Fig. 3 The known relationships between statements (a–e), with references

6.1 Subvarieties with Many Preperiodic Points and Preperiodic Subvarieties

Historically, one of the first of these properties to be studied was property (d), as a special case of the following more general question in the field of unlikely intersections: For an algebraic variety X with a self-map $\Phi : X \to X$, which subvarieties Y of X contain a Zariski dense subset of preperiodic points for Φ?

One plausible necessary and sufficient condition would be for Y to be preperiodic for Φ, but it is easy to construct counterexamples to this equivalence in its full generality. However, Zhang [38] conjectured that this equivalence holds under the condition that the pair (X, Φ) is polarized.

Definition 6.1 A polarization of a pair (X, Φ), where X is a projective variety and $\Phi : X \to X$ is a self-map, is a line bundle $\mathcal{L}$ such that $\Phi^*\mathcal{L} = d\mathcal{L}$ for some integer $d > 1$ (which can be thought of as the degree of Φ).

This conjecture is a generalization of the Manin–Mumford conjecture (proved by Raynaud [32, 33]) on subvarieties of abelian varieties containing infinitely many torsion points, and so Zhang in [39] calls this the Dynamical Manin–Mumford Conjecture.

Conjecture 6.2 (Dynamical Manin–Mumford Conjecture) Let (X, Φ) be a pair of a projective variety X and a dominant map $\Phi : X \to X$ that admits a polarization. Then a subvariety Y of X contains a Zariski dense subset of preperiodic points if and only if Y is preperiodic.

In terms of our diagram, this corresponds to the implication (d) $\iff$ (c). (Note that although $\mathbb{A}^n$ is not a projective variety, we can complete $\mathbb{A}^n$ to a projective variety X to which the map $\Phi = f \times \cdots \times f$ extends in such a way that (X, Φ) admits a polarization; for more details, see Sect. 6.2.) However most of the study of this conjecture has been focused on the implication (d) $\Longrightarrow$ (c), which has generally been the harder direction.

The Dynamical Manin–Mumford Conjecture has been studied in various contexts and is now known not to hold in full generality as originally stated (counterexamples, and a refined statement, have been given in [17]). However, in the case of interest for our application, it is known to be true:

Theorem 6.3 (Ghioca, Nguyen, Ye [15, 16]) *If $\Phi : (\mathbb{P}^1)^n \to (\mathbb{P}^1)^n$ is of the form $f_1 \times \cdots \times f_n$, where each f_i is a non-exceptional rational map (not conjugate to a power map or plus or minus a Chebyshev polynomial), at least one of them is not a Lattès map, and they have all the same degree, then the Dynamical Manin–Mumford conjecture holds for the pair $((\mathbb{P}^1)^n, \Phi)$.*

Note that although this theorem is for $(\mathbb{P}^1)^n$, the result also holds for the restriction to $\mathbb{A}^n$, since $\mathbb{A}^n$ is Zariski dense in $(\mathbb{P}^1)^n$. Dujardin and Favre [11] have shown a related result: that Dynamical Manin–Mumford holds for $(\mathbb{A}^2, \Phi)$, where Φ is any automorphism of Hénon type.

The set of invariant subvarieties for maps $\Phi : \mathbb{A}^n \to \mathbb{A}^n$ of the form $(x_1, \dots, x_n) \mapsto (f_1(x_1), \dots, f_n(x))$ was first determined by Medvedev and Scanlon [27], and their work can be extended to give all preperiodic subvarieties. We are only interested in the case where $f_1 = \cdots = f_n$ and of preperiodic hypersurfaces. (However, it turns out that $f_1 = \cdots = f_n$ is the most interesting case, and also that all lower-dimensional preperiodic subvarieties are generated as intersections of preperiodic hypersurfaces.) We compile the relevant results in the theorem statement below:

Theorem 6.4 (Medvedev, Scanlon [27]) *If $\Phi : \mathbb{A}^n \to \mathbb{A}^n$ is of the form $f \times \cdots \times f$, where f is a non-exceptional polynomial map (not conjugate to a power map or plus or minus a Chebyshev polynomial), then any hypersurface in $\mathbb{A}^n$ that is preperiodic for Φ is of the form $\{P = 0\}$, where $P(x)$ divides some polynomial in $\mathbb{C}[x_1, \dots, x_n]$, which is a product of factors of the form $\tilde{f}^n(x_i) - L(\tilde{f}^m(x_j))$.*

In terms of Fig. 3, Theorem 6.4 says that (c) implies (e1).

The proof of Dynamical Manin–Mumford by Ghioca, Nguyen, and Ye similarly proceeds by explicitly describing the subvarieties of $(\mathbb{P}^1)^n$ that have infinitely many preperiodic points for Φ, proving that (d) implies (e1) in Fig. 3. They combine this with an argument for (c) implies (d) to give an independent proof of (c) implies (e1).

6.2 The Connection with Heights

From Eqs. (1) and (2), we see that single-variable Mahler measure is directly related to heights of points in $\mathbb{P}^n(\overline{\mathbb{Q}})$. It is natural to ask if multivariate Mahler measure is also given by a height.

A very strong candidate for such a height would be a dynamical height of subvarieties of the type constructed by Zhang in [38] in the setting where X is a projective variety and $\Phi : X \to X$ is a map such that (X, Φ) has a polarization $\mathcal{L}$ (see Definition 6.1). Zhang showed that this dynamical height vanishes on both preperiodic subvarieties of X and on subvarieties containing infinitely many preperiodic points of X, and he conjectured that the converse holds. Like the Dynamical Manin–Mumford conjecture, this conjecture does not hold in full generality, and the same counterexamples apply.

Since dynamical Mahler measure also detects polynomials whose zero locus is preperiodic or has infinitely many preperiodic points, it is natural to conjecture that $\mathrm{m}_f(P)$ is equal to the dynamical height of the hypersurface $\{P = 0\}$ with respect to the map $f \times \cdots \times f$. There is an issue with making this precise, though: the hypersurface $\{P(x_1, \ldots, x_n) = 0\}$ is most naturally viewed as a subset of the affine space $\mathbb{A}^n$, and in order to use Zhang's construction, we must first complete $\mathbb{A}^n$ to a projective variety X to which the map $\Phi = f \times \cdots \times f$ extends (for instance, $X = \mathbb{P}^n$ or $(\mathbb{P}^1)^n$) and then choose a line bundle on X. There are multiple ways of doing this, which will lead to different heights, not all of which can equal the Mahler measure.

We conjecture that the correct choice is to take $X = \mathbb{P}^n$, with the map Φ given by $[x_0, \ldots, x_n] \mapsto [x_0^{\deg f}, x_0^{\deg f} f(x_1/x_0), \ldots, x_n^{\deg f} f(x_n/x_0)]$ and $\mathcal{L} = \mathcal{O}(1)$ (so $d = \deg f$).

This conjecture is supported by the work of Chambert-Loir and Thuillier [10] showing that ordinary multivariate Mahler measure agrees with the dynamical height for hypersurfaces in $\mathbb{P}^n$ under the map $f \times \cdots \times f$, where f is a power map.

A promising direction for future work would be to prove this relationship between dynamical Mahler measure and dynamical height, which would provide an additional connection between (a) and (b) in Fig. 3. Zhang's height is too technical to define here, but we give a general overview of the main ideas.

Let X be a projective variety with a line bundle $\mathcal{L}$, and $\Phi : X \to X$ a dominant endomorphism (that acts compatibly with the line bundle: $\Phi^*\mathcal{L} \cong \mathcal{L}^d$, where d should be thought of as the degree of Φ). Zhang defined a height $h_{\Phi,\mathcal{L}}$ on subvarieties of X (or more generally, cycles on X) that, like the dynamical height we saw earlier, behaves nicely under pushforward: $h_{\Phi,\mathcal{L}}(\Phi(Y)) = dh_{\Phi}(Y)$.

The dynamical height of a subvariety is always nonnegative. If Y is preperiodic, then a formal consequence of the compatibility with pushforward is that $h_{\Phi,\mathcal{L}}(Y) = 0$ ([38, Theorem 2.4(b)]). The converse implication is conjectured [38, Conjecture 2.5]. Zhang also showed that if $h_{\Phi,\mathcal{L}}(Y) > 0$, then there is a Zariski open subset $U \subseteq Y$ that contains no preperiodic subvarieties, hence in particular no preperiodic points. So the preperiodic points of Y are not Zariski dense. Taking the contrapositive, we have the following result:

Proposition 6.5 (Zhang [38]) *If the preperiodic points of Y are Zariski dense, then* $h_{\Phi,\mathcal{L}}(Y) = 0$.

In our situation, we are interested in polynomials $P(x_1, \ldots, x_n) \in \mathbb{C}[x_1, \ldots, x_n]$. These polynomials naturally cut out hypersurfaces in $\mathbb{A}^n$, which is not a projective variety. However, we can solve this problem by completing $\mathbb{A}^n$ to a projective variety X to which the map $\Phi = f \times \cdots \times f$ extends, and choosing a compatible line bundle $\mathcal{L}$. This can be done either by taking $X = \mathbb{P}^n$ (as in the work of Chambert-Loir and Thullier [10]), in which case the natural choice for $\mathcal{L}$ is $\mathcal{O}(1)$, or by taking $X = (\mathbb{P}^1)^n$ (as in the work of Ghioca, Nguyen, and Ye [15, 16]), in which case there are many choices for $\mathcal{L}$ (some of these heights for subvarieties of $(\mathbb{P}^1)^n$ are used in [15, 16]).

In this setup, the results and conjectures of Zhang can be stated in terms of Fig. 3: [38, Theorem 2.4(b)] gives the implication (c) implies (b), [38, Conjecture 2.5] would give (b) implies (c), and Proposition 6.5 gives (d) implies (b).

It is natural to hope that one of these heights is in fact equal to our Dynamical Mahler measure. We believe that the most likely possibility is, as in the work of Chambert-Loir and Thuillier, to take $X = \mathbb{P}^n$ and $\mathcal{L} = \mathcal{O}(1)$. If this were the case it would give a direct connection between statement (a) and the case of statement (b) with $X = \mathbb{P}^n$.

6.3 New Results

Here we highlight the equivalences and implications from Fig. 3 proved in [9] and in Sects. 7 and 8 of the current work.

(a) implies (d): This result for two-variable polynomials, conditional on the Dynamical Lehmer's Conjecture, is the main result of [9]. (See Theorem 4.4 here for a complete statement.) The proof proceeds in two main steps, first obtaining (a) implies (d) conditional on the Dynamical Lehmer's Conjecture, and then using the two-variable case of Theorem 6.3 from [16] for the (d) implies (e) step. More specifically, in [9, Proposition 7.6] we show that (a) implies (d) conditional on Dynamical Lehmer's Conjecture for pairs (P, f) satisfying a technical condition that we call the bounded orders property, which we then show holds in all cases.

In Sect. 9 of this chapter, we strengthen the two-variable result by showing that (a) implies (d) without Dynamical Lehmer's Conjecture for polynomials f such that $\mathrm{PrePer}(f) \subseteq \mathcal{J}(f)$. Again, combining this with the implication (d) implies (e) from [16] gives us a two-variable Kronecker's Lemma for these polynomials. The implication (a) implies (d) for polynomials in more than two variables is still open.

(e2) implies (a): This was shown in [9, Corollary 6.4]. The strategy for the proof consists of noticing that $\mathrm{m}_f(x-y) = 0$ for arbitrary $f \in \mathbb{Z}[x]$ monic, and then using the fact that the Mahler measure is invariant under composition with any polynomial commuting with f.

(e1) implies (e2): This is shown in Sect. 7 by studying integrality of polynomials that satisfy certain commutative properties. A key step is to prove that a polynomial commuting with some monic $f \in \overline{\mathbb{Z}}[x]$ (and satisfying certain technical conditions) must have coefficients in $\overline{\mathbb{Z}}[x]$. (See Proposition 7.7.)

7 An Integrality Property of Commuting Polynomials

To show that (e1) implies (e2) in Fig. 3, we need to know that we can choose our product of factors of the form $\tilde{f}^n(x_i) - L(\tilde{f}^m(x_j))$ to have integer coefficients. We will do this by showing that $\tilde{f}$ and L have algebraic integer coefficients, hence the product also has algebraic integer coefficients, and the coefficients of the product can be then assumed to be rational integers by enlarging the set of factors, if necessary, to be stable under the Galois action.

Let $\overline{\mathbb{Z}}$ be the ring of algebraic integers, which has fraction field $\overline{\mathbb{Q}}$ and unit group $\overline{\mathbb{Z}}^\times$. First we show that $\tilde{f} \in \overline{\mathbb{Z}}[x]$.

Lemma 7.1 *If $g, h \in \overline{\mathbb{Q}}[x]$ are polynomials of positive degree with leading coefficients in $\overline{\mathbb{Z}}^\times$ such that $f = g \circ h \in \overline{\mathbb{Z}}[x]$, then the polynomials $g(x + h(0))$ and $h(x) - h(0)$ both lie in $\overline{\mathbb{Z}}[x]$.*

Proof We follow the method of proof of [18, Theorem 2.1].

By replacing $g(x)$ and $h(x)$ with $g(x + h(0))$ and $h(x) - h(0)$, respectively, we may assume that $h(0) = 0$. In this case, it suffices to show that $g(x), h(x) \in \overline{\mathbb{Z}}[x]$.

Write $g(x) = a \prod_i (x - \alpha_i)$ and $h(x) = b \prod_j (x - \beta_j)$. By assumption $a, b \in \overline{\mathbb{Z}}^\times$. Note that the polynomial $f = g \circ h$ has leading coefficient equal to $a \cdot b^{\deg g} \in \overline{\mathbb{Z}}^\times$. Hence, the roots of f are all algebraic integers, and we can factor $f(x)$ in $\overline{\mathbb{Z}}[x]$ as $f(x) = a \cdot b^{\deg g} \prod_k (x - \gamma_k)$ with $\gamma_k \in \overline{\mathbb{Z}}$. We have another factorization:

$$f(x) = g(h(x)) = a \prod_i (h(x) - \alpha_i).$$

By unique factorization, we must have $h(x) - \alpha_i = b \prod_{k \in S_i} (x - \gamma_k)$ for some subset S_i of the γ_k, and so in particular $h(x) - \alpha_i \in \overline{\mathbb{Z}}[x]$ has algebraic integer coefficients. Since we have assumed that $h(0) = 0$, we conclude that $\alpha_i \in \overline{\mathbb{Z}}$ and $h(x) \in \overline{\mathbb{Z}}[x]$. Then also $g(x) = a \prod_i (x - \alpha_i) \in \overline{\mathbb{Z}}[x]$, and this concludes the proof. □

Proposition 7.2 *Suppose that $f \in \overline{\mathbb{Q}}[x]$ has degree > 1 and that the n-fold iterate $f^n = f \circ \cdots \circ f$ is a monic polynomial in $\overline{\mathbb{Z}}[x]$. Then in fact $f \in \overline{\mathbb{Z}}[x]$.*

Proof Since the leading coefficient of f^n is 1, we see that f must have leading coefficient in $\overline{\mathbb{Z}}^\times$ (in fact, it must be a root of unity, but we do not need this). The same is true for any iterate of f.

Write $c = f(0)$. We now apply Lemma 7.1 to the composition $f^n = (f^{n-1}) \circ f$ and conclude that $f(x) - c \in \overline{\mathbb{Z}}[x]$.

Hence, if we write $f(x) = a_d x^d + \cdots + a_1 x + c$, we have shown that all a_j with $j > 0$ are algebraic integers. It remains to show that also c is an algebraic integer. For this, we note that

$$f^n(c) - c = f^{n+1}(0) - c = f(f^n(0)) - c = a_d(f^n(0))^d + \cdots + a_1(f^n(0)), \quad (13)$$

where the constant terms cancel out. The right-hand side lies in $\overline{\mathbb{Z}}$ because all the a_i do, as does $f^n(0)$ since $f^n \in \overline{\mathbb{Z}}[x]$.

Hence, c is a root of a polynomial of the form $f^n(x) - x - A = 0$, where $A \in \overline{\mathbb{Z}}$ is the right-hand side of (13). Since f has degree > 1, this is a monic polynomial with algebraic integer coefficients; hence, $c \in \overline{\mathbb{Z}}$ also, as desired. □

Before proving our next statement, we need the following result:

Theorem 7.3 ([21, 34]) *If two polynomials f and g commute under composition, then up to conjugation with the same linear polynomial, either both are power functions, both are plus or minus Chebyshev polynomials, or an iterate of one is equal to an iterate of the other.*

Corollary 7.4 *If $f \in \overline{\mathbb{Z}}[x]$ is a nonlinear, monic polynomial and $g \in \overline{\mathbb{Q}}[x]$ is a nonlinear polynomial that commutes with some iterate of f, then in fact $g \in \overline{\mathbb{Z}}[x]$.*

Proof Suppose that f has degree n and g has degree m. Guided by Theorem 7.3, suppose first that f is conjugate via $L \in \mathbb{C}[x]$ to a power function; write $L(x) = ax + b$, so that $f^L(x) = x^n$. By [5, Theorem 4], $g^L(x) = uz^m$ for u a $(n-1)^{\text{st}}$ root of unity; in particular, $g^L \in \overline{\mathbb{Z}}[x]$. Now, as f and f^L are both monic, the coefficient a of L must be a root of unity, and by examining the constant term of $f^L = L^{-1} \circ f \circ L$, we find

$$0 = a^{-1}(f(b) - b),$$

which is to say that b is a root of the monic polynomial $f(x) - x \in \overline{\mathbb{Z}}[x]$. Thus L has coefficients in $\overline{\mathbb{Z}}$, and so g does also.

Next, suppose that f is conjugate via L to a Chebyshev polynomial, $f^L = T_n$. By [5, Theorem 4], $g^L(x) = uT_m(x)$ for $u = \pm 1$; in particular, $g^L \in \overline{\mathbb{Z}}[x]$. Once again the coefficient a of L must be a root of unity, and b satisfies

$$T_n(0) = a^{-1}(f(b) - b),$$

so that b is a root of the monic polynomial $f(x) - x - aT_n(0) \in \overline{\mathbb{Z}}[x]$. So L has coefficients in $\overline{\mathbb{Z}}$, and thus g does also.

Finally, assume g commutes with f^k for some k. It follows from Theorem 7.3 that $g^a = (f^k)^b$ for some positive integers a and b. Hence, $g^a \in \overline{\mathbb{Z}}[x]$ and is monic. By Proposition 7.2, we conclude that $g \in \overline{\mathbb{Z}}[x]$. □

We now show that the commuting linear function L has coefficients in the algebraic integers.

Proposition 7.5 *If f in $\overline{\mathbb{Z}}[x]$ is monic of degree > 1, and $L \in \overline{\mathbb{Q}}[x]$ is a linear polynomial that commutes with f, then $L \in \overline{\mathbb{Z}}[x]$.*

Proof Write $f(x) = x^n + c_{n-1}x^{n-1} + \cdots + c_1x + c_0$, and $L(x) = ax + b$. First look at the leading coefficient of $L \circ f = f \circ L$:

$$a = a^n,$$

so a is a root of unity; hence in $\overline{\mathbb{Z}}$.

Now look at the constant coefficient of $L \circ f = f \circ L$:

$$ac_0 + b = f(b),$$

so b is a root of the equation $f(x) - x - ac_0 \in \overline{\mathbb{Z}}[x]$. So L has algebraic integer coefficients. □

Lemma 7.6 *If f in $\overline{\mathbb{Q}}[x]$ has degree > 1, then any polynomial $g \in \mathbb{C}[x]$ commuting with f has coefficients in $\overline{\mathbb{Q}}[x]$.*

Proof Since $f \in \overline{\mathbb{Q}}[x]$, the requirement that $f \circ g = g \circ f$ gives algebraic conditions on the coefficients of g. From Boyce [5], we know that there are finitely many g of a fixed degree commuting with f, so this combined with the algebraic conditions on the coefficients ensures that $g \in \overline{\mathbb{Q}}[x]$. More specifically, the algebraic condition imposed by $f \circ g = g \circ f$ guarantees that the lead coefficient of g is in $\overline{\mathbb{Q}}$, and Boyce describes an algorithm due to Jacobsthal [19] for computing all of the coefficients of g from this lead coefficient. □

Combining the above, we have the following result.

Proposition 7.7 *If $f \in \overline{\mathbb{Z}}[x]$ is monic of degree > 1, then any polynomial $g \in \mathbb{C}[x]$ commuting with f has coefficients in $\overline{\mathbb{Z}}[x]$.*

We now prove our desired theorem.

Theorem 7.8 *Let $f \in \mathbb{Z}[x]$ be a monic polynomial with integer coefficients.*

Let $P \in \mathbb{Z}[x_1, \ldots, x_n]$ be a primitive polynomial with integer coefficients. Suppose that, working in the ring $\mathbb{C}[x_1, \ldots, x_n]$, P divides some polynomial Q in $\mathbb{C}[x_1, \ldots, x_n]$ that is a product of factors of the form $\tilde{f}^n(x_i) - L(\tilde{f}^m(x_j))$ where $m, n \geq 0$ are integers, $L \in \mathbb{C}[x]$ is a linear polynomial commuting with an iterate of f, and $\tilde{f} \in \mathbb{C}[x]$ is a nonlinear polynomial of minimal degree commuting with an iterate of f (with possibly different choices of L, $\tilde{f}$, n, and m for each factor).

Then, working in the ring $\mathbb{Z}[x_1, \ldots, x_n]$, $P(x)$ divides some polynomial R in $\mathbb{Z}[x_1, \ldots x_n]$ that is a product of factors of the form $\tilde{f}^n(x_i) - L(\tilde{f}^m(x_j))$, where $m, n, \tilde{f}$, and L are as above.

Proof Let $Q = Q_1, \ldots, Q_n$ be all the Galois conjugates of Q. Since the property of commuting with the polynomial f, which has $\mathbb{Q}$-coefficients, is preserved by the action of $\mathrm{Gal}(\overline{\mathbb{Q}}/\mathbb{Q})$, the Q_i all have factorizations of our desired form.

Now let $R = \prod_i Q_i$, which is also a product of factors of this form. Since all these factors have coefficients in $\overline{\mathbb{Z}}$, so does R. But also R is invariant under the Galois action by construction, so $R \in \mathbb{Q}[x]$; therefore, $R \in \mathbb{Z}[x]$.

Finally, we need to check that P divides R in the ring $\mathbb{Z}[x_1, \ldots, x_n]$. From the above, we know that, inside the ring $\mathbb{C}[x_1, \ldots, x_n]$, P divides Q, so also R. Since both P and R have rational coefficients, the quotient R/P does also, and P divides R in $\mathbb{Q}[x_1, \ldots, x_n]$. Since additionally P and R have integer coefficients, and P is primitive, by Gauss's Lemma P divides R in $\mathbb{Z}[x_1, \ldots, x_n]$, as desired. □

8 Dynamical Kronecker's Lemma in Some Two-Variable Cases

In this section we prove that (a) implies (d) in Fig. 3 for polynomials f that satisfy a certain condition. More precisely, we give an alternate proof of the following key step in the proof of two-variable Dynamical Kronecker, which replaces the assumption of the Dynamical Lehmer's Conjecture with the assumption that the preperiodic points of f belong to its Julia set:

Theorem 8.1 *Assume that* $\mathrm{PrePer}(f) \subseteq \mathcal{J}_f$. *If* $P(x, y) \in \mathbb{Z}[x, y]$ *with* $\mathrm{m}_f(P) = 0$, *then the graph of* $P(x, y) = 0$ *passes through infinitely many points* (α, β) *for which* α *and* β *are both preperiodic for* P.

Remark 8.2 The property $\mathrm{PrePer}(f) \subseteq \mathcal{J}_f$ is a strong assumption. In Sect. 9, we discuss some conditions that guarantee that this property is satisfied, and so Dynamical Kronecker's Lemma holds unconditionally for such f.

Before proving Theorem 8.1, we consider two lemmas.

Lemma 8.3 *Assume that* $\mathrm{PrePer}(f) \subseteq \mathcal{J}_f$. *If*

$$P(x, y) = a(x)y^k + (\textit{lower order terms in } y) \in \mathbb{Z}[x, y]$$

is a two-variable polynomial with $\mathrm{m}_f(P) = 0$, *then:*

(i) $\mathrm{m}_f(a) = 0$.
(ii) The polynomial $a(x) \in \mathbb{Z}[x]$ *is primitive (the gcd of its coefficients is 1).*
(iii) $a(x)$ *divides the polynomial* $P(x, y)$ *(in* $\mathbb{Z}[x, y]$*).*

Proof

(i) This result follows from the equality case of Proposition 5.6. In particular, from Eq. (12), we see that $\mathrm{m}_f(P)$ can be zero if and only if the two (nonnegative) summands are both zero, one of which corresponds to $\mathrm{m}_f(a)$ in this two-variable case.
(ii) This follows from (i) and the fact that non-primitive polynomials have positive dynamical Mahler measure.
(iii) It is enough to show that $a(x)$ divides $P(x, y)$ in $\mathbb{C}[x, y]$, since if this is the case, the polynomial $P(x, y)/a(x)$ must have rational coefficients, and by Gauss's Lemma, the coefficients must also be integers.

We argue by contradiction, somewhat in the style of [12, Lemma 3.20]. Suppose not: Then there is a root α of $a(x)$ such that $P(\alpha, y)$ is not identically 0. By the single-variable Dynamical Kronecker's Lemma (Lemma 4.1), since $\mathrm{m}_f(a) = 0$, the roots of $a(x)$ are in $\mathrm{PrePer}(f)$. By assumption, they are in $\mathcal{J}_f$.

As in the proof of Proposition 5.6, we have a factorization

$$P(x, y) = a(x) \prod_{i=1}^{k} (y - g_i(x)),$$

where the g_i are algebraic functions that may have branch cuts or singularities. In particular, plugging in the α above, we see that some $g_i(x)$ must have a pole at $x = \alpha$. We will show this leads to a contradiction.

We can decompose the Mahler measure of P as

$$\mathrm{m}_f(P) = \mathrm{m}_f(a) + \sum_{i=1}^{k} \int_{\mathcal{J}_f} p_\mu(g_i(x)) d\mu_f(x).$$

Since all summands are nonnegative, in order to have $\mathrm{m}_f(P) = 0$, we must have $\int_{\mathcal{J}_f} p_\mu(g_i(x)) d\mu_f(x) = 0$ for each i.

On the other hand, if g_i has a pole at α of order $r \in \mathbb{Q}$, we have a power series expansion

$$g_i(x) = (x - \alpha)^{-r} + \cdots ,$$

where the first term dominates near $x = \alpha$, so there must be some neighborhood U of α such that $p_\mu(g_i(x)) > 1$ for $x \in U$.

Then,

$$\begin{aligned} 0 = \int_{\mathcal{J}_f} p_\mu(g_i(x)) d\mu_f(x) &\geq \int_{U \cap \mathcal{J}_f} p_\mu(g_i(x)) d\mu_f(x) \\ &\geq \int_{U \cap \mathcal{J}_f} d\mu_f(x) = \mu_f(U \cap \mathcal{J}_f) > 0, \end{aligned}$$

which gives the desired contradiction. (The last step uses the fact that α lies in the support $\mathcal{J}_f$ of μ.) □

After factoring out the leading term, we are reduced to considering the case when

$$P(x, y) = y^k + (\text{lower order terms in } y)$$

as a monic polynomial in y.

Lemma 8.4 *If $P(x, y) = y^k +$(lower order terms in y) $\in \mathbb{Z}[x, y]$ is a two-variable polynomial, monic in y, with* $\mathrm{m}_f(P) = 0$*, then for any* $\alpha \in \mathcal{J}_f$ *the polynomial* $P_\alpha \in \mathbb{C}[y]$ *given by* $P_\alpha(y) = P(\alpha, y)$ *satisfies* $\mathrm{m}_f(P_\alpha) = 0$.

Proof The polynomial $P_\alpha \in \mathbb{C}[y]$ is monic, so $\mathrm{m}_f(P_\alpha) \geq 0$ for all α.
However,

$$0 = \mathrm{m}_f(P) = \int_{\mathcal{J}_f} \mathrm{m}_f(P_\alpha) d\mu_f(\alpha), \tag{14}$$

from which we can immediately deduce that $\mathrm{m}_f(P_\alpha) = 0$ for almost all $\alpha \in \mathcal{J}_f$ (that is, except possibly on a set of invariant measure 0).

Next we prove that $\mathrm{m}_f(P_\alpha)$ is a continuous function of α. To do this, write $P(\alpha, y) = \prod_i (y - g_i(\alpha)) \in \mathbb{C}[y]$, where, as above, the g_i are algebraic functions. By Jensen's formula,

$$\mathrm{m}_f(P_\alpha) = \sum_i p_{\mu_f}(g_i(\alpha)).$$

By Proposition 5.4, p_{μ_f} is continuous, and therefore, $\mathrm{m}_f(P_\alpha)$ is continuous as a function of $\alpha \in \mathbb{C}$.

Since the support of μ_f is exactly $\mathcal{J}_f$ (as discussed in the proof of [31, Theorem 6.5.8] or [36, Theorem 2, page 169]), Eq. (14) then implies that $\mathrm{m}_f(P_\alpha) = 0$ for every $\alpha \in \mathcal{J}_f$. □

Proof of Theorem 8.1 Write

$$P(x, y) = a(x)y^k + (\text{lower order terms in y}) \in \mathbb{Z}[x, y]$$

Case 1: $a(x)$ is not constant. By the single-variable Dynamical Kronecker's Lemma (Lemma 4.1), the roots of $a(x)$ are preperiodic for f, and by Part (iii) of Lemma 8.3, $P(x, y)$ contains the vertical lines through those preperiodic x-coordinates.

Case 2: $a(x)$ is constant, so it equals ± 1 by Part (i) of Lemma 8.3. By flipping the sign as necessary, we may assume that $a(x) = 1$, so we are in the situation of Lemma 8.4.

Now let α be any preperiodic point for f, and let its Galois conjugates be $\alpha = \alpha_1, \alpha_2, \ldots, \alpha_n$. Consider the polynomial $P_{\alpha_1} P_{\alpha_2} \cdots P_{\alpha_n} \in \mathbb{C}[y]$: This has algebraic integer coefficients since α is an algebraic integer (here we are crucially using the fact that f is monic) and $P \in \mathbb{Z}[x, y]$, and in fact rational integer coefficients since it is invariant under the Galois action. Since $\alpha_1, \ldots, \alpha_n \in \mathrm{PrePer}(f)$, by our assumption they are also in $\mathcal{J}_f$, and by Lemma 8.4, $\mathrm{m}_f(P_{\alpha_1} P_{\alpha_2} \cdots P_{\alpha_n}) = 0$.

From the single-variable Dynamical Kronecker's Lemma, we conclude that all roots of $P_{\alpha_1} P_{\alpha_2} \cdots P_{\alpha_n}$ are preperiodic for f. Hence, any point on the curve

$P(x, y) = 0$ with x-coordinate $\alpha = \alpha_1$ also has preperiodic y-coordinate. Since $P(x, y)$ is monic in y, the polynomial $P(\alpha, y)$ is nonzero for any such α, and thus there is some β for which $P(\alpha, \beta) = 0$. Since there are infinitely many choices for the periodic point α, we get infinitely many points with both coordinates preperiodic. □

9 Conditions for the Preperiodic Points of f to Lie in the Julia Set $\mathcal{J}_f$

Given the results of Sect. 8, it is natural to ask how restrictive the hypothesis is that $\mathrm{PrePer}(f) \subseteq \mathcal{J}_f$. This seems to be a delicate question in general. In the case of unicritical polynomials—polynomials with a unique critical point $\gamma \in \mathbb{C}$—we can answer the question completely. We begin with some background on the connection between Julia sets, filled Julia sets, and periodic points.

For $f \in \mathbb{C}[z]$, a fixed point z_0 is:

- **Repelling** if $|f'(z_0)| > 1$
- **Neutral** if $|f'(z_0)| = 1$
- **Attracting** if $|f'(z_0)| < 1$

If $|f'(z_0)| > 1$, then the image under f of a small neighborhood around z_0 expands so that z_0 "repels" nearby points. If $|f'(z_0)| < 1$, the image under f of small neighborhood around z_0 shrinks, so that z_0 "attracts" nearby points. The number $f'(z_0)$ is called the **multiplier** of the fixed point. These ideas generalize to n-cycles by considering points on the cycle as fixed points of the iterated polynomial f^n. So to study an n-cycle containing the point z_0 and determine whether it is repelling, neutral, or attracting, we consider the absolute value of

$$\left.\frac{df^n}{dz}\right|_{z=z_0} = \prod_{z_i \text{ on the cycle}} f'(z_i).$$

The equality comes from applying the chain rule to the derivative of f^n.

As noted in Sect. 2.2, all preperiodic points of a polynomial f lie in the filled Julia set $\mathcal{K}_f$. But in fact more is true: The Julia set $\mathcal{J}_f$ is the closure of the repelling periodic points of f [2, Theorem 6.9.2]. Since the Julia set $\mathcal{J}_f$ is completely invariant under f—meaning that $f(\mathcal{J}_f) = \mathcal{J}_f = f^{-1}(\mathcal{J}_f)$ [2, Theorem 3.2.4]—we need only to determine when the nonrepelling cycles lie in the Julia set.

Notice that it suffices to check this when $\mathcal{J}_f \subsetneq \mathcal{K}_f$, since in the case of $\mathcal{J}_f = \mathcal{K}_f$, the preperiodic points lie in $\mathcal{J}_f$ trivially. Observe that the condition $\mathcal{J}_f = \mathcal{K}_f$ is guaranteed if $\mathcal{J}_f$ is totally disconnected.

The following two results will be useful in the sequel.

Theorem 9.1 ([35, Theorem 1.35 (a)]) *Let $f(z) \in \mathbb{C}[z]$ be a polynomial of degree $d \geq 2$. Then f has at most $d - 1$ nonrepelling periodic cycles in $\mathbb{C}$.*

Fig. 4 The Mandelbrot set

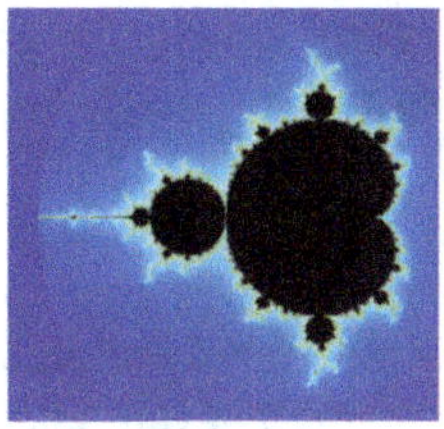

Theorem 9.2 ([37, Corollary 8.2]) *If z_0 is a periodic point with multiplier $\left.\frac{df^n}{dz}\right|_{z=z_0} = \lambda$ a root of unity, then $z_0 \in \mathcal{J}_f$.*

9.1 *The Degree-2 Case*

We start by considering the case of $\deg(f) = 2$. In this case, we can completely classify monic integer polynomials that fail to have all preperiodic points in the Julia set.

Proposition 9.3 *Let $f \in \mathbb{Z}[z]$ be monic and quadratic. Then* $\mathrm{PrePer}(f) \subseteq \mathcal{J}_f$ *unless f is affine conjugate over $\mathbb{Z}$ to either z^2 or $z^2 - 1$.*

To prove Proposition 9.3, we need a couple of tools. First, note that any quadratic polynomial $f \in \mathbb{C}[z]$ is affine conjugate over $\mathbb{C}$ to a polynomial of the form $z^2 + c$ with $c \in \mathbb{C}$. To see this, choose a conjugating function $L(z) = \frac{z}{a} - \gamma$ where a is the leading coefficient of f and γ is the unique critical point of f. Then f^L will be monic and have a critical point at zero, which gives it the desired form.

Let $f_c(z) = z^2 + c$. The Mandelbrot set (Fig. 4) is defined as

$$\mathcal{M}_2 = \left\{c \in \mathbb{C} : \sup |f_c^n(0)| < \infty\right\}.$$

The Mandelbrot set is contained in the disk of radius 2; furthermore, $\mathcal{M}_2$ satisfies $\mathcal{M}_2 \cap \mathbb{R} = \left[-2, \frac{1}{4}\right]$ (see [8, Chapter VIII, Theorem 1.2]). It follows from the discussion of Julia sets in Sect. 2.2 that for $c \notin \mathcal{M}_2$, the Julia set for f_c is totally disconnected and $\mathrm{PrePer}(f) \subseteq \mathcal{J}_f$.

The Mandelbrot set completely classifies conjugacy classes of complex quadratic polynomials. In fact, the conjugacy described above preserves the field of definition of a quadratic polynomial provided the field does not have characteristic 2. However, to prove Proposition 9.3, we need to understand conjugacy classes of monic integral quadratic polynomials, which is slightly more delicate.

Lemma 9.4 *Let $f \in \mathbb{Z}[z]$ be monic and quadratic. Then f is affine conjugate over $\mathbb{Z}$ to an integral polynomial of the form $z^2 + c$ or $z^2 + z + c$.*

Proof Let $f(z) = z^2+\alpha z+\beta \in \mathbb{Z}[z]$. We will find a linear polynomial $L(z) \in \mathbb{Z}[z]$ such that

$$f = g^L = L^{-1} \circ g \circ L$$

with g a polynomial as in the statement. We have two cases:

If α is even, take $\alpha_1 \in \mathbb{Z}$ such that $\alpha = 2\alpha_1$. Then for $L = z + \alpha_1$ we have $g(z) = z^2 + \beta - \alpha_1^2 + \alpha_1$.

If α is odd, take $\alpha_1 \in \mathbb{Z}$ such that $\alpha = 2\alpha_1 + 1$. Then for $L = z + \alpha_1$ we have $g(z) = z^2 + z + \beta - \alpha_1^2$. □

Proof of Proposition 9.3 By Theorem 9.1, a degree-2 polynomial has at most one nonrepelling periodic cycle in $\mathbb{C}$. It suffices to find one such cycle in each case.

First consider the case in which $f(z)$ is conjugate to $g(z) = z^2 + c$. Since $\mathcal{M}_2 \cap \mathbb{R} = \left[-2, \frac{1}{4}\right]$, it suffices to consider the cases of $c = 0, -1, -2$.

When $c = 0$, we immediately see that $z_0 = 0$ is an attracting fixed point in $\mathcal{K}_f \setminus \mathcal{J}_f$.

When $c = -2$, we have $g = T_2$ (the second Chebyshev polynomial), and $\mathcal{J}_f = [-2, 2] = \mathcal{K}_f$.

Finally, when $c = -1$, we find that $\{-1, 0\}$ is an attracting cycle. Indeed $g^2(z) = z^4 - 2z^2$ and $\frac{dg^2}{dz} = 4z^3 - 4z$. This gives

$$\left.\frac{dg^2}{dz}\right|_{z=0} = \left.\frac{dg^2}{dz}\right|_{z=-1} = 0.$$

(This is a general phenomenon: When a critical point f is strictly periodic, one can show that the multiplier of the cycle will be 0, and the cycle is called **superattracting**.)

Now we consider the case in which f is conjugate to $g(z) = z^2 + z + c$. Letting $L(z) = z - \frac{1}{2}$, we find that $g^L = z^2 + c + \frac{1}{4} \in \mathbb{Q}[z]$. Again using the fact that $\mathcal{M}_2 \cap \mathbb{R} = \left[-2, \frac{1}{4}\right]$, we have $\mathcal{K}_f = \mathcal{J}_f$ for $c < -\frac{9}{4}$ and $c > 0$, and we only need to check $c = 0, -1, -2$.

When $c = 0$, we have $g(z) = z^2 + z$, and $z_0 = 0$ is a neutral fixed point since $g'(z) = 1+2z$ and $g'(0) = 1$. Since the multiplier is 1, Theorem 9.2 implies $0 \in \mathcal{J}_f$.

When $c = -1$, we have $g(z) = z^2 + z - 1$ and $z_0 = -1$ is a neutral fixed point since $g'(z) = 1 + 2z$ and $f'(-1) = -1$. Since the multiplier is -1, Theorem 9.2 implies $-1 \in \mathcal{J}_f$.

Finally, when $c = -2$, we have $g(z) = z^2 + z - 2$. We claim that the roots of $z^3 + 2z^2 - z - 1$ give a neutral 3-cycle with multiplier 1. First notice that

$$g^3(z) - z = (z^2 - 2)(z^3 + 2z^2 - z - 1)^2. \tag{15}$$

The roots of the first factor in (15) are (repelling) fixed points. Let z_0 be a root of $z^3+2z^2-z-1=0$. Since z^3+2z^2-z-1 has exponent 2 in the factorization of $g^3(z)-z$, it must be a factor of the derivative $\frac{d(g^3(z)-z)}{dz}$, but this implies that

$$\left.\frac{d(g^3(z)-z)}{dz}\right|_{z=z_0}=0,$$

from which we get $\left.\frac{dg^3}{dz}\right|_{z=z_0}=1$. Since the multiplier is 1, Theorem 9.2 implies $z_0\in\mathcal{J}_f$. □

9.2 The Family z^d+c

A polynomial of the form $f_{d,c}(z)=z^d+c$ has only one critical point, namely $z_0=0$. Therefore, in this family, we have the dichotomy between connected and totally disconnected Julia sets, and we can define the Mandelbrot or Multibrot set (Fig. 5)

$$\mathcal{M}_d=\left\{c\in\mathbb{C}:\sup|f_{d,c}^n(0)|<\infty\right\}.$$

When d is odd and $d>1$, Parisé and Rochon [28] proved

$$\mathcal{M}_d\cap\mathbb{R}=\left[-\frac{d-1}{d^{\frac{d}{d-1}}},\frac{d-1}{d^{\frac{d}{d-1}}}\right],\tag{16}$$

and we remark that this implies $\mathcal{M}_d\cap\mathbb{Z}=\{0\}$ for $d>1$.

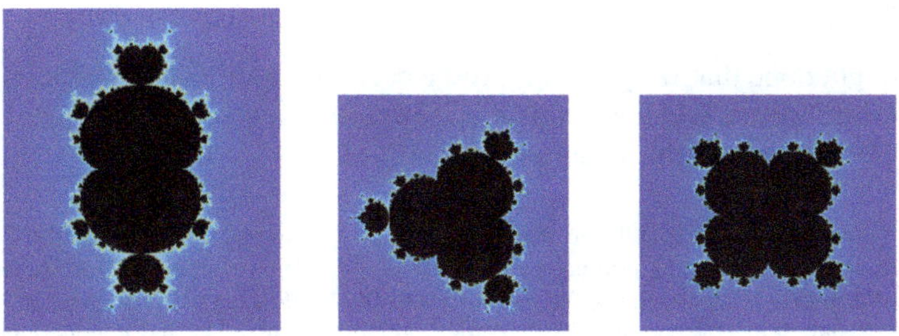

(A) (B) (C)

Fig. 5 Some Multibrot sets. (**a**) Multibrot set for $d=3$. (**b**) Multibrot set for $d=4$. (**c**) Multibrot set for $d=5$

When d is even, Parisé, Ransford, and Rochon [29] proved

$$\mathcal{M}_d \cap \mathbb{R} = \left[-2^{\frac{1}{d-1}}, \frac{d-1}{d^{\frac{d}{d-1}}}\right], \tag{17}$$

and we remark that this implies $\mathcal{M}_d \cap \mathbb{Z} = \{-1, 0\}$ for $d > 2$.

Theorem 9.5 *Let $f \in \mathbb{Z}[z]$ be a monic polynomial that is affine conjugate over $\mathbb{C}$ to a polynomial of the form $z^d + c$ with $d > 2$. Then* $\mathrm{PrePer}(f) \not\subseteq \mathcal{J}_f$ *if and only if either $c = 0$ or d is even and $c = -1$.*

Proof If $f \in \mathbb{Z}[x]$ is affine conjugate over $\mathbb{C}$ to $z^d + c$, then f has a unique critical point $\gamma \in \mathbb{C}$. Since f is monic, the derivative factors over $\mathbb{C}$ as $f'(z) = d(z-\gamma)^{d-1}$. Integrating, we get $f(z) = (z-\gamma)^d + b$ where $\gamma, b \in \mathbb{C}$, but we also know that $f(z) \in \mathbb{Z}[z]$. Of course, the derivative $f'(z) = d(z-\gamma)^{d-1} \in \mathbb{Z}[z]$ as well. The coefficient of z^{d-2} in $f'(z)$ is $-d(d-1)\gamma$, so we see that $\gamma \in \mathbb{Q}$. We claim that when $d > 2$, in fact $\gamma \in \mathbb{Z}$.

Looking at the coefficient of z in $f(z)$, we have $d\gamma^{d-1} \in \mathbb{Z}$. Let p be a prime dividing the denominator of γ. Since $d\gamma^{d-1} \in \mathbb{Z}$, we must have $p^{d-1} \mid d$. But this is impossible since $p^{d-1} \geq 2^{d-1} > d$ when $d > 2$. Now consider the constant term of f, which is $(-\gamma)^d + b$. Since $\gamma \in \mathbb{Z}$, we have $b \in \mathbb{Z}$ as well.

Choose $L(z) = z + \gamma$, and we see that $f^L(z) = z^d + b - \gamma \in \mathbb{Z}[z]$. So f is conjugate to a polynomial of the form $z^d + c$ with $c \in \mathbb{Z}$.

We know that $c \notin \mathcal{M}_d$ implies $\mathrm{PrePer}(f) \subseteq \mathcal{J}_f$. From Eqs. (16) and (17), we deduce that for $c \in \mathbb{Z}$, $c \notin \mathcal{M}_d$ iff $c \neq 0$ for d odd and $c \neq 0, -1$ for d even. Thus, it suffices to consider the exceptional cases $c = 0$ and $c = -1$.

We already know that for the power function $z^d + 0$, the point $z_0 = 0$ is an attracting fixed point in $\mathcal{K}_f \setminus \mathcal{J}_f$.

Now consider $g(z) = z^d - 1$ with d even. In this case we find that $\{-1, 0\}$ is an attracting cycle. Indeed $g^2(z) = (z^d - 1)^d - 1$ and $\frac{dg^2}{dz} = d^2 z^{d-1}(z^d - 1)^{d-1}$. This gives $\left.\frac{dg^2}{dz}\right|_{z=0} = \left.\frac{dg^2}{dz}\right|_{z=-1} = 0$. □

One might hope that for $f \in \mathbb{Z}[z]$, if the coefficients of f are sufficiently large, then all critical points will have unbounded orbit. In this case, the Julia set would be totally disconnected, so that again $\mathrm{PrePer}(f) \subseteq \mathcal{J}_f$.

Acknowledgments We are grateful to Patrick Ingram for suggesting that we study the dynamical Mahler measure of multivariate polynomials. Thanks to Patrick Ingram and to Rob Benedetto for many helpful discussions. We thank the organizers of the BIRS workshop "Women in Numbers 5"—Alina Bucur, Wei Ho, and Renate Scheidler—for their leadership and encouragement that extended for the whole duration of this project. We are also grateful to the anonymous referees for helpful comments that greatly improved the article. This work has been partially supported by the Natural Sciences and Engineering Research Council of Canada (Discovery Grant 355412-2013 to ML), the Fonds de recherche du Québec - Nature et technologies (Projets de recherche en équipe 256442 and 300951 to ML), the Simons Foundation (grant number 359721 to MM), and the National Science Foundation (grant DMS-1844206 supporting AC).

References

1. P. Atela, J. Hu, Commuting polynomials and polynomials with same Julia set. Int. J. Bifurcation Chaos Appl. Sci. Eng. **6**(12A), 2427–2432 (1996). MR 1445904
2. A.F. Beardon, *Iteration of Rational Functions: Complex Analytic Dynamical Systems*, vol. 132 (Springer, Berlin, 2000)
3. R. Benedetto, P. Ingram, R. Jones, M. Manes, J.H. Silverman, T.J. Tucker, Current trends and open problems in arithmetic dynamics. Bull. Am. Math. Soc. (N.S.) **56**(4), 611–685 (2019). MR 4007163
4. M.-J. Bertin, M. Lalín, *Mahler Measure of Multivariable Polynomials*. Women in Numbers 2: Research Directions in Number Theory. Contemporary Mathematics, vol. 606 (American Mathematical Society, Providence, 2013), pp. 125–147. MR 3204296
5. W.M. Boyce, On polynomials which commute with a given polynomial. Proc. Am. Math. Soc. **33**, 229–234 (1972). MR 291138
6. D.W. Boyd, Speculations concerning the range of Mahler's measure. Can. Math. Bull. **24**(4), 453–469 (1981). MR 644535
7. F. Brunault, W. Zudilin, *Many variations of Mahler measures: a lasting symphony*. Australian Mathematical Society Lecture Series, 28 (Cambridge University Press, Cambridge, 2020), xv+167 pp. MR 4382435
8. L. Carleson, T.W. Gamelin, *Complex Dynamics*. Universitext: Tracts in Mathematics (Springer, New York, 1993), x+175 pp. MR 1230383
9. A. Carter, M. Lalín, M. Manes, A.B. Miller, L. Mocz, Two-variable polynomials with dynamical Mahler measure zero. Res. Number Theory **8**(2), Paper No. 25, 22 pp. (2022). MR 4404457
10. A. Chambert-Loir, A. Thuillier, Mesures de Mahler et équidistribution logarithmique. Ann. Inst. Fourier (Grenoble) **59**(3), 977–1014 (2009). MR 2543659
11. R. Dujardin, C. Favre, The dynamical Manin-Mumford problem for plane polynomial automorphisms. J. Eur. Math. Soc. (JEMS) **19**(11), 3421–3465 (2017). MR 3713045
12. G. Everest, T. Ward, *Heights of Polynomials and Entropy in Algebraic Dynamics*. Universitext (Springer, London, 1999). MR 1700272
13. P. Fatou, Sur les équations fonctionnelles. Bull. Soc. Math. France **47**(48), 161–271, 33–94, 208–314 (1919, 1920)
14. A. Freire, A. Lopes, and R. Mañé, An invariant measure for rational maps. Bol. Soc. Brasil. Mat. **14**(1), 45–62 (1983). MR 736568
15. D. Ghioca, K.D. Nguyen, H. Ye, The dynamical Manin-Mumford conjecture and the dynamical Bogomolov conjecture for endomorphisms of $(\mathbb{P}^1)^n$. Comp. Math. **154**(7), 1441–1472 (2018). MR 3826461
16. D. Ghioca, K.D. Nguyen, H. Ye, The dynamical Manin-Mumford conjecture and the dynamical Bogomolov conjecture for split rational maps. J. Eur. Math. Soc. **21**(5), 1571–1594 (2019). MR 3941498
17. D. Ghioca, T.J. Tucker, S. Zhang, Towards a dynamical Manin-Mumford conjecture. Int. Math. Res. Not. IMRN **201122**, 5109–5122 (2011). MR 2854724
18. I. Gusić, On decomposition of polynomials over rings. Glas. Mat. Ser. III **43(63)**(1), 7–12 (2008). MR 2426659
19. E. Jacobsthal, Über vertauschbare Polynome. Math. Z. **63**, 243–276 (1955). MR 74373
20. J.L.W.V. Jensen, Sur un nouvel et important théorème de la théorie des fonctions. Acta Math. **22**(1), 359–364 (1899). MR 1554908
21. G. Julia, Mémoire sur la permutabilité des fractions rationnelles. Ann. Sci. École Norm. Sup. (3) **39**, 131–215 (1922). MR 1509242
22. L. Kronecker, Zwei Sätze über Gleichungen mit ganzzahligen Coefficienten. J. Reine Angew. Math. **53**, 173–175 (1857). MR 1578994
23. W.M. Lawton, A problem of Boyd concerning geometric means of polynomials. J. Number Theory **16**(3), 356–362 (1983). MR 707608

24. D.H. Lehmer, Factorization of certain cyclotomic functions. Ann. Math. (2) **34**(3), 461–479 (1933). MR 1503118
25. M.J. Lyubich, Entropy properties of rational endomorphisms of the Riemann sphere. Ergod. Theory Dyn. Syst. **3**(3), 351–385 (1983). MR 741393
26. K. Mahler, On some inequalities for polynomials in several variables. J. Lond. Math. Soc. **37**, 341–344 (1962). MR 0138593
27. A. Medvedev, T. Scanlon, Invariant varieties for polynomial dynamical systems. Ann. Math. (2) **179**(1), 81–177 (2014). MR 3126567
28. P.-O. Parisé, D. Rochon, Tricomplex dynamical systems generated by polynomials of odd degree. Fractals **25**(3), 1750026, 11 pp. (2017). MR 3654403
29. P.-O. Parisé, T. Ransford, D. Rochon, Tricomplex dynamical systems generated by polynomials of even degree. Chaotic Model. Simul. **1**, 37–48 (2017)
30. J. Pineiro, L. Szpiro, T.J. Tucker, Mahler measure for dynamical systems on $\mathbb{P}^1$ and intersection theory on a singular arithmetic surface, in *Geometric Methods in Algebra and Number Theory*. Progress in Mathematics, vol. 235 (Birkhäuser, Boston, 2005), pp. 219–250. MR 2166086
31. T. Ransford, Potential theory in the complex plane, in *London Mathematical Society Student Texts*, vol. 28 (Cambridge University, Cambridge, 1995). MR 1334766
32. M. Raynaud, Courbes sur une variété abélienne et points de torsion. Invent. Math. **71**(1), 207–233 (1983). MR 688265
33. M. Raynaud, Sous-variétés d'une variété abélienne et points de torsion, in *Arithmetic and Geometry*, vol. I. Progress in Mathematics, vol. 35 (Birkhäuser, Boston, 1983), pp. 327–352. MR 717600
34. J.F. Ritt, Permutable rational functions. Trans. Am. Math. Soc. **25**(3), 399–448 (1923). MR 1501252
35. J.H. Silverman, The arithmetic of dynamical systems, in *Graduate Texts in Mathematics*, vol. 241 (Springer, New York, 2007). MR 2316407
36. N. Steinmetz, Rational iteration, in *De Gruyter Studies in Mathematics*, vol. 16 (Walter de Gruyte, Berlin, 1993). Complex Analytic Dynamical Systems. MR 1224235
37. S. Sutherland, An introduction to Julia and Fatou sets, in *Fractals, Wavelets, and their Applications*. Springer Proceedings and Mathematical Statistics, vol. 92 (Springer, Cham, 2014), pp. 37–60. MR 3280213
38. S. Zhang, Small points and adelic metrics. J. Algebraic Geom. **4**(2), 281–300 (1995). MR 1311351
39. S.-W. Zhang, Distributions in algebraic dynamics, in *Surveys in Differential Geometry*, vol. X. Surveys in Differential Geometry, vol. 10 (International Press, Somerville, 2006), pp. 381–430. MR 2408228

Geometric Decomposition of Abelian Varieties of Order 1

Toren D'Nelly-Warady and Kiran S. Kedlaya

1 Introduction

We say that an abelian variety A over a finite field $\mathbb{F}_q$ has *order* 1 if $\#A(\mathbb{F}_q) = 1$. The simple abelian varieties of order 1 were studied by Madan–Pal [11], who showed that there are none for $q \geq 5$, one isogeny class (of genus 1) for each of $q = 3$ and $q = 4$, and infinitely many isogeny classes for $q = 2$. More precisely, Madan–Pal produced a family of abelian varieties[1] A_n over $\mathbb{F}_2$ of order 1 with the property that every simple abelian variety over $\mathbb{F}_2$ of order 1 occurs as an isogeny factor of A_n for exactly one value of n. They also gave a partial analysis of the isogeny factors of the A_n, which was completed by Robinson [15]. We recall some more details in Theorem 4.1.

The main result of this paper is to determine how the simple abelian varieties over $\mathbb{F}_2$ of order 1 decompose into isogeny factors after base extension to an algebraic closure $\overline{\mathbb{F}}_2$ of $\mathbb{F}_2$. (See Sect. 8 for the proof.)

Theorem 1.1 *Let A be a simple abelian variety over $\mathbb{F}_2$ of order 1. Let n be the unique positive integer for which A is an isogeny factor of the Madan–Pal variety A_n. Then $A_{\overline{\mathbb{F}}_2}$ is isogenous to B^f for some simple abelian variety B over $\overline{\mathbb{F}}_2$, where*

[1] Note that the A_n are only specified by Madan–Pal up to isogeny, not up to isomorphism.

D'Nelly-Warady was supported by the UC LEADS program. Kedlaya was supported by NSF (grants DMS-1802161, DMS-2053473) and UC San Diego (Warschawski Professorship).

T. D'Nelly-Warady · K. S. Kedlaya (✉)
Department of Mathematics, University of California San Diego, La Jolla, CA, USA
e-mail: tdnellywarady@ucsd.edu; kedlaya@ucsd.edu

A. Bucur et al. (eds.), *Research Directions in Number Theory*, Association for Women in Mathematics Series 33, https://doi.org/10.1007/978-3-031-51677-1_8

$$f = \begin{cases} 1 & n \text{ is a power of } 2 \text{ and } n \neq 4 \\ 2 & n \text{ is not a power of } 2 \text{ and } n \neq 7, 30 \\ 2 & n = 4 \\ 3 & n = 7 \\ 4 & n = 30. \end{cases}$$

The main step between the Madan–Pal–Robinson classification and Theorem 1.1 is to identify ways that a ratio of two Frobenius eigenvalues can equal a nontrivial root of unity. This reduces to finding solutions to a certain polynomial equation in three roots of unity; to solve this problem, we apply the paradigm used by Kedlaya–Kolpakov–Poonen–Rubinstein [8] to classify tetrahedra with rational dihedral angles, incorporating two improvements. First, the classification of minimal additive relations among roots of unity has been extended from weight 12 (by Poonen–Rubinstein [14, Theorem 3.1]) to weight 20 (by Christie–Dykema–Klep [3, Theorem 4.3]; see Theorem 6.1). Second, the parallel classification of relations mod 2 has been extended from weight 12 (see [8, Theorem 6.8]) to weight 18 (see Theorem 6.4).

This analysis leaves a handful of exceptional cases, all involving abelian varieties of dimension at most 6. For these we may appeal directly to the L-Functions and Modular Forms Database (LMFDB) [10]: a single query returns all simple abelian varieties over $\mathbb{F}_2$ of order 1 of dimension at most 6, and we may simply look up the geometric decompositions of these.

As a byproduct, we also determine when two different abelian varieties in the Madan–Pal classification are geometrically isogenous to each other.

Corollary 1.2 *Let B_1, B_2 be simple abelian varieties over $\mathbb{F}_2$ of order* 1. *Then* $\mathrm{Hom}(B_{1,\overline{\mathbb{F}}_q}, B_{2,\overline{\mathbb{F}}_q}) \neq 0$ *if and only if either B_1 and B_2 are isogenous or there exists a pair (n_1, n_2) with*

$$\{n_1, n_2\} \in \{\{1, 2\}, \{1, 4\}, \{2, 4\}, \{3, 30\}, \{6, 7\}, \{7\}, \{30\}\}$$

such that B_i is an isogeny factor of A_{n_i} for $i = 1, 2$.

It is natural to ask about other properties of simple abelian varieties of order 1; for instance, how often are they ordinary, or principally polarizable? We discuss some such questions in Sect. 9 and leave the rest to the interested reader.

Some of our arguments depend on computer calculations made in SAGEMATH [16]. We have collected these in a series of Jupyter notebooks and made them available via a GitHub repository [5]. We also use the code accompanying [8] for computing torsion closures of ideals in Laurent polynomial ideals; see [9].

2 Weil Polynomials

Throughout this section and the next, let $\mathbb{F}_q$ be a finite field of characteristic p. (We will eventually take $p = q = 2$, but we do not impose this restriction yet.)

By a *Weil polynomial* (or more precisely a *q-Weil polynomial* if we need to specify q), we will mean a monic integer polynomial $Q(x)$ of some even degree $2g$ such that every root of $Q(x)$ in $\mathbb{C}$ has absolute value $q^{1/2}$. Note that this implies that

$$Q(x) = \pm q^{-g} x^{2g} Q(q/x). \tag{2.1}$$

By a *real Weil polynomial*, we will mean a monic integer polynomial $R(x)$ with roots in the interval $[-2\sqrt{q}, 2\sqrt{q}]$. For every real Weil polynomial, the formula

$$Q(x) = x^{\deg R(x)} R(x + qx^{-1}) \tag{2.2}$$

defines a Weil polynomial $Q(x)$; the Weil polynomials that occur in this way are precisely the ones for which the plus sign occurs in (2.1).

Now let A be an abelian variety over $\mathbb{F}_q$. By results of Weil, there is a Weil polynomial $Q(x)$ which equals the characteristic polynomial of Frobenius on the ℓ-adic Tate module for every prime ℓ not dividing q. Moreover, the plus sign occurs in (2.1), so we may associate to A both a Weil polynomial and a real Weil polynomial.

We say that A is *simple* if it is nonzero and not isogenous to the product of two nonzero abelian varieties. Since multiplication of Weil polynomials corresponds to taking products of abelian varieties, any abelian variety with irreducible Weil polynomial is simple. The converse is not quite true because of the Honda–Tate theorem; see Sect. 3.

By the *Newton polygon* of A, we will mean the Newton polygon of its associated Weil polynomial $Q(x)$ with respect to the p-adic valuation v_q for the normalization $v_q(q) = 1$. It can also be computed from the Newton polygon of the real Weil polynomial $R(x)$ (for the same normalized valuation): by (2.2), all slopes in $[0, 1/2)$ have the same multiplicity in both polygons. (Explicitly, if $v_q(\alpha) \in [0, 1/2)$, then $v_q(q\alpha^{-1}) \in (1/2, 1]$ and so $v_q(\alpha + q\alpha^{-1}) = v_q(\alpha)$.) For each $s \in [0, 1)$, the multiplicity m_s of s as a slope in the Newton polygon has the property that $m_s s \in \mathbb{Z}$, e.g., from Manin's description of the formal group of A in terms of the quantities $m_s s$ [12, Theorem 4.1]; in other words, the Newton polygon of A has integral vertices. We say that A is *ordinary* if the slopes of its Newton polygon are all equal to 0 or 1.

For n a positive integer, let $A_{\mathbb{F}_{q^n}}$ denote the base extension of A from $\mathbb{F}_q$ to $\mathbb{F}_{q^n}$. From the definition of the Weil polynomial, we see that the Weil polynomial $Q_n(x)$ of $A_{\mathbb{F}_{q^n}}$ has roots which are the n-th powers of the roots of $Q(x)$; that is,

$$Q_n(x) = \mathrm{Res}_y(Q(y), y^n - x),$$

where Res denotes the resultant.

We say that A is *geometrically simple* if $A_{\mathbb{F}_{q^n}}$ is simple for each positive integer n, or equivalently if $A_{\overline{\mathbb{F}}_q}$ is simple.

Lemma 2.1 *Let A be a simple abelian variety over $\mathbb{F}_q$ with irreducible Weil polynomial $Q(x)$. If no two roots of $Q(x)$ have ratio equal to a nontrivial root of unity, then A is geometrically simple.*

Proof Let $\pi \in \overline{\mathbb{Q}}$ be a root of $Q(x)$. Let f be the integer $[\mathbb{Q}(\pi) : \mathbb{Q}(\pi^n)]$. If $\pi_1, \dots, \pi_d$ are the conjugates of π listed without repetition, then $d = [\mathbb{Q}(\pi) : \mathbb{Q}]$; $\pi_1^n, \dots, \pi_d^n$ are the conjugates of π^n; and the number of distinct entries in this list is $[\mathbb{Q}(\pi^n) : \mathbb{Q}]$. It follows that $Q_n(x)$ is the f-th power of an irreducible polynomial.

In the above notation, the condition on $Q(x)$ implies that $f = 1$ for all n, so $Q_n(x)$ is irreducible for all n. □

Lemma 2.2 *Let A be an abelian variety over $\mathbb{F}_q$ of dimension $g > 2$ with Weil polynomial $Q(x)$. Suppose that the Newton polygon of A consists of the slopes $1/g$ and $1 - 1/g$, each with multiplicity g. Then A is geometrically simple.*

Proof Since the condition on the Newton polygon is stable under base extension, it suffices to check that A is simple. For this, note that there is no way to separate the slopes of the Newton polygon into two sets of slopes which each satisfy the conditions on the Newton polygon of an abelian variety (namely, integer vertices and symmetry of slopes under $s \mapsto 1 - s$). □

3 The Honda–Tate Theorem

We summarize [17, Theorem 8, Theorem 9].

Theorem 3.1 (Honda–Tate) *There is a one-to-one correspondence between isogeny classes of simple abelian varieties over $\mathbb{F}_q$ and irreducible q-Weil polynomials, in which a simple abelian variety A corresponds to an irreducible q-Weil polynomial $Q(x)$ with the property that the Weil polynomial of A equals $Q(x)^e$ for some positive integer e. More precisely, e is the least common denominator of the following rational numbers indexed by places v of the number field $\mathbb{Q}(\pi) = \mathbb{Q}[x]/(Q(x))$:*

- *If v is real:* $\frac{1}{2}$
- *If v is complex:* 0
- *If v lies over a finite prime not equal to p:* 0
- *If v lies over p:* $\frac{\mathrm{ord}_v(\pi)}{\mathrm{ord}_v(q)}[\mathbb{Q}(\pi)_v : \mathbb{Q}_p]$

More precisely, the endomorphism algebra of A is a division algebra with center $\mathbb{Q}(\pi)$, and the listed quantities are the Brauer invariants of this division algebra.

Corollary 3.2 *For every simple abelian variety A over $\mathbb{F}_p$, the Weil polynomial of A is irreducible* unless *it equals* $(x^2 - p)^2$.

Proof In Theorem 3.1, the invariant associated to every place above p is an integer (namely the degree of the residue field of $\mathbb{Q}(\pi)_v$ over $\mathbb{F}_p$). Hence the only way to get a nontrivial denominator is for $\mathbb{Q}(\pi)$ to have a real place, which occurs if and only if $Q(x) = x^2 - p$. □

Corollary 3.3 *For every* ordinary *simple abelian variety A over $\mathbb{F}_q$, the Weil polynomial of A is irreducible.*

Proof In Theorem 3.1, the invariant associated to every place above p is an integer because $\operatorname{ord}_v(\pi)/\operatorname{ord}_v(q) \in \{0, 1\}$. Hence the only way to get a nontrivial denominator is for $\mathbb{Q}(\pi)$ to have a real place, which occurs if and only if $Q(x) = x^2 - q$ or $Q(x) = x \pm \sqrt{q}$; but these cases are not ordinary. □

4 The Madan–Pal–Robinson Classification

We describe the classification of abelian varieties over $\mathbb{F}_2$ of order 1 in terms of their associated real Weil polynomials. This is due to Madan–Pal [11] modulo the computation of certain irreducible factors, which was completed by Robinson [15].

For n a positive integer, define the integer polynomial

$$P_n(x) = \prod_{0 \le k \le n/2, \gcd(n,k)=1} \left(x^2 - (4 + 2\cos \frac{2\pi k}{n})x + 1 \right) \tag{4.1}$$

of degree $\max\{2, \phi(n)\}$. For $n = 2, 7, 30$, $P_n(x)$ is reducible:

$$P_2(x) = (x-1)^2$$
$$P_7(x) = (x^3 - 5x^2 + 6x - 1)(x^3 - 6x^2 + 5x - 1)$$
$$P_{30}(x) = (x^4 - 8x^3 + 14x^2 - 7x + 1)(x^4 - 7x^3 + 14x^2 - 8x + 1).$$

Theorem 4.1 (Madan–Pal, Robinson) *The real Weil polynomials associated to simple abelian varieties over $\mathbb{F}_2$ of order 1 are precisely those of the form $P(3-x)$, where $P(x)$ equals either $P_n(x)$ for some positive integer $n \neq 2, 7, 30$ or an irreducible factor of one of $P_2(x)$, $P_7(x)$, $P_{30}(x)$.*

Proof We first check that Corollary 3.2 applies to show that $P_n(3-x)$ is the real Weil polynomial of some abelian variety A_n over $\mathbb{F}_2$ of order 1. For this, it suffices to verify that $x^2 - 2$ never occurs as a factor of $P_n(3-x)$, or equivalently that $(3-x)^2 - 2 = x^2 - 6x + 7$ never occurs as a factor of $P_n(x)$; this is apparent because $P_n(0) = 1$.

By [11, Theorem 4], every simple abelian variety over $\mathbb{F}_2$ of order 1 is a factor of A_n for a unique value of n. It thus remains to compute the factorizations of the polynomials $P_n(x)$; this is established in [15] based on partial results from [11]. □

We next compute the Newton polygons of these abelian varieties.

Lemma 4.2 *For $n = 2^m$ with $m \geq 2$, $P_n(1) = (-1)^{n/4}2$.*

Proof Let ζ be a primitive n-th root of unity. We may then write

$$\begin{aligned} P_n(1) &= \prod_{0 \leq k \leq n/2, \gcd(n,k)=1} \left(-2 - 2\cos\frac{2\pi k}{n}\right) \\ &= (-1)^{n/4} \prod_{0 \leq k \leq n/2, \gcd(n,k)=1} (2 + \zeta^k + \zeta^{-k}) \\ &= (-1)^{n/4} \prod_{0 \leq k \leq n/2, \gcd(n,k)=1} (1 + \zeta^k)(1 + \zeta^{-k}) \\ &= (-1)^{n/4} \Phi_n(-1), \end{aligned}$$

where $\Phi_n(x) = x^{n/2} + 1$ is the n-th cyclotomic polynomial. Since evidently $\Phi_n(-1) = 2$, this proves the claim. □

Lemma 4.3 *Let A_n be the abelian variety with real Weil polynomial $P_n(3 - x)$:*

(a) *If n is not a power of 2, then A_n is ordinary.*
(b) *If n is a power of 2, then the Newton polygon of A_n has all slopes equal to $1/2^m$ or $1 - 1/2^m$ where $m = \max\{1, \log_2(n) - 1\}$.*

Proof Write $n = 2^m j$ with j odd. Over the ring of algebraic integers, (4.1) implies

$$P_n(x) \equiv \prod_{0 \leq k \leq n/2, \gcd(n,k)=1} \left(x^2 - (2\cos\frac{2\pi k}{n})x + 1\right) \equiv \Phi_n(x) \pmod 2; \tag{4.2}$$

in particular, each root of $P_n(x)$ is congruent modulo a prime above 2 to a primitive j-th root of unity. In case (a), this implies that none of the roots of $P_n(3 - x)$ have positive valuation, as each root is congruent modulo a prime above 2 to $1 - \eta$, where η is a primitive j-th root of unity for some $j > 1$; consequently, A_n is ordinary. In case (b), it implies that all of the roots of $P_n(3 - x)$ have positive valuation; this automatically implies the claim for $m = 0, 1$. For $m \geq 2$, we also need to know that $P_n(3) \equiv 2 \pmod 4$ in order to deduce that $P_n(3 - x)$ is an Eisenstein polynomial at 2; this follows by applying Lemma 4.2 and (4.2) to obtain

$$P_n(3) \equiv P_n(1) + 2P_n'(1) \equiv P_n(1) + 2\Phi_n'(1) \equiv 2 \pmod 4.$$

□

5 Reduction to an Equation in Roots of Unity

Let A_n be an abelian variety over $\mathbb{F}_2$ with real Weil polynomial $P_n(3-x)$. In light of Lemma 2.1, the key step in the proof of Theorem 1.1 will be to determine when the ratio of two Frobenius eigenvalues of A_n can equal a nontrivial root of unity. The following lemmas reduce this to a tractable problem, which we solve in the remainder of the paper.

Lemma 5.1 *For any positive integer n, the Frobenius eigenvalues of A_n are precisely the numbers α satisfying an equation of the form*

$$\alpha - 2\eta\alpha^{-1} + \eta - 1 = 0 \tag{5.1}$$

for some primitive n-th root of unity η.

Proof By definition, α is a Frobenius eigenvalue of A_n if and only if $\alpha + 2\alpha^{-1}$ is a root of the real Weil polynomial $P_n(3-x)$. From (4.1), the latter condition means that there exists a primitive n-th root of unity η such that

$$(3 - \alpha - 2\alpha^{-1}) + (3 - \alpha - 2\alpha^{-1})^{-1} = 4 + \eta + \eta^{-1}.$$

This equation simplifies to

$$(\alpha - 2\eta\alpha^{-1} + \eta - 1)(\alpha - 2\eta^{-1}\alpha^{-1} + \eta^{-1} - 1) = 0$$

and so holds if and only if one of the two factors on the left vanishes. If it is the first factor, then (5.1) holds; otherwise, (5.1) becomes true after we replace η with η^{-1}. □

Lemma 5.2 *Let α_1, α_2 be distinct Frobenius eigenvalues of A_n. If $\eta_3 = \alpha_1/\alpha_2$ is a root of unity (of any order), then there exist roots of unity η_1, η_2 of order n such that $g(\eta_1, \eta_2, \eta_3) = 0$, where*

$$\begin{aligned} g(z_1, z_2, z_3) = {} & z_1 + z_1^{-1} + z_2 + z_2^{-1} \\ & + z_3 + z_3^{-1} - z_1 z_3^{-1} - z_1^{-1} z_3 - z_2 z_3^{-1} - z_2^{-1} z_3 \\ & + z_1 z_2 z_3^{-1} + z_1^{-1} z_2^{-1} z_3 \\ & - 2 z_1 z_2 z_3^{-2} - 2 z_1^{-1} z_2^{-1} z_3^2 \end{aligned} \tag{5.2}$$

as an element of the Laurent polynomial ring $R = \mathbb{Z}[z_1^{\pm}, z_2^{\pm}, z_3^{\pm}]$.

Proof By Lemma 5.1, there exist roots of unity η_1, η_2 of order n such that

$$\alpha_1 - 2\eta_1\alpha_1^{-1} + \eta_1 - 1 = \alpha_2 - 2\eta_2^{-1}\alpha_2^{-1} + \eta_2^{-1} - 1 = 0.$$

We obtain (5.2) by substituting $\alpha_1 = \alpha_2\eta_3$ in the first equation to obtain

$$0 = \alpha_2^2\eta_3^2 + \alpha_2\eta_3(\eta_1 - 1) - 2\eta_1$$
$$0 = \alpha_2^2 + \alpha_2(\eta_2^{-1} - 1) - 2\eta_2^{-1},$$

then eliminating α_2. This is an easy resultant computation in SAGEMATH but can also be done by hand: Subtract η_3^2 times the second equation from the first to obtain

$$(\eta_3 + \eta_2^{-1}\eta_3^2 - \eta_3^2 - \eta_1\eta_3)\alpha_2 = 2\eta_2^{-1}\eta_3^2 - 2\eta_1,$$

and then substitute for α_2. □

6 Additive Relations Among Roots of Unity

Our approach to solving the equation $g(\eta_1, \eta_2, \eta_3) = 0$ from Lemma 5.2 is to treat it as an additive relation among 16 roots of unity. A strategy for analyzing such relations was described by Conway–Jones [4] and used thereafter by numerous authors; a recent example is [8], whose notation and terminology we follow (with some improvements as noted in the introduction).

Let μ be the multiplicative group of roots of unity in $\mathbb{C}$ and let $\mathbb{Z}[\mu] \subset \mathbb{C}$ be the ring of cyclotomic integers. For N a positive integer, let μ_N be the subgroup of μ generated by $\zeta_N = e^{2\pi i/N}$. By a *cyclotomic relation*, we will mean a multisubset (subset with multiplicity) of μ with sum zero; this corresponds to the notion of a *sorou* (acronym for *sum of roots of unity*) in [3]. By a *mod* 2 *cyclotomic relation*, we will mean a multisubset of μ with sum divisible by 2 in $\mathbb{Z}[\mu]$.

A cyclotomic relation (resp., a mod 2 cyclotomic relation) is *indecomposable* if it is nonempty and cannot be partitioned into two nonempty cyclotomic relations (resp., mod 2 cyclotomic relations). The cardinality of a cyclotomic relation (resp., a mod 2 cyclotomic relation) is also called its *weight*.

Theorem 6.1 *Every indecomposable cyclotomic relation of weight at most* 20 *is a subset of* μ*; that is, its elements are pairwise distinct. (The bound is best possible; see Remark 6.5.)*

Proof This follows from the explicit classification of minimal cyclotomic relations of weight at most 20 [3, Theorem 4.3]. For the case of weight at most 12, see also [14, Theorem 3.1]. □

For smaller weight, we will also need the explicit forms of such sequences.

Theorem 6.2 *Let* S *be an indecomposable cyclotomic relation of weight at most* 8*. Then then there exists* $\zeta \in \mu$ *such that* $\zeta^{-1}S = \{\zeta^{-1}\eta : \eta \in S\}$ *appears in Table 1.*

Table 1 Indecomposable cyclotomic relations of weight at most 8

n	Type	Relation
2	R_2	$1, -1$
3	R_3	$1, \zeta_3, \zeta_3^2$
5	R_5	$1, \zeta_5, \zeta_5^2, \zeta_5^3, \zeta_5^4$
6	$(R_5 : R_3)$	$\zeta_5, \zeta_5^2, \zeta_5^3, \zeta_5^4, -\zeta_3, -\zeta_3^2$
7	R_7	$1, \zeta_7, \zeta_7^2, \zeta_7^3, \zeta_7^4, \zeta_7^5, \zeta_7^6$
7	$(R_5 : 2R_3)$	$1, \zeta_5^2, \zeta_5^3, -\zeta_3\zeta_5, -\zeta_3^2\zeta_5, -\zeta_3\zeta_5^4, -\zeta_3^2\zeta_5^4$
7	$(R_5 : 2R_3)$	$1, \zeta_5, \zeta_5^4, -\zeta_3\zeta_5^2, -\zeta_3^2\zeta_5^2, -\zeta_3\zeta_5^3, -\zeta_3^2\zeta_5^3$
8	$(R_5 : 3R_3)$	$\zeta_5^2, \zeta_5^3, -\zeta_3, -\zeta_3^2, -\zeta_3\zeta_5, -\zeta_3^2\zeta_5, -\zeta_3\zeta_5^4, -\zeta_3^2\zeta_5^4$
8	$(R_5 : 3R_3)$	$\zeta_5, \zeta_5^4, -\zeta_3, -\zeta_3^2, -\zeta_3\zeta_5^2, -\zeta_3^2\zeta_5^2, -\zeta_3\zeta_5^3, -\zeta_3^2\zeta_5^3$
8	$(R_7 : R_3)$	$\zeta_7, \zeta_7^2, \zeta_7^3, \zeta_7^4, \zeta_7^5, \zeta_7^6, -\zeta_3, -\zeta_3^2$

Proof The result is originally due to Włodarski [18] and was extended to weight 9 by Conway–Jones [4]. Our notation follows [14, Table 3.1]. □

In many cases, one is particularly interested in cyclotomic relations which are stable under complex conjugation. Any such relation can be partitioned into indecomposable relations, but not uniquely; hence it is not automatic that there is such a partition which is stable under complex conjugation. We first address this question in the easier mod 2 setting, recalling [8, Lemma 6.9]; see Lemma 6.7 for an analogous statement for genuine cyclotomic relations.

Lemma 6.3 *Let S be a mod 2 cyclotomic relation (of any weight) which is stable under complex conjugation. Then S can be partitioned into indecomposable mod 2 cyclotomic relations in a manner which is itself stable under complex conjugation; that is, the conjugate of each part is also a part.*

Proof We induct on the weight of S. Let T be an indecomposable mod 2 cyclotomic relation contained in S. Let $\overline{T}$ be the complex conjugate of T. If either $T = \overline{T}$ or $T \cap \overline{T} = \emptyset$, we apply the induction hypothesis to the complement of $T \cup \overline{T}$ in S; otherwise, we apply it to the symmetric difference $T \oplus \overline{T}$ and to its complement in S. □

The following is an extension of [8, Theorem 6.7, Corollary 6.8]. The bound is best possible; see again Remark 6.5.

Theorem 6.4 *Let $S = \{\eta_1, \ldots, \eta_m\}$ be a mod 2 cyclotomic relation of weight at most 18. Then there exist signs $\sigma_1, \ldots, \sigma_m \in \{\pm 1\}$ such that $\sigma_1\eta_1 + \cdots + \sigma_m\eta_m = 0$; that is, S can be "lifted" to a cyclotomic relation.*

Proof We proceed by induction on m, using [8, Theorem 6.7] to treat the cases $m \leq 12$ as base cases. For the induction step, we may assume that S is indecomposable and $m \geq 13$. Let N be the *level* of S in the sense of [8, §6], i.e., the smallest positive integer for which $S \subseteq \mu_N$. We may assume that S is *minimal* in the sense that no rotation of S (i.e., the product of S with a root of unity) has level strictly less than N; in this case, N is odd and squarefree [8, Lemma 6.1]. Let p be the largest prime factor of N; since $m > 10$, we must have $p \geq 7$. We can write S as a disjoint union $\bigsqcup_{i=0}^{p-1} \zeta_p^i T_i$, where T_i is a nonempty set of roots of unity whose orders are not divisible by p; the sums of the T_i are pairwise congruent modulo 2 (see the proof of [8, Lemma 6.4]).

If some T_i is empty, then each T_i is itself a mod 2 cyclotomic relation; this gives a contradiction against either the hypothesis that S is indecomposable (if there is more than one nonempty T_i) or the definition of the level (if only one T_i is nonempty).

If some T_i is a singleton, we may assume without loss of generality that $T_0 = \{1\}$. Then for each $i > 0$, $T_i \cup \{1\}$ is itself a mod 2 cyclotomic relation of weight at most $n-p+1 < n$, so by the induction hypothesis we can find a function $f_i : T_i \to \{\pm 1\}$ such that

$$1 + \sum_{\eta \in T_i} f_i(\eta)\eta = 0.$$

Combining these, we obtain a lift of S to a cyclotomic relation.

We may thus assume that $\#T_i \geq 2$ for each i. This rules out $p \geq 11$, as this would imply $m \geq \sum_i \#T_i \geq 22$; we thus have $p = 7$. We also must have $\#T_i \leq 6$ for each i, as otherwise $m \geq \sum_i \#T_i \geq 7 + 6 \times 2 = 21$. We may assume without loss of generality that $T_0 = \{1, \zeta_N\}$, where $N \in \{3, 5, 15\}$. Then each T_i consists of powers of ζ_{15}, has sum congruent to $1 + \zeta_N$ mod 2, and contains no mod 2 cyclotomic relation; it is straightforward to enumerate all sets satisfying these conditions (of cardinality at most 6) using SAGEMATH.

It now suffices to show that there exist functions $f_i : T_i \to \{\pm 1\}$ for $i = 1, \dots, 6$ such that the sums $\sum_{\eta \in T_i} f_i(\eta)\eta$ are either all equal to $1 + \zeta_N$ or all equal to $1 - \zeta_N$. From the enumeration, we see that for $N \in \{3, 5\}$, we can always achieve $\sum_{\eta \in T_i} f_i(\eta)\eta = 1 + \zeta_N$. When $N = 15$, there do exist indices $i \neq j \in \{1, \dots, 6\}$ such that for any functions $f_i : T_i \to \{\pm 1\}$, $f_j : T_j \to \{\pm 1\}$,

$$\sum_{\eta \in T_i} f_i(\eta)\eta \neq 1 + \zeta_N, \qquad \sum_{\eta \in T_j} f_j(\eta)\eta \neq 1 - \zeta_N;$$

however, for all such indices, we have

$$\#T_i \geq 5, \qquad \#T_j \geq 4,$$

and this cannot occur when $n \leq 18$. □

Remark 6.5 In [3, §2.2], an example is given of an indecomposable cyclotomic relation of weight 21 with one pair of repeated elements. Omitting this pair gives a mod 2 cyclotomic relation of weight 19 which cannot be lifted to a cyclotomic relation; one can extend the proof of Theorem 6.4 to give a complete classification of such relations.

Remark 6.6 In Theorem 6.4, if S is indecomposable (mod 2), then the choice of the σ_i is unique up to multiplying them all by -1, by the following reasoning. Suppose that $\sigma'_1, \dots, \sigma'_m \in \{\pm 1\}$ is a second choice for which $\sum_i \sigma'_i \eta_i = 0$. Define the partition $\{1, \dots, m\} = J_1 \sqcup J_2$ by

$$J_1 = \{i \in \{1, \dots, m\} : \sigma_i = \sigma'_i\}, \qquad J_2 = \{i \in \{1, \dots, m\} : \sigma_i \neq \sigma'_i\}.$$

Then both $\{\sigma_i \eta_i : i \in J_1\}$ and $\{\sigma_i \eta_i : i \in J_2\}$ are cyclotomic relations, and hence mod 2 cyclotomic relations. Since S is indecomposable as a mod 2 cyclotomic relation, this is only possible if $J_1 = \emptyset$ or $J_2 = \emptyset$.

The following is an extension of [14, Lemma 4.1], but with a different proof. See also Remark 6.8.

Lemma 6.7 *Let S be a cyclotomic relation of weight at most* 18 *which is stable under complex conjugation. Then S can be partitioned into indecomposable cyclotomic relations in a manner which is itself stable under complex conjugation; that is, the conjugate of each part is also a part.*

Proof We induct on the weight of S. By Lemma 6.3, S admits a conjugation-stable partition as a mod 2 cyclotomic relation. Apply Theorem 6.4 to lift each part to a genuine cyclotomic relation; by Remark 6.6, each of these lifts is unique up to an overall sign, so we can ensure that conjugate parts lift to conjugate relations.

This yields a conjugation-equivariant function $f : S \to \{\pm 1\}$ such that $\sum_{\eta \in S} f(\eta)\eta = 0$. If f is constant, we have a partition of S of the desired form; otherwise as in Remark 6.6 the level sets of f partition S nontrivially into two cyclotomic relations, each stable under complex conjugation. □

Remark 6.8 In this paper, we will only need Lemma 6.7 for weight at most 16. We describe an alternate proof of this restricted result in the style of the original proof of [14, Lemma 4.1].

Let T be an indecomposable cyclotomic relation of minimal weight contained in S. Let $\overline{T}$ be the complex conjugate of T; it will suffice to check that either $T = \overline{T}$ or $T \cap \overline{T} = \emptyset$, as then we can remove $T \cup \overline{T}$ from S and then apply the induction hypothesis to conclude.

If T is of type R_p for some prime p in the sense of Table 1 (i.e., a collection of p equally spaced roots of unity), then $\overline{T}$ is a rotation of T and so the claim is evident. In particular, this covers all cases where T has weight at most 5; since S has weight at most 17, this also covers all cases where S can be partitioned into three or more indecomposable cyclotomic relations.

The remaining case is where the complement T' of T in S is itself an indecomposable cyclotomic relation and both T and T' have weights in the range $\{6, 7, 8\}$. By Theorem 6.2, there exists $\zeta \in \mu$ such that $\zeta^{-1}T$ is listed in Table 1. Since $T \cap \overline{T} \neq \emptyset$, there must be two (not necessarily distinct) entries η_i, η_j in the listed sequence such that $\zeta \eta_i = (\zeta \eta_j)^{-1}$, yielding

$$\zeta^2 = \eta_i^{-1}\eta_j^{-1}.$$

This limits T to a computable finite list of options, and similarly for T'.

Write T as the disjoint union of T_1 and T_2, where $T_1 = T \cap \overline{T}$, and similarly for T'. Then each of T_2 and T_2' is disjoint from its complex conjugate, and yet together they are stable under complex conjugation; they must therefore be conjugates of each other. In particular,

$$\sum T_1 = -\sum T_2 = -\overline{\sum T_2'} = \overline{\sum T_1'} = \sum T_1'.$$

We may thus check the claim by enumerating the candidates for T: for each T, computing the partition $T_1 \sqcup T_2$ and the sum $t = \sum T_1$, finding pairs $T \neq T'$ with matching values of t, and checking that for each such pair, the union $T \cup T'$ contains a cyclotomic relation which is either stable under complex conjugation or disjoint from its complex conjugate. This computation takes a few minutes in SAGEMATH.

Remark 6.9 For $\eta_1, \eta_2 \in \mu$, if $\eta_1 - \eta_2$ has 2-adic valuation greater than 1 with respect to some prime above 2 in $\mathbb{Z}[\mu]$, then $\eta_1 = \eta_2$. To see this, suppose by way of contradiction that η_1/η_2 has order $m > 1$. If $m = \ell^e$ for some prime ℓ and some positive integer e, then by [7, Theorem 10.1] the 2-adic valuation of $\eta_1 - \eta_2$ is 0 if $\ell \neq 2$ or $1/\phi(m) \leq 1$ if $\ell = 2$. Otherwise, $\eta_1 - \eta_2$ is a unit in $\mathbb{Z}[\mu]$ [7, §I.10, Exercise 2].

If instead $\eta_1 - \eta_2$ has 2-adic valuation greater than $1/2^n$ for some positive integer n, then $\eta_1^2 - \eta_2^2 = (\eta_1 - \eta_2)(\eta_1 - \eta_2 + 2\eta_2)$ has 2-adic valuation at least $1/2^{n-1}$. By induction, we deduce that $\eta_1^{2^n} = \eta_2^{2^n}$.

7 Solving an Equation in Roots of Unity

We now apply the results from Sect. 6, plus the computation of torsion closures described in [8, §7], to solve the equation $g(\eta_1, \eta_2, \eta_3) = 0$ from Lemma 5.2.

We first note that g is invariant under the group G generated by the substitutions

$$(z_1, z_2, z_3) \mapsto (z_1^{-1}, z_2^{-1}, z_3^{-1}), (z_2, z_1, z_3), (z_1, z_2^{-1}, -z_1/z_3). \tag{7.1}$$

We next apply the results of Sect. 6.

Lemma 7.1 *In the ring R, define the element $u = -z_1 z_2 z_3^{-2}$ and the subset*

$$S = \{z_1^{\pm 1}, z_2^{\pm 1}, z_3^{\pm 1}, -(z_1/z_3)^{\pm 1}, -(z_2/z_3)^{\pm 1}, (z_1 z_2/z_3)^{\pm 1}\}.$$

Then for any $\eta_1, \eta_2, \eta_3 \in \mu$ with $g(\eta_1, \eta_2, \eta_3) = 0$, there exists an element $h \in R$ with $h(\eta_1, \eta_2, \eta_3) = 0$ of one of the following forms:

(a) $h = \eta \pm 1$ *for some* $\eta \in S \cup \{u, u^{-1}\}$; *or*
(b) $h = u + u^{-1} + \sum_{\eta \in T} \eta$ *for some subset* T *of* S *with* $\#T \leq 6$, *which is invariant under the substitution* $z_i \mapsto z_i^{-1}$

Proof We apply Theorem 6.1 to the cyclotomic relation of weight 16 obtained from g by separating into monomials with coefficients ± 1, taking u and u^{-1} twice each, then evaluating at $z_i = \eta_i$. If any of the 16 terms is equal to its own conjugate, then (a) holds; we may thus assume that this does not occur.

By Theorem 6.1, this cyclotomic relation admits a partition into indecomposable cyclotomic relations; by Lemma 6.7, this partition can further be taken to be stable under complex conjugation. If we group together conjugate pairs, the two evaluations of u must end up in separate parts; hence one of these parts has weight at most 8, and so (b) holds. □

We finally proceed to solve the desired equation.

Lemma 7.2 *If* $\eta_1, \eta_2, \eta_3 \in \mu$ *satisfy* $g(\eta_1, \eta_2, \eta_3) = 0$, *then either there exists* $\zeta \in \mu$ *with*

$$\eta_1 = \eta_2 = \zeta, \eta_3 = 1 \quad or \quad \eta_1 = \eta_2 = \zeta, \eta_3 = -\zeta, \tag{7.2}$$

or $\eta_1, \eta_2, \eta_3 \in \mathbb{Q}(\mu_N)$ *for* $N \in \{15, 21, 24\}$; *more precisely, the orders of* η_1, η_2, η_3 *must fit one of the patterns listed in Table 2.*

Proof Consider the elements $h \in R$ of the forms given by Lemma 7.1 as the vertices of a Cayley graph for the action of the group G and the generators specified above, also adding an edge from h to $g - h$ when both occur as vertices. Using SAGEMATH, we compute a set U of representatives of the connected components of this graph; we have $\#U = 16$. By Lemma 7.1, our original triple (η_1, η_2, η_3) is G-equivalent to one which is a zero of some $h \in U$.

For each $h \in U$, we compute the *torsion closure* of the ideal (g, h) of R in the sense of [8, §7], i.e., the maximal ideal of R whose support contains every solution to $g(\eta_1, \eta_2, \eta_3) = h(\eta_1, \eta_2, \eta_3) = 0$ with $\eta_1, \eta_2, \eta_3 \in \mu$. This computation in SAGEMATH (for all h) takes under 5 minutes on a laptop, using the algorithm described in [8, §7] as implemented in [9].

Of the associated primes of the resulting torsion closures, two are the one-dimensional ideals corresponding to the parametric solutions (7.2). The rest are zero-dimensional ideals corresponding to Galois orbits of sporadic solutions in $\mathbb{Q}(\mu_N)$ for $N \in \{15, 21, 24\}$; we construct Table 2 by direct inspection. □

Remark 7.3 We sketch an alternate proof of Lemma 7.2 that can in principle be carried out by hand, although we do not do this completely here.

Suppose first that $h \in U$ arises from case (a) of Lemma 7.1. Modulo the ideal $(h, 2)$ of R, we may write g as indicated and then deduce the indicated consequences from Remark 6.9. (The listed values of h cover all G-orbits.)

Table 2 Orders of elements of sporadic solutions in Lemma 7.2

Order of η_1	Order of η_2	Orders of η_3
1	2	8
1	4	24
2	2	4
2	4	6,12
3	30	10, 15, 30
4	4	3, 12
6	7	21
7	7	7, 14
30	30	5, 6, 10, 15, 30

h	g modulo $(h, 2)$	Consequence
$z_1 \pm 1$	$z_2 + z_2^{-1}$	$\eta_2^4 = 1$
$z_1 z_2 z_3^{-1} \pm 1$	$z_3 + z_3^{-1}$	$\eta_3^4 = 1$
$z_1 z_2 z_3^{-2} \pm 1$	$z_1^{-2}(z_1 + z_2)(z_1 + z_2^{-1})$	$\eta_1^4 = \eta_2^{\pm 4}$

Similarly, if h arises from case (b) with $T = \emptyset$, then $\eta_1 \eta_2 \eta_3^{-2} = \pm i$ and $g \equiv z_1^{-2}(z_1 + z_2)(z_1 + z_2^{-1}) \pmod{h, 1+i}$, so by Remark 6.9 we have $\eta_1^8 = \eta_2^{\pm 8}$. With further calculation, this yields the parametric solutions (7.2) plus solutions with $\eta_1, \eta_2, \eta_3 \in \mathbb{Q}(\mu_N)$ for $N = 24$.

Suppose next that $h \in U$ arises from case (b) of Lemma 7.1. We have already treated the case $\#T = 0$. If $\#T = 2$, then $T = \{\mu, \mu^{-1}\}$ for some $\mu \in S$; the cases $\mu = z_2, -z_2 z_3^{-1}$ cover all G-orbits. Since Table 1 contains no entries with $n = 4$, either $u + \mu$ or $u + \mu^{-1}$ evaluates to 0. With further calculation, this yields the parametric solutions (7.2) plus solutions with $\eta_1, \eta_2, \eta_3 \in \mathbb{Q}(\mu_N)$ for $N \in \{7, 12, 15\}$.

If $\#T = 4$, then up to G-action we may take T to be one of four options. One of these is

$$T = \{(-z_1 z_3^{-1})^{\pm 1}, (-z_2 z_3^{-1})^{\pm 1}\},$$

in which we have the algebraic relation $(-z_1 z_3^{-1})(-z_2 z_3^{-1}) = -u$. In this case, we first solve the equation $h(\eta_1, \eta_2, \eta_3) = 0$ in roots of unity; this yields solutions with $\eta_1 \eta_3^{-1}, \eta_2 \eta_3^{-1} \in \mathbb{Q}(\mu_N)$ with $N = 24$. We then substitute into g and solve further to confirm that $\eta_1, \eta_2, \eta_3 \in \mathbb{Q}(\mu_N)$.

The remaining options for T with $\#T = 4$ are

$$\{z_1^{\pm 1}, z_2^{\pm 1}\}, \{z_2^{\pm 1}, (-z_2 z_3^{-1})^{\pm 1}\}, \{z_3^{\pm 1}, (-z_2 z_3^{-1})^{\pm 1}\}.$$

By Lemma 6.7, either h corresponds to the unique sequence listed in Table 1 with $n = 6$ rotated by ± 1, or h splits into two conjugate relations of type R_3. In the first case, we must have $\eta_1, \eta_2, \eta_3 \in \mathbb{Q}(\mu_N)$ for $N = 15$ without further calculation.

(When $T = \{z_1^{\pm 1}, z_2^{\pm 1}\}$ solving for η_3 requires a square root but ends up not forcing an increase in N.) In the second case, u equals a cube root of unity times one of $z_2^{\pm 1}, (-z_2 z_3^{-1})^{\pm 1}$; with further calculation, this yields solutions with $\eta_1, \eta_2, \eta_3 \in \mathbb{Q}(\mu_N)$ for $N \in \{15, 24\}$.

If $\#T = 6$, then both h and $g - h$ correspond to sequences listed in Table 1 rotated by ± 1. (Note that h cannot be the sum of two disjoint conjugate minimal sequences because Table 1 contains no entries with $n = 4$.) In particular, since z_1, z_2, z_3 each appear in either h or $g - h$, we must have $\eta_1, \eta_2, \eta_3 \in \mathbb{Q}(\mu_N)$ for $N \in \{15, 21\}$ without further calculation.

8 Geometric Decompositions

With Lemma 7.2 in hand, we now proceed to the proof of Theorem 1.1 and Corollary 1.2.

Proof of Theorem 1.1 As noted earlier, the LMFDB asserts this result in all cases where $\dim(A) \leq 6$, so we may ignore those cases in what follows. Since the cases $n = 1, 2, 7, 30$ are excluded by our dimension bound, Theorem 4.1 implies that the real Weil polynomial associated to A equals $P_n(3 - x)$.

Suppose that α_1, α_2 are distinct Frobenius eigenvalues of A such that $\eta_3 = \alpha_1/\alpha_2$ is a root of unity. By Lemma 5.2, the roots of unity η_1, η_2 of order n corresponding to α_1, α_2 via (5.1) satisfy $g(\eta_1, \eta_2, \eta_3) = 0$. Note that each of the exceptional cases listed in Table 2 includes a now-excluded value of n and that the first family listed in (7.2) is inconsistent with the hypothesis that $\alpha_1 \neq \alpha_2$. Consequently, Lemma 7.2 implies that η_1, η_2, η_3 must belong to the second family listed in (7.2); that is, α_1 and α_2 are the two roots of (5.1) for some common value of η.

From the previous calculation, it follows that for any positive integer m, the Weil polynomial associated to $A_{\mathbb{F}_{2^m}}$ is either irreducible or the square of an irreducible Weil polynomial. Moreover, the latter outcome occurs when $m = n$: For each root of unity η of order n, both roots of (5.1) occur as Frobenius eigenvalues of A_n and their ratio is $-\eta$. We deduce that the Weil polynomial associated to $A_{\mathbb{F}_{q^n}}$ is the square of an irreducible Weil polynomial $Q(x)$, and moreover, no two of the roots of $Q(x)$ have ratio equal to a nontrivial root of unity.

If n is not a power of 2, then A is ordinary by Lemma 4.3, as then is $A_{\mathbb{F}_{q^n}}$. Hence Corollary 3.3 implies that $Q(x)$ is the Weil polynomial associated to an abelian variety over $\mathbb{F}_{q^n}$, so we may apply Lemma 2.1 to deduce that $A_{\overline{\mathbb{F}}_q}$ is isogenous to the square of an abelian variety.

If n is a power of 2, then we may use our dimension bound to reduce to the case $n \geq 8$. Lemma 4.3 implies that the slopes of the Newton polygon of A are all equal to $2/n$ or $1 - 2/n$; we may thus apply Lemma 2.2 to deduce that A is geometrically simple. □

Proof of Corollary 1.2 By Theorem 4.1, there exists a unique integer n_i such that B_i is an isogeny factor of A_{n_i}. Let α_i be a Frobenius eigenvalue of B_i. By Theorem 3.1, $\mathrm{Hom}(B_{1,\overline{\mathbb{F}}_q}, B_{2,\overline{\mathbb{F}}_q}) \neq 0$ if and only if some conjugate of α_2 equals α_1 times a root of unity. Using this criterion, it is straightforward to verify the "if" implication.

To check the "only if" implication, we may assume that B_1 and B_2 are not isogenous and that α_1/α_2 is a root of unity. If $n_1 = n_2$, then by Theorem 4.1 the common value is in $\{7, 30\}$. If $n_1 \neq n_2$, we may choose a root of unity η_i of order n_i associated to α_i via Lemma 5.1; by Lemma 7.2, n_1 and n_2 must appear as the first two entries in some row of Table 2. □

9 Additional Properties

In [2], it is shown for any fixed q, any sufficiently large positive integer is the order of some abelian variety over $\mathbb{F}_q$ which is simultaneously ordinary, geometrically simple, and principally polarizable. It is thus natural to ask how these conditions interact for simple abelian varieties of order 1 over $\mathbb{F}_2$.

We first consider the combination of the ordinary and geometrically simple conditions. From the proof of Theorem 1.1 and our calculation of Newton polygons (Lemma 4.3), we obtain the following.

Corollary 9.1 *Let A be a simple abelian variety over $\mathbb{F}_2$ of order* 1. *Then A cannot be both ordinary and geometrically simple. In addition, if* $\dim(A) \geq 3$, *then A is either ordinary or geometrically simple (but not both).*

As for the principally polarizable condition, we have the following partial result.

Theorem 9.2 *Let n be a power of an odd prime p. Then A_n is isogenous to a principally polarizable abelian variety.*

Proof Let α be a Frobenius eigenvalue of A_n and choose $\eta \in \mu$ of order n for which (5.1) holds; then $\eta \in \mathbb{Q}(\alpha)$. Define

$$\beta = \alpha + 2\alpha^{-1}, \qquad \xi = \eta + \eta^{-1};$$

we derive from (5.1) the equation

$$\beta^2 + (\xi - 2)\beta - (2 + 3\xi) = 0,$$

which as a quadratic equation in β has discriminant

$$(\xi - 2)^2 + 4(2 + 3\xi) = (\xi + 2)(\xi + 6).$$

We now have the following field diagram:

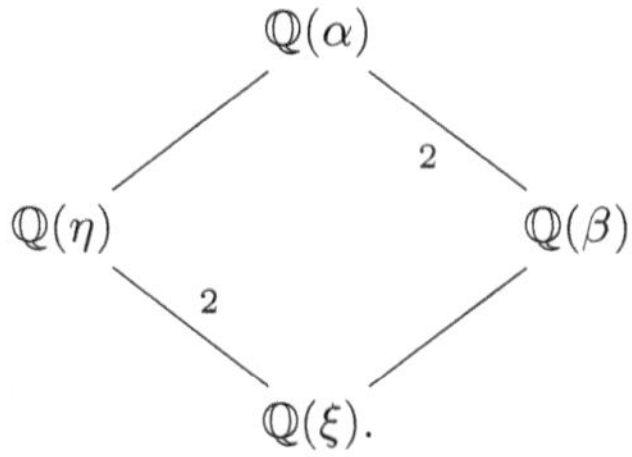

Let v be a prime of $\mathbb{Q}(\xi)$ lying above p; then modulo v, ξ is congruent to 2 and so $(\xi+2)(\xi+6)$ is congruent to 32. As a result, v does not ramify in $\mathbb{Q}(\beta)$; since v does ramify in $\mathbb{Q}(\eta)$, every prime of $\mathbb{Q}(\beta)$ above v ramifies in $\mathbb{Q}(\alpha)$. We may thus apply [6, Theorem 1.1] to deduce that A_n is isogenous to a principally polarizable abelian variety. □

Remark 9.3 For positive integers n such that the simple isogeny factors of A_n are of dimension at most 6, LMFDB reports the following as to whether or not these factors are isogenous to a principally polarizable abelian variety:

- Yes: $n = 1, 2, 3, 4, 5, 7, 9$.
- No: $n = 6, 10, 12, 14, 18, 30$.
- Unknown: $n = 8$.

It may be possible to collect more data using the methods of [13] and [1], especially when A_n is ordinary (which by Lemma 4.3 occurs when n is not a power of 2).

References

1. J. Bergström, V. Karemaker, S. Marseglia, Polarizations of abelian varieties over finite fields via canonical liftings (2021). arXiv:2101.05531v3
2. R. van Bommel, E. Costa, W. Li, B. Poonen, A. Smith, Abelian varieties of prescribed order over finite fields (2021). arXiv:2106.13651v1
3. L. Christie, K.J. Dykema, I. Klep, Classifying minimal vanishing sums of roots of unity (2020). arXiv:2008.11268v1
4. J.H. Conway, A.J. Jones, Trigonometric Diophantine equations (On vanishing sums of roots of unity). Acta Arith. **30**, 229–240 (1976)
5. T. D'Nelly-Warady, K.S. Kedlaya, GitHub repository. https://github.com/mwarady22/Geometric-decomposition-of-abelian-varieties-of-order-1
6. E.W. Howe, Kernels of polarizations of abelian varieties over finite fields. J. Algebraic Geom. **5**, 583–608 (1996)
7. G.J. Janusz, *Algebraic Number Fields*, 2nd edn. Graduate Studies in Math. 7, (American Mathematical Society, Providence, 1996)
8. K.S. Kedlaya, A. Kolpakov, B. Poonen, M. Rubinstein, Space vectors forming rational angles (2020). arXiv:2011.14232v1
9. K.S. Kedlaya, A. Kolpakov, B. Poonen, M. Rubinstein, GitHub repository. https://github.com/kedlaya/tetrahedra/
10. LMFDB Collaboration, The L-Functions and Modular Forms Database. https://lmfdb.org (Retrieved July 2022)

11. M.L. Madan, S. Pal, Abelian varieties and a conjecture of R. M. Robinson. J. Reine Angew. Math. **291**, 78–91 (1977)
12. Yu.I. Manin, The theory of commutative formal groups over fields of finite characteristic [Russian]. Usp. Math. **18**, 3–90 (1963); English translation, Russ. Math. Surveys **18**, 1–80 (1963)
13. S. Marseglia, Computing square-free polarized abelian varieties over finite fields. Math. Comp. **90**, 953–971 (2021)
14. B. Poonen, M. Rubinstein, The number of intersection points made by the diagonals of a regular polygon. SIAM J. Discrete Math. **11**, 135–156 (1998)
15. R.M. Robinson, Conjugate algebraic units in a special interval. Math. Z. **154**, 31–40 (1977)
16. Sage Developers, The SageMath version 9.6, 2022. https://www.sagemath.org
17. W.C. Waterhouse, J.S. Milne, *Abelian Varieties Over Finite Fields*. 1969 Number Theory Institute (Proceedings of Symposia in Pure Mathematics, vol. XX, State Universities, New York, Stony Brook, 1969, pp. 53–64; American Mathematical Society, Providence, 1971)
18. Ł. Włodarski, On the equation $\cos\alpha_1+\cos\alpha_2+\cos\alpha_3+\cos\alpha_4 = 0$. Ann. Univ. Sci. Budapest. Eötvös Sect. Math. **12**, 147–155 (1969)

On Markoff Type Surfaces over Number Fields and the Arithmetic of Markoff Numbers

Seoyoung Kim, Damaris Schindler, and Jyothsnaa Sivaraman

1 Introduction

The main goal of this chapter is to study the arithmetic of Markoff equations and generalized Markoff equations. The classical Markoff equation [18] is the affine cubic equation over the rational integers

$$x^2 + y^2 + z^2 - 3xyz = 0.$$

There are infinitely many integer solutions to this equation, and by ordering them into a Markoff tree, Zagier [21] has given an asymptotic for the number of integer solutions of bounded naive height, refining previous results of Gurwood [14]. More recently, Mirzakhani [19] has given a yet different proof as special case of her much more general work, using a dynamical point of view. For generalizations of Zagier's work, see work of Baragar [2] and Gamburd, Magee and Ronan [12]. It is a natural question to deform this equation and say for $k \in \mathbb{Z}$ study the solubility of

$$x^2 + y^2 + z^2 - 3xyz = k \tag{1}$$

over the rational integers. If there is a solution with coordinates in $\mathbb{Z}$, then there is also a solution in $\mathbb{R}$ and solutions in $\mathbb{Z}_p$ for every prime p. If the existence of

S. Kim · D. Schindler
Mathematisches Institut, Georg-August Universität Göttingen, Göttingen, Germany
e-mail: seoyoung.kim@mathematik.uni-goettingen.de;
damaris.schindler@mathematik.uni-goettingen.de

J. Sivaraman (✉)
IISER Thiruvananthapuram, Vithura, Kerala, India
e-mail: jyothsnaa.s@gmail.com

A. Bucur et al. (eds.), *Research Directions in Number Theory*, Association for Women in Mathematics Series 33, https://doi.org/10.1007/978-3-031-51677-1_9

these local solutions is already sufficient to deduce the existence of a solution over $\mathbb{Z}$, then we say that the affine Hasse principle holds. The question, for which k the affine Hasse principle holds, has been taken up among other questions in the work of Ghosh and Sarnak [13]. First they find congruence conditions for $k \in \mathbb{Z}$ such that Eq. (1) has solutions in all of $\mathbb{Z}_p$ for every prime p. Those values of k are called admissible. Moreover, an admissible k is called generic if there is no solution to (1) with one coordinate in the set $\{0, \pm 1, \pm 2\}$. For generic admissible k, they then describe fundamental domains for the integer solutions to (1) modulo the action of Vieta involutions, permutations of coordinates, and sign changes of two of the coordinates. These fundamental domains, which are in this case compact, are then used to find infinite families of counterexamples to the affine Hasse principle. Note that not all of the counterexamples to the affine Hasse principle for Eq. (1) can be explained by a Brauer–Manin obstruction, see [17] and [10] for a study of the associated Brauer groups and Brauer–Manin sets. Once the Hasse principle holds for a certain member in the family, it is an interesting question when strong approximation holds. For the original Markoff equation, this has been studied in the work of Bourgain, Gamburd and Sarnak [5] and has recently seen a breakthrough by Chen [9], who showed that the classical Markoff equation satisfies a strong approximation property at all but finitely many primes.

One goal of this chapter is to study the arithmetic of homogeneous and inhomogeneous Markoff equations over number fields. Let K be a number field and $\mathcal{O}_K$ its ring of integers. For $a, k \in K$ consider the equation

$$\mathcal{M}_{a,k} : x^2 + y^2 + z^2 - axyz = k,$$

and more generally for $n \in \mathbb{N}_{\geq 3}$, the affine hypersurface

$$\mathcal{M}_{a,k,n} : x_1^2 + \ldots + x_n^2 - ax_1x_2 \ldots x_n = k.$$

For $k = 0$, these have already been studied in 1907 by Hurwitz [16] and are now also known as Markoff–Hurwitz equations. Here again we can ask if a local to global principle holds for solutions in the ring of integers $\mathcal{O}_K$. First one notices that one has again n Vieta involutions mapping solutions over $\mathcal{O}_K$ to solutions over $\mathcal{O}_K$ assuming that a and k are integral. It is then a natural question to ask if there are finitely many orbits under the group Γ generated by all n Vieta involutions, permutations of variables, and sign changes of two of the variables. Hurwitz [16] showed that over the rational integers the set of integer solutions falls into finitely many orbits under Γ. The question of existence of solutions has, for example, been studied by Herzberg in [15] for the homogeneous case and in the work of Fine et al. [11] for more general cases.

Over other rings of integers of number fields, one may not at all expect that the set of solutions is a union of only finitely many orbits under Γ. Silverman in [20] showed in the case of the homogeneous Markoff equation in 3 variables and

$a \neq 0$ that one has infinitely many orbits as soon as the group of units in the ring of integers is infinite and the equation $u^2 + v^2 + 1 \equiv 0 \bmod a\mathcal{O}_K$ is soluble in units. The question of finite generation of all integral solutions under the group Γ has been taken up for the homogeneous Markoff–Hurwitz equation by Baragar [3]. He showed that for $k = 0$, $a \neq 0$, and a number field $K \neq \mathbb{Q}$ that has a real embedding, the set of solutions is either empty or the set of orbits under the group Γ is infinite. The strategy of proof is to show that every orbit of Γ is discrete after embedding the solutions into $\mathbb{R}$ and that the solution set over $\mathcal{O}_K$ is either empty or has a cluster point. We study the question on finiteness of fundamental solutions in the case of the inhomogeneous Markoff–Hurwitz equation. In Sect. 2, we show that for $k < 0$ and $a > 0$ every orbit under Γ is discrete. Our first main goal of this chapter is then to generalize the work of Silverman and Baragar to inhomogeneous Markoff equations, where in the case of three variables, we obtain the following result.

Theorem 1.1 *Let $K/\mathbb{Q}$ be a number field and $a \in \mathcal{O}_K$ with $a \neq 0$. Suppose that K has a real embedding and that $K \neq \mathbb{Q}$. Then consider the equation*

$$x^2 + y^2 + z^2 = axyz + k \tag{2}$$

with $a, k \in \mathcal{O}_K$. We assume that $\tau(a) > 0$ and $\tau(k) < 0$ for all real embeddings τ of K. Then the number of orbits is either 0 or infinite.

Next we focus on the situation of real quadratic fields and explicitly work out domains for minimal solutions in orbits under Γ inspired by the fundamental domains that appear in the work of Ghosh and Sarnak [13]. For this we consider the case of three variables and allow k to be positive or negative.

We already know that the family of inhomogeneous Markoff equations (1) with $a = 3$ in general does not satisfy the integral Hasse principle over the rational integers. What happens over number fields? Our second main goal of this chapter is the study of local solubility (see Sect. 5) as well as global solubility (see Sect. 6) and the construction of counterexamples to the Hasse principle over number fields. For the violation to the integral Hasse principle over number fields, we obtain the following example with elementary methods. To produce further examples and work over more general number fields, one could study the Brauer–Manin obstruction and evaluation of Brauer classes as in [17] and [10].

Theorem 1.2 *Let K be the subfield of degree 3 of the field $\mathbb{Q}(\zeta_{31})$. Consider the equation*

$$x_1^2 + x_2^2 + x_3^2 - 3x_1x_2x_3 = 118. \tag{3}$$

This has no global solutions but has local solutions modulo each power of a prime ideal of K.

In another direction, one can study the arithmetic of Markoff equations in fixing one member of the families described above and try to learn more about the arithmetic properties of the solutions. Considering $\mathcal{M}_{3,0}$, we find all Markoff numbers as coordinates of positive solutions. It is not much known about arithmetic properties of Markoff numbers. By a deep result of Bourgain, Gamburd and Sarnak [5], we know that almost all Markoff numbers are composite. For more background on Markoff numbers, we refer the reader to the book of Aigner [1] (see also work of Bombieri [4]).

What is the typical size of the largest prime factor of Markoff numbers? In Sect. 7, we study the linear recurrence sequences occurring as branches of the Markoff tree. Combining this with the work of Bugeaud [6], we give a lower bound on the largest prime factor of Markoff numbers that appear in certain branches of the Markoff tree.

2 Inhomogeneous Markoff Equations over Number Fields with a Real Embedding

Let K be a number field with ring of integers $\mathcal{O}_K$ and at least one real embedding. For the rest of this section, fix a real embedding $K \subset \mathbb{R}$ and let

$$\mathcal{M}_{a,k} : x_1^2 + x_2^2 + x_3^2 = ax_1x_2x_3 + k,$$

where $a, k \in \mathcal{O}_K$ and $a > 0, k < 0$. Further $\mathcal{M}_{a,k}(\mathcal{O}_K)$ shall be used to denote the set of all solutions of $\mathcal{M}_{a,k}$ with coordinates in $\mathcal{O}_K$. Let ϕ_i for $i \in \{1, 2, 3\}$ be used to denote the following involutions:

$$\begin{aligned} \phi_1(x_1, x_2, x_3) &= (x_2, x_3, ax_2x_3 - x_1), \\ \phi_2(x_1, x_2, x_3) &= (x_1, x_3, ax_1x_3 - x_2) \\ \text{and} \quad \phi_3(x_1, x_2, x_3) &= (x_1, x_2, ax_1x_2 - x_3). \end{aligned}$$

These involutions are called the Vieta involutions. We denote by Γ the group generated by the involutions ϕ_i, the automorphisms of the $\mathcal{M}_{a,k}(\mathcal{O}_K)$ given by permuting the coordinates of each solution, and the automorphisms given by sign change of the first two entries in a solution tuple.

We define more generally the following form of equations:

$$\mathcal{M}_{a,k,n} : x_1^2 + x_2^2 + \cdots + x_n^2 = ax_1x_2\cdots x_n + k, \quad \text{for } a > 0, k < 0,$$

which was introduced and studied in [3]. That is to say, $\mathcal{M}_{a,k,3} = \mathcal{M}_{a,k}$. Similarly, the idea of n Vieta involutions

$$\phi_i(x_1, x_2, \ldots, x_n) = (x_1, \ldots, \hat{x}_i, \ldots, x_n, \left(\frac{ax_1\cdots x_n}{x_i}\right) - x_i), \quad i = 1, \ldots, n, \tag{4}$$

where the hat denotes that the component is omitted, and possible permutations along with sign changes can be perused to study the set $\mathcal{M}_{a,k,n}(\mathcal{O}_K)$ of integral solutions of $\mathcal{M}_{a,k,n}$. In case that $x_i = 0$, we read the expression above as $\frac{ax_1 \ldots x_n}{x_i} = ax_1 \ldots x_{i-1}x_{i+1} \ldots x_n$.

In this section, we assume that **the set $\mathcal{M}_{a,k,n}(\mathcal{O}_K)$ is non-empty**.

Definition 1 For a point P in $\mathcal{M}_{a,k,n}(\mathcal{O}_K)$, we define $h(P) = \max_i |x_i|$.

Definition 2 If $h(\phi(P)) \geq h(P)$ for every Vieta involution ϕ, then $P \in \mathcal{M}_{a,k}(\mathcal{O}_K)$ will be called a fundamental solution.

We say that a solution $(x_1, x_2, \ldots, x_n)$ is a positive ordered solution in $\mathcal{M}_{a,k,n}$ if $0 < x_1 \leq x_2 \leq \cdots \leq x_n$. We say that two positive ordered solutions are connected if there exists a Vieta involution that maps one to the other. In this case, we say that the two positive ordered solutions are neighbors. The following lemma illustrates the relation between neighbors when the height of the image of ϕ_n is strictly less than its preimage.

Lemma 2.1 *Let $\mathbf{x} = (x_1, x_2, \ldots, x_n)$ be a positive ordered solution in $\mathcal{O}_K$ to the equation*

$$\mathcal{M}_{a,k,n} : x_1^2 + x_2^2 + \cdots + x_n^2 = ax_1x_2\cdots x_n + k, \quad \text{for } a > 0, k < 0,$$

such that $h(\phi_n(\mathbf{x})) < h(\mathbf{x})$. Then either $\phi_n(\mathbf{x})$ is a fundamental solution, or we have

$$h(\mathbf{x}) > h(\phi_n(\mathbf{x})) - \frac{k}{h(\mathbf{x})} + \frac{n-2}{2h(\mathbf{x})}\left(\frac{2}{a}\right)^{2/(n-2)}.$$

Proof Since the proof is similar to the proof of Lemma 1.1 in [3], we will give a rough sketch of the proof. As we assume $k < 0$, we have the $ax_1x_2\cdots x_{n-2} > 2$. This implies that for $i \leq n-1$ we have

$$ax_1 \cdots \hat{x}_i \cdots x_{n-1} > 2.$$

Therefore,

$$ax_1 \cdots \hat{x}_i \cdots x_n - x_i > 2x_n - x_i > x_n.$$

Therefore, if $\mathbf{x} = (x_1, x_2, \ldots, x_n)$ is a positive ordered solution, then $\phi_i(\mathbf{x})$ is also a positive ordered solution for $i = 1, \ldots, n-1$. In other words, ϕ_n is the only involution that can (potentially) decrease the height of a given solution in $\mathcal{M}_{a,k,n}(\mathcal{O}_K)$. Now the arithmetic–geometric inequality gives

$$x_1^2 + \cdots + x_{n-2}^2 \geq (n-2)\left(\frac{2}{a}\right)^{2/(n-2)}.$$

Under the assumption $h(\phi_n(\mathbf{x})) < h(\mathbf{x})$, the above inequality implies

$$\frac{n-2}{2x_n}\left(\frac{2}{a}\right)^{2/(n-2)} - \frac{k}{x_n} < x_n - x_{n-1}.$$

If $x_n' = ax_1 \cdots x_{n-1} - x_n > x_{n-1}$, then $\phi_n(\mathbf{x})$ is a positive ordered solution. This implies that $\phi_n(\mathbf{x})$ is a fundamental solution. Otherwise, we have $x_n' \leq x_{n-1}$ and $h(\phi_n(\mathbf{x})) = x_{n-1}$. Thus, we obtain the result. □

Lemma 2.2 *Any orbit in the set of ordered solutions in $\mathcal{M}_{a,k,n}(\mathcal{O}_K)$ is discrete in $\mathbb{R}^n$.*

Proof The proof is similar to that of Baragar [3] section 1. For simplicity, we consider the case $n = 3$. A similar argument can easily be reproduced for the case $n \geq 3$ using Lemma 2.1. We assume the following ordering on the solution $P = (x, y, z) \in \mathcal{M}_{a,k}(\mathcal{O}_K)$

$$0 \leq |x| \leq |y| \leq |z|.$$

Given a solution to the Markoff equation, either all the entries are non-negative or exactly two are negative. We first work with the non-negative solutions in the orbit. That is to say, the solutions (x, y, z) with $0 \leq x \leq y \leq z$ to the equation $\mathcal{M}_{a,k} : x^2 + y^2 + z^2 = axyz + k$ satisfy

$$x^2 + (y-z)^2 = (ax-2)yz + k.$$

It follows that since $k < 0$, $ax > 2$. If for a point P, $|axy - z| > z$, then consider the Vieta involution ϕ given by

$$(x, y, z) \to (x, y, axy - z).$$

We have $\phi \cdot \phi(P) = P$, and therefore,

$$h(\phi \cdot \phi(P)) < h(\phi(P)).$$

Since this is the only involution for which we may obtain a Markoff triple with lower height, it follows that P is a fundamental solution. If for a positive ordered solution P, $|axy - z| \leq z$, it is still true that $axy - z > 0$ since two entries of this triple are already positive. Therefore, we have $axyz \leq 2z^2$, and it now follows that

$$x^2 + y^2 + z^2 - k \leq 2z^2$$

by the Markoff equation. This further implies that

$$x^2 + (z-y)^2 - k \leq 2z(z-y).$$

Since $ax > 2$, we get

$$(z - y) > \frac{\left(\frac{2}{a}\right)^2 - k}{2z}.$$

In this case, if $|axy - z| > |y|$, we use the same argument as before. We get

$$\phi \cdot \phi(P) = P$$
$$\text{but } h(\phi \cdot \phi(P)) \geq h(\phi(P)).$$

This implies that $\phi(P)$ is a fundamental solution. Further, if we consider a sequence of n consecutive non-negative solutions in this orbit from P_1, culminating in P, we get

$$2h(P) > \left(\left(\frac{2}{a}\right)^2 - k\right) \cdot \left(\frac{1}{h(P_1)} + \frac{1}{h(P_2)} + \ldots + \frac{1}{h(P_{n-1})}\right).$$

This gives the following upper bound on n

$$\frac{2h(P)^2}{\left(\frac{2}{a}\right)^2 - k} + 1 > n.$$

Finally, if Q is a cluster point for the non-negative solutions in this orbit, then there are infinitely many non-negative solutions R in the orbit with $h(R) < h(Q) + 1$. But from the above bound on n, the number of non-negative solutions in the orbit with $h(R) < h(Q) + 1$ is finite. This implies that the set of non-negative solutions in the orbit is discrete in $\mathbb{R}$.

If we consider the negative solutions in the orbit, they can be obtained by changing the signs of two of the entries in each triple of the positive solutions. It follows that the orbit is discrete. □

Remark 2.1 *In due course of the above proof, we have shown that since $k < 0$, we have $ax > 2$ for any ordered positive solution $(x, y, z) \in \mathcal{O}_K^3$. This implies that if we consider the involutions*

$$(x, y, z) \to (x, axz - y, z) \quad \textit{and} \quad (x, y, z) \to (ayz - x, y, z),$$

we have

$$axz - y > 2z - y > z \quad \textit{and} \quad ayz - x > axz - x > 2z - x > z.$$

Therefore, by rearranging the new solutions in ascending order and applying the above Vieta involutions, we observe that there are infinitely many solutions in each orbit. If the set of z coordinates of the solutions in an orbit is bounded above, then the set of all solutions in the orbit must lie in a compact set and therefore have a limit point. This contradicts Lemma 2.2. However, if the set of x coordinates is bounded, then the above implies that the set of z coordinates is also bounded.

3 Inhomogeneous Markoff Equations over Arbitrary Number Fields

Let K be an arbitrary number field. We now consider the Markoff–Hurwitz equation

$$x^2 + y^2 + z^2 = axyz + k$$

for some $a, k \in \mathcal{O}_K$. Suppose we have a solution (p_1, p_2, p_3) in $\mathcal{O}_K$. We now consider the equation

$$x^2 + y^2 + p_3^2 = ap_3xy + k. \tag{5}$$

Let $A = ap_3$ and $B = p_3^2 - k$. This brings us to the equation

$$x^2 - Ax + 1 = 0.$$

Let the roots be ω and ω'. We can now rewrite Eq. (5) as

$$(x - \omega y)(x - \omega' y) = -B.$$

Suppose further that $\omega \notin K$. Then the left-hand side is simply

$$N_{L/K}(x - \omega y) = -B,$$

where $L = K(\omega)$. However, we now note that if we multiply any solution with $z = p_3$ by a unit u of norm one, we get another solution. Let $S^* = \mathcal{O}_L^*$, the group of units of $\mathcal{O}_L^*$, and

$$S_1^* = \{u \in S^* : N_{L/K}(u) = 1\}.$$

We now recall two theorems of [3].

Theorem 3.1 ([3, Theorem 2.1]) *Suppose that R is an order in a number field K, and K is neither $\mathbb{Q}$ nor imaginary quadratic. Suppose also that*

$$\omega^2 - A\omega + 1 = 0$$

for $A \in R$ *and* $\omega \notin K$. *Set* $L = K(\omega)$ *and* $S = R \bigoplus \omega R$. *Finally suppose that every real embedding* τ *of* K *in* $\mathbb{C}$ *satisfies* $|\tau(A)| > 2$. *Then*

$$rank(S_1^*) \geq 2.$$

The second theorem talks about the existence of a cluster point in $\mathcal{O}_K^2$ for the set of solutions of the equation

$$x^2 + y^2 - Axy = -B,$$

where A, B are in $\mathcal{O}_K$.

Theorem 3.2 ([3, Theorem 2.2]) *Suppose that* R *is an order in a number field* K *and* R^* *is infinite. Further suppose that* $A \in R$ *and* $\tau(A) > 2$ *for all real embeddings* τ *of* K. *Let* ω *satisfy*

$$\omega^2 - A\omega + 1 = 0$$

and suppose that $\omega \notin K$. *Finally, if* $(p_1, p_2) \in R^2$ *is a solution to*

$$x^2 - Axy + y^2 = -B, \tag{6}$$

where $B \in R$, *then* (p_1, p_2) *is a cluster point for the solutions of* (6) *in* $\mathbb{C}^2$.

For convenience of the reader, we now state a version of Kronecker's approximation theorem, which we need in the next lemma.

Theorem 3.3 *For any* $\alpha, \beta, \epsilon \in \mathbb{R}$ *with* $\alpha \notin \mathbb{Q}$, *and* $\epsilon > 0$, *there exist integers* p *and* q *such that*

$$|p\alpha - q - \beta| < \epsilon.$$

Two intermediate lemmas we will require are the following.

Lemma 3.4 *Let* $K \subset \mathbb{R}$ *and* $S \subseteq \mathcal{O}_K^*$ *be a subgroup. If the rank of* S *is at least* 2, *then* S *is dense at* 1 *in* $\mathbb{C}$. *That is to say,* 1 *is a cluster point of* S.

Proof Let v_1 and v_2 be two positive multiplicatively independent units in S. It follows that $\log v_1$ and $\log v_2$ are $\mathbb{Q}$ linearly independent. Therefore, $\log_{v_1} v_2$ is irrational. By Theorem 3.3, the $\mathbb{Z}-$ span of 1 and $\log_{v_2} v_1$ is dense in $\mathbb{R}$. Therefore, the subgroup S is dense at 1. □

Before we proceed to the next lemma, we make the following observation. Going back to the Markoff–Hurwitz equation

$$x^2 + y^2 + z^2 = axyz + k,$$

suppose we have a solution (p_1, p_2, p_3) in $\mathcal{O}_K^3$. Now set $A = ap_3$ and $B = p_3^2 - k$ and consider the equation

$$x^2 + y^2 - Axy = -B$$

with $A, B \in \mathcal{O}_K$. For any real embedding τ of K into $\mathbb{C}$, we have

$$\tau(p_1)^2 + \tau(p_2)^2 - \tau(A)\tau(p_1 p_2) = -\tau(B).$$

We may assume that $\tau(p_1 p_2) > 0$. Suppose this were false, we may choose another solution by multiplying p_1 and p_3 by -1. This equation can be rewritten as

$$(\tau(p_1) - \tau(p_2))^2 = (\tau(A) - 2)\tau(p_1 p_2) - \tau(B).$$

If $\tau(B) \geq 0$ and if the solution (p_1, p_2, p_3) is non-trivial, then it follows that $\tau(A) > 2$.

Lemma 3.5 *If the number field $K \subset \mathbb{R}$ is such that the rank of $\mathcal{O}_K^*$ is 1, and $(p_1, p_2, p_3) \in \mathcal{O}_K^3$ is a positive solution to the equation*

$$x^2 + y^2 + z^2 = axyz + k, \tag{7}$$

with $k < 0$ and $a > 0$, then the quadratic equation

$$x^2 - ap_3 x + 1 = 0$$

has no solution in K.

Proof The proof of this lemma is already contained in the second half of the proof of [3, Theorem 2.3], and we repeat the arguments here for convenience of the reader.

Consider the equations $x^2 - ap_i x + 1 = 0$, for $i \in \{1, 2, 3\}$. If any of these equations have a solution $\omega \in K$, then it follows that $\omega \in \mathcal{O}_K^*$. Since the rank of $\mathcal{O}_K^*$ is 1, we have the existence of a positive fundamental unit u, and therefore, $\omega = \pm u^{k_z}$ for some $k_z \in \mathbb{Z}$. Since $k < 0$ and $a > 0$, without loss of generality, we may assume that the $p_i \geq 0$. Therefore, we have the following expressions for the p_i,

$$p_1 = \frac{u^{k_x} + u^{-k_x}}{a}, \qquad p_2 = \frac{u^{k_y} + u^{-k_y}}{a}, \qquad p_3 = \frac{u^{k_z} + u^{-k_z}}{a}$$

for integers k_x, k_y and k_z. Without loss of generality, we assume that u^{k_x}, u^{k_y} and u^{k_z} are greater than 1. Further we may assume that $u^{k_x} < u^{k_y} < u^{k_z}$. By Remark 2.1, we may assume that, given a positive constant M, if we traverse far enough along an orbit, $\frac{M+1}{a} < p_1 < p_2 < p_3$. Therefore, $u^{k_x} > M + 1 - u^{-k_x} \geq M$. We also have from the usual Vieta involutions that (p_1', p_2, p_3) is also a solution, where

$p_1' = ap_2p_3 - p_1$. Further, by the same argument as above, we have an expression for p_1',

$$p_1' = \frac{u^{k_x'} + u^{-k_x'}}{a}.$$

From the two expressions for p_1', we get

$$\frac{u^{k_x'} + u^{-k_x'}}{a} + \frac{u^{k_x} + u^{-k_x}}{a} = a \cdot \frac{u^{k_y} + u^{-k_y}}{a} \cdot \frac{u^{k_z} + u^{-k_z}}{a}.$$

This implies

$$u^{k_x'-k_y-k_z} + u^{-k_x'-k_y-k_z} + u^{k_x-k_y-k_z} + u^{-k_x-k_y-k_z}$$
$$= 1 + u^{-2k_z} + u^{-2k_y} + u^{-2k_y-2k_z}.$$

Since $u^{k_x-k_y} < 1$, we have $u^{k_x-k_y-k_z} < u^{-k_z} < \frac{1}{M}$. Therefore,

$$1 - \frac{3}{M} < u^{k_x'-k_y-k_z} < 1 + \frac{3}{M} \implies k_x' - k_y - k_z = 0$$

for M sufficiently large. Since $p_1' = ap_2p_3 - p_1'$, we get

$$u^{k_x} + u^{-k_x} = u^{k_y-k_z} + u^{k_z-k_y}.$$

Substituting in (7), we get $4 - k(a)^2 = 0$, which is a contradiction since $k < 0$. □

Proof of Theorem 1.1 We have already proved in Lemma 2.2 that each orbit in the set of solutions is discrete. Therefore, to show infinitude of the number of orbits, it suffices to show that the set of solutions has a cluster point. If (p_1, p_2, p_3) is a solution to (2), then setting $A = ap_3$ and $B = p_3^2 - k$, we consider the equation

$$x^2 - Axy + y^2 = -B. \tag{8}$$

Let ω be a solution to $\omega^2 - A\omega + 1 = 0$.
Case 1: If $\omega \notin K$ applying Theorem 3.2, we get that (p_1, p_2) is a cluster point in $\mathbb{C}^2$ of the solutions in $\mathcal{O}_K$ to the equation

$$x^2 - Axy + y^2 = -B, \tag{9}$$

if a solution ω to $\omega^2 - A\omega + 1 = 0$ is not in K. Therefore, if $\omega \notin K$, (p_1, p_2, p_3) is a cluster point in $\mathbb{C}^3$ in the set of all solutions to Eq. (2) by Theorem 3.2.
Case 2: If $\omega \in K$, we first identify ω, K, p_1, p_2, p_3, A and B with their images under a fixed real embedding. Then in K we have

$$p_1^2 + p_2^2 - Ap_1p_2 = (p_1 - \omega p_2)(p_1 - \omega^{-1}p_2) = -B.$$

Let $\beta_1 = p_1 - \omega p_2$ and $\beta_2 = p_1 - \omega^{-1}p_2$. For any $u \in \mathcal{O}_K^*$, if we set $x - \omega y = u\beta_1$ and $x - \omega^{-1}y = u^{-1}\beta_2$, then solving for x and y, we get

$$x = \frac{\omega u^{-1}\beta_2 - \omega^{-1}u\beta_1}{\omega - \omega^{-1}} \quad \text{and} \quad y = \frac{u^{-1}\beta_2 - u\beta_1}{\omega - \omega^{-1}}.$$

It follows immediately that $x, y \in \mathcal{O}_K$ whenever $u \equiv 1 \bmod (\omega - \omega^{-1})$.

Subcase 1: Suppose that the rank of $\mathcal{O}_K^*$ is greater than or equal to 2. Since the subgroup of units given by

$$S^* = \{u \in \mathcal{O}_K^* : u \equiv 1 \bmod (\omega - \omega^{-1})\}$$

is of finite index in $\mathcal{O}_K^*$, by Lemma 3.4, 1 is a cluster point for S^*. Now consider for $u \in S^*$,

$$\begin{aligned}
|x - p_1| &= \left|\frac{\omega\beta_2}{\omega - \omega^{-1}}(u^{-1} - 1) + \frac{\omega^{-1}\beta_1}{\omega - \omega^{-1}}(1 - u)\right| \\
&\leq \left|\frac{\omega\beta_2}{\omega - \omega^{-1}}(u^{-1} - 1)\right| + \left|\frac{\omega^{-1}\beta_1}{\omega - \omega^{-1}}(1 - u)\right|. \\
|y - p_2| &= \left|\frac{\beta_2}{\omega - \omega^{-1}}(u^{-1} - 1) + \frac{\beta_1}{\omega - \omega^{-1}}(1 - u)\right| \\
&\leq \left|\frac{\beta_2}{\omega - \omega^{-1}}(u^{-1} - 1)\right| + \left|\frac{\beta_1}{\omega - \omega^{-1}}(1 - u)\right|.
\end{aligned}$$

Since β_1, β_2, ω and ω^{-1} are fixed by p_1 and p_2, it follows by Lemma 3.4 that (p_1, p_2) is a cluster point for the solutions of (8) in $\mathcal{O}_K$. Therefore, once again we have that (p_1, p_2, p_3) is a cluster point in $\mathbb{C}^3$ for the set of all solutions to Eq. (2).

Subcase 2: As for the case when the rank of $\mathcal{O}_K^*$ is 1, up to an isomorphism we may assume that it is imbedded in the reals. It now follows from Lemma 3.5, the comment before Lemma 3.5 and Theorem 3.2 that if there is a solution (p_1, p_2, p_3), then it is a cluster point to the set of solutions to (2).

□

4 Fundamental Domain

Throughout this section, we assume that the field K is a real quadratic field.

4.1 Case $k < 0$ and $a > 1$

In this section (following [13]), we consider the equation for $k < 0$ and $a > 1$

$$x_1^2 + x_2^2 + x_3^2 = ax_1x_2x_3 + k. \tag{10}$$

We would like to construct the fundamental domain for the set of solutions under the action of the group Γ generated by the Vieta involutions, permutations, and double sign change in the ring of integers of a real quadratic field. Treating the above equation as a quadratic in x_3, we get

$$2x_3 = ax_1x_2 \pm \sigma \text{ where } \sigma = \sqrt{(ax_1x_2)^2 - 4(x_1^2 + x_2^2 - k)}. \tag{11}$$

Since for an ordered solution $x_1 \leq x_2 \leq x_3$, $ax_1x_2 - 2x_3 \leq x_2(ax_1 - 2)$, we have

$$|\sigma|^2 \leq (x_2(ax_1 - 2))^2.$$

Simplifying we get

$$x_2^2(ax_1 - 2) \leq x_1^2 - k. \tag{12}$$

Considering the points for which $x_3 \leq ax_1x_2 - x_3$, we get $x_1 \leq x_2 \leq x_3 \leq \frac{a}{2}x_1x_2$.
If $x_1 \leq \frac{2}{a}$, then from Eq. (10) we get

$$x_1^2 + (x_2 - x_3)^2 \leq k.$$

This implies that for a solution with $x_1 \geq 0$, $\frac{2}{a} < x_1$. Further there is no solution for which only one $x_i \leq 0$ and no solution with $x_i \leq 0$ for all i. This leaves the case where $x_1 \leq x_2 < 0 < x_3$. However, this solution is equivalent by double sign change to the one given by $|x_1|$, $|x_2|$ and x_3.

Lemma 4.1 *Every solution to the equation given by* (10) *in the real quadratic field K is Γ equivalent to a point in*

$$G_a^-(k) = \left\{(x_1, x_2, x_3) \,:\, \frac{2}{a} < x_1 \leq x_2 \leq x_3 \leq \frac{a}{2}x_1x_2\right\}.$$

Let us assume $\frac{2}{a} < x_1 \leq x_2 \leq x_3 \leq \frac{a}{2}x_1x_2$. If $x_1 \geq \frac{4}{a}$, then we obtain from Eq. (12) that

$$2x_2^2 \leq x_1^2 + |k| \implies x_2 \leq \sqrt{|k|} \implies x_3 \leq \frac{a|k|}{2}.$$

For $\frac{2}{a} < x_1 < \frac{4}{a}$, then from Eq. (12), we get

$$x_2^2 \leq \frac{x_1^2 + |k|}{ax_1 - 2}.$$

Therefore, we have the following lemma.

Lemma 4.2 *Every solution to the equation given by* (10) *in the real quadratic field* K *is* Γ *equivalent to a point in*

$$G_1 = I_1 \bigcup I_2,$$

where

$$I_1 = \left\{ (x_1, x_2, x_3) \ : \ \frac{4}{a} \leq x_1 \leq x_2 \leq \sqrt{|k|}, \frac{4}{a} \leq x_3 \leq \frac{a|k|}{2} \right\}$$

and

$$I_2 = \left\{ (x_1, x_2, x_3) \ : \ \frac{2}{a} < x_1 < \frac{4}{a}, \ x_1 \leq x_2 \leq \sqrt{\frac{x_1^2 + |k|}{ax_1 - 2}}, \right.$$
$$\left. x_2 \leq x_3 \leq \frac{ax_1}{2} \sqrt{\frac{x_1^2 + |k|}{ax_1 - 2}} \right\}.$$

4.2 Case k ≥ 0 and a > 1

We now describe the fundamental domain of Eq. (10)

$$x_1^2 + x_2^2 + x_3^2 = ax_1x_2x_3 + k$$

for $k \geq 0$ and $a > 1$. In particular, we obtain the following Lemma.

Lemma 4.3 *For* $k > \frac{4}{a^2}$, *every solution to the equation given by* (10) *in the real quadratic field* K *is* Γ *equivalent to a point in the compact set* $G^+(k) \cup \mathfrak{T}(k)$, *where* $\mathfrak{T}(k) \subset \mathbb{Z}^3$ *is the finite set of solutions having coordinates either* 0 *or* 1 *and*

$$G^+(k) = \left\{ (-x_1, x_2, x_3) \in O_K^3 : \frac{2}{a} < x_1 \leq x_2 \leq x_3, \right.$$
$$\left. x_1^2 + x_2^2 + x_3^2 - ax_1x_2x_3 = k \right\}, \tag{13}$$

or if not, then it is equivalent to a triple $(x_1, ax_1x_2 - x_3, x_2)$ *with* $\frac{2}{a} < ax_1x_2 - x_3 < x_2 \leq x_3$ *and* $x_1 > \frac{2}{a}$.

Proof When $k \geq 0$, there are only finitely many triples $(x_1, x_2, x_3) \in \{0, 1\}^3$ satisfying Eq. (10). We denote the finite set by $\mathfrak{T}(k)$. Now we consider the Vieta involution on (10)

$$(x_1, x_2, x_3) \to (x_1, x_2, ax_1x_2 - x_3).$$

If $ax_1x_2 - x_3 < 0$, then (x_1, x_2, x_3) is equivalent to a solution of $G_a^+(k)$. Hence, we consider the case $ax_1x_2 - x_3 \geq 0$. Now, we assume that $ax_1x_2 - x_3 \geq x_3$. In this case, we have $2x_3 \leq ax_1x_2$. Similar to the description (11), solving for x_3 of Eq. (10) gives

$$2x_3 = ax_1x_2 \pm \sigma \text{ where } \sigma = \sqrt{(ax_1x_2)^2 - 4(x_1^2 + x_2^2 - k)},$$

and this further implies $\sigma = |ax_1x_2 - 2x_3| = ax_1x_2 - 2x_3$ and (12) for an ordered solution. If $x_1 < \frac{2}{a}$, we conclude $k \leq \frac{4}{a^2}$. Thus for all $k > \frac{4}{a^2}$, we have a contradiction to the assumption $ax_1x_2 - x_3 \geq x_3$. So, when $k > \frac{4}{a^2}$, we have

$$0 \leq ax_1x_2 - x_3 < x_3.$$

Moreover, we can prove $\frac{2}{a} < ax_1x_2 - x_3 < x_2 \leq x_3$: If $x_2 \leq ax_1x_2 - x_3 < x_3$, then we have $ax_1x_2 < 2x_3 \leq 2x_2(ax_1 - 1)$. This gives

$$\sigma = 2x_3 - ax_1x_2 \leq 2x_2(ax_1 - 1) - ax_1x_2 = ax_1x_2 - 2x_2 = (ax_1 - 2)x_2,$$

and the same argument gives a contradiction. Hence, we proved the desired result. □

5 Finding Local Solutions

Let K be a number field and let $\mathfrak{p}$ be a prime ideal of $\mathcal{O}_K$ above a prime $p \in \mathbb{Z}$. We denote by $\mathbf{F} = \mathcal{O}_K/\mathfrak{p}$ the residue field of size $|\mathbb{F}| = \mathfrak{N}(\mathfrak{p})$. In order to show the existence of local solutions of the Markoff–Hurwitz equation, we need the following result of Carlitz [7, 8].

Theorem 5.1 *Given a number field K and a prime ideal $\mathfrak{p}$ that does not lie above 2, the number of solutions $N_\mathfrak{p}$ modulo $\mathfrak{p}$ to the equation, with an arbitrary e and $abcd \neq 0$,*

$$ax^2 + by^2 + cz^2 = 2dxyz + e$$

is determined by

$$N_{\mathfrak{p}} = \mathfrak{N}(\mathfrak{p})^2 + 1 + \mathfrak{N}(\mathfrak{p})\{\psi(a) + \psi(b) + \psi(c) + \psi(e)\}\psi(d^2 - abc),$$

where ψ is the quadratic symbol taking value 1 *for squares,* -1 *for non-squares, and* 0 *at* 0.

Proof The proof can be found in [7]. □

Theorem 5.1 implies the following local property of the Markoff–Hurwitz equation, that is to say, when the parameters are $a = b = c = 1$ and suitable d.

Corollary 5.2 *Given a number field K and a prime ideal $\mathfrak{p}$ that does not lie above 2, the number of solutions $N_{\mathfrak{p}}$ modulo $\mathfrak{p}$ to the equation*

$$x_1^2 + x_2^2 + x_3^2 = \alpha x_1 x_2 x_3 + \gamma \quad \textit{with} \quad \alpha, \gamma \in \mathbb{Z}$$

is strictly greater than 0. *In fact*

$$N_{\mathfrak{p}} \geq \mathfrak{N}(\mathfrak{p})^2 - 4\mathfrak{N}(\mathfrak{p}).$$

We now prove a lemma on finding solutions modulo $\mathfrak{p}^n$.

Lemma 5.3 *Let K be a number field. Further let $\mathfrak{p}$ be a prime ideal such that $\mathfrak{N}(\mathfrak{p}) > 4$ and such that $\mathfrak{p}$ does not lie above* 2. *We consider the equation*

$$x_1^2 + x_2^2 + x_3^2 = \alpha x_1 x_2 x_3 + \gamma \quad \textit{with} \quad \alpha, \gamma \in \mathbb{Z}. \tag{14}$$

It has a solution modulo $\mathfrak{p}^n$ for all $n \geq 1$.

Proof By Corollary 5.2, we know that the equation has solutions modulo $\mathfrak{p}$. We now break the proof into three cases. Let

$$f = x_1^2 + x_2^2 + x_3^2 - \alpha x_1 x_2 x_3;$$

then f' takes one of the following three forms: $2x_i - \alpha x_j x_k$ for $\{i, j, k\} = \{1, 2, 3\}$. If all three forms are 0 mod $\mathfrak{p}$ for (a, b, c) with entries in $\mathcal{O}_K$ where (a, b, c) is a solution mod $\mathfrak{p}$, then we call the solution singular.
Case 1: If we have a non-singular solution, then by Hensel's lemma we have solutions modulo $\mathfrak{p}^n$ for all $n \geq 1$.
Case 2: If a given solution is singular, we further suppose that $\mathfrak{p} \nmid (\gamma)$ and $\mathfrak{p} \nmid (abc)$. Since the solution is singular, we have

$$2x_i \equiv \alpha x_j x_k \bmod \mathfrak{p} \quad ,$$

where x_i, x_j, x_k is a permutation of a, b, c. Since $\mathfrak{p}$ does not lie above 2, we have

$$\alpha abc \equiv 2a^2 \bmod \mathfrak{p},$$

and therefore,

$$a^2 \equiv b^2 \equiv c^2 \bmod \mathfrak{p}.$$

Substituting in (14), we get

$$0 \not\equiv a^2 \equiv \gamma \bmod \mathfrak{p}.$$

It follows that $(a, 0, 0)$ is a solution modulo $\mathfrak{p}$ to (14). Further, since $2a - \alpha bc = 2a$ and $\mathfrak{p}$ does not lie above 2, it follows that this solution is non-singular, thereby leading back to case 1.

Case 3: If a given solution is singular and $\mathfrak{p} \mid x_1x_2x_3$, we have that $\gamma \in \mathfrak{p}^2$. However, again using the fact that the solution is singular and that $\mathfrak{p}$ does not lie above 2, we get that the solution is $(0, 0, 0) \bmod \mathfrak{p}$. On the other hand from Corollary 5.2, we observe that there is at least one other solution that reduces the problem to one of the above two cases. □

5.1 Primes Above 3

If $\alpha \in \mathfrak{p}$ and $\gamma - 1 \in \mathfrak{p}$, where $\mathfrak{p}$ lies above 3, then $(1, 0, 0)$ is a non-singular solution modulo $\mathfrak{p}$. If $\alpha \in \mathfrak{p}$ and $\gamma \in \mathfrak{p}^2$ where $\mathfrak{p}$ lies above 3, then $(3, 0, 0)$ is a non-singular solution modulo $\mathfrak{p}^2$ and $(0, 0, 0)$ is a solution modulo $\mathfrak{p}$.

5.2 Primes Above 2

If $\alpha - 1, \gamma \in \mathfrak{p}$ where $\mathfrak{p}$ where $\mathfrak{p}$ lies above 2, then $(1, 1, 1)$ is a non-singular solution modulo $\mathfrak{p}$.

In all the above two cases, we have by Hensel's lemma solutions modulo $\mathfrak{p}^n$ for $n \geq 1$.

6 Exhibiting Failure of the Hasse Principle

The following lemma is a formulation of the well-known supplement to the law of quadratic reciprocity in number fields.

Lemma 6.1 *Given a number field K and a prime ideal $\mathfrak{p}$ not lying above 2, we have*

$$\left(\frac{2}{\mathfrak{p}}\right) = (-1)^{\frac{\mathfrak{N}(\mathfrak{p})^2-1}{8}}.$$

Here $(\cdot)$ is used to indicate the 2-power residue symbol modulo $\mathfrak{p}$, and the norm is with respect to $K/\mathbb{Q}$.

Proof By definition, we have

$$\left(\frac{2}{\mathfrak{p}}\right) \equiv 2^{\frac{\mathfrak{N}(\mathfrak{p})-1}{2}} \bmod \mathfrak{p}\mathcal{O}_K.$$

Consider the field given by $K(i)$, denoted L. We now have

$$(2i)^{\frac{\mathfrak{N}(\mathfrak{p})-1}{2}} \cdot (1+i) \equiv (1+i)^{\mathfrak{N}(\mathfrak{p})} \equiv 1 + i^{\mathfrak{N}(\mathfrak{p})} \equiv 1 + i \cdot (-1)^{\frac{\mathfrak{N}(\mathfrak{p})-1}{2}} \bmod \mathfrak{p}\mathcal{O}_L.$$

This implies that

$$\left(\frac{2}{\mathfrak{p}}\right) \equiv \frac{1 + i \cdot (-1)^{\frac{\mathfrak{N}(\mathfrak{p})-1}{2}}}{i^{\frac{\mathfrak{N}(\mathfrak{p})-1}{2}}(1+i)} \equiv \frac{(1 + i \cdot (-1)^{\frac{\mathfrak{N}(\mathfrak{p})-1}{2}})(1-i)}{2i^{\frac{\mathfrak{N}(\mathfrak{p})-1}{2}}} \bmod \mathfrak{p}\mathcal{O}_L.$$

Therefore,

$$\left(\frac{2}{\mathfrak{p}}\right) \equiv \begin{cases} (-1)^{\frac{\mathfrak{N}(\mathfrak{p})-1}{4}} \bmod \mathfrak{p}\mathcal{O}_L, & \frac{\mathfrak{N}(\mathfrak{p})-1}{2} \text{ is even} \\ (-1)^{\frac{\mathfrak{N}(\mathfrak{p})+1}{4}} \bmod \mathfrak{p}\mathcal{O}_L, & \frac{\mathfrak{N}(\mathfrak{p})-1}{2} \text{ is odd.} \end{cases}$$

Combining both

$$\left(\frac{2}{\mathfrak{p}}\right) \equiv (-1)^{\frac{\mathfrak{N}(\mathfrak{p})^2-1}{8}} \bmod \mathfrak{p}\mathcal{O}_L \implies \left(\frac{2}{\mathfrak{p}}\right) \equiv (-1)^{\frac{\mathfrak{N}(\mathfrak{p})^2-1}{8}} \bmod \mathfrak{p}\mathcal{O}_K.$$

Hence the lemma. □

Lemma 6.2 *Let K be a Galois number field of degree 3 such that 2 splits in K. Further let $v \neq 0$ be a rational integer such that all its prime divisors in $\mathbb{Z}$ are in $\{\pm 1 \bmod 8\}$. Then the set of solutions in $\mathcal{O}_K$ to the equation*

$$x_1^2 + x_2^2 + x_3^2 - ax_1x_2x_3 = a^{-2}(4 + 2v^2), \quad \textit{with } a \in \mathcal{O}_K$$

is empty.

Proof It suffices to show that the equation

$$x_1^2 + x_2^2 + x_3^2 - x_1x_2x_3 = 4 + 2v^2 \tag{15}$$

has no solutions in $\mathcal{O}_K$. Suppose otherwise and let $(x_1, x_2, x_3) \in \mathcal{O}_K^3$ be a solution. Let $w = 2x_3 - x_1x_2$; we can now rewrite the equation as

$$w^2 - 8v^2 = (x_1^2 - 4)(x_2^2 - 4). \tag{16}$$

Since all the prime divisors of v are in $\{\pm 1 \bmod 8\}$, they must all be odd. Therefore, $4 + 2v^2$ is not divisible by 4. Let $\mathfrak{p}$ be a prime ideal in $\mathcal{O}_K$ above $2\mathbb{Z}$. We now have $4 + 2v^2 \notin \mathfrak{p}^2 \cap \mathbb{Z}$, since $v_{\mathfrak{p}}(2) = 1$. So for each $\mathfrak{p}$ above $2\mathbb{Z}$ at least one $x_i \notin \mathfrak{p}$. If $\mathfrak{p}_1, \mathfrak{p}_2, \mathfrak{p}_3$ are the primes above $2\mathbb{Z}$ in K, we claim that there exists a solution (x_1, x_2, x_3) such that at least one $x_i \notin \mathfrak{p}_j$ for all $j \in \{1, 2, 3\}$. It is easy to see from (15) that if $x_1, x_2 \in \mathfrak{p}_1$, then so is x_3. So let us assume this does not happen and assume that $x_1 \in \mathfrak{p}_1$, $x_2 \in \mathfrak{p}_2$, and $x_3 \in \mathfrak{p}_3$. We now look at the solution given by

$$(x_1, x_2, x_1x_2 - x_3).$$

Since $x_1 \in \mathfrak{p}_1$ and $x_2 \in \mathfrak{p}_2$, we have that $x_1x_2 - x_3 \notin \mathfrak{p}_1 \cup \mathfrak{p}_2$. But the fact that $x_3 \in \mathfrak{p}_3$ implies that $x_1x_2 - x_3 \notin \mathfrak{p}_i$ for all i. This proves our claim. Now let (x_1, x_2, x_3) be a solution such that $x_1 \notin \mathfrak{p}$ for all primes $\mathfrak{p} \mid 2\mathcal{O}_K$. We now look at the ring homomorphism

$$\begin{aligned} \phi : \mathbb{Z} &\to \mathcal{O}_K/\mathfrak{p}^3 \\ a &\to a \bmod \mathfrak{p}^3. \end{aligned}$$

The kernel is $\mathfrak{p}^3 \cap \mathbb{Z} \supseteq 8\mathbb{Z}$, and since $v_{\mathfrak{p}}(4) = 2$, we have $Ker(\phi) = 8\mathbb{Z}$. We now have the injective map

$$\begin{aligned} \psi : \mathbb{Z}/8\mathbb{Z} &\to \mathcal{O}_K/\mathfrak{p}^3 \\ a &\to a \bmod \mathfrak{p}^3. \end{aligned}$$

Finally, since the norm of $\mathfrak{p}^3$ is 8, the map is surjective. This implies that since invertible classes must go to invertible classes through ψ,

$$x_1^2 \equiv 1 \bmod \mathfrak{p}^3.$$

If $\mathfrak{p}_1, \mathfrak{p}_2, \mathfrak{p}_3$ are the primes above $2\mathbb{Z}$ in K and $x_1 \notin \mathfrak{p}_i$ for all i, we have

$$x_1^2 \equiv 1 \bmod \mathfrak{p}_i^3$$

for all i. By the Chinese remainder theorem

$$x_1^2 \equiv 1 \bmod 8\mathcal{O}_K \implies x_1^2 - 4 \equiv 5 \bmod 8\mathcal{O}_K.$$

Therefore, if the degree of K over $\mathbb{Q}$ is 3

$$\begin{aligned} \mathfrak{N}(x_1^2 - 4) \equiv 5^3 \bmod 8\mathcal{O}_K &\implies \mathfrak{N}(x_1^2 - 4) \equiv 5^3 \bmod 8\mathbb{Z} \\ &\implies \mathfrak{N}(x_1^2 - 4) \equiv 5 \bmod 8\mathbb{Z}. \end{aligned}$$

Therefore, there exists a rational prime p such that

$$p \equiv \pm 5 \bmod 8\mathbb{Z} \text{ such that } p \mid \mathfrak{N}(x_1^2 - 4).$$

By our hypothesis $p \nmid v$. Let $\mathfrak{q}_1$ be a prime in $\mathcal{O}_K$ that lies above p with $\mathfrak{N}(\mathfrak{q}_1) = p^f$ for some natural number f and $\mathfrak{q}_1 \mid (x_1^2 - 4)\mathcal{O}_K$. By (16), 2 is a quadratic residue modulo $\mathfrak{q}_1$. Since the number field is Galois and of degree 3, the exponent f is odd. We now have $p = \pm 5 + 8k$ and therefore

$$p^{2f} = (25 \pm 80k + 64k^2)^f \implies p^{2f} - 1 \equiv 25^f - 1 \bmod 16$$
$$\implies \frac{p^{2f} - 1}{8} \equiv \frac{25^f - 1}{8} \bmod 2.$$

However,

$$\frac{p^{2f} - 1}{8} \equiv \frac{25^f - 1}{8} \bmod 2 \implies \frac{p^{2f} - 1}{8} \equiv 3 \cdot (25^{f-1} + \cdots + 1) \bmod 2 \equiv 1 \bmod 2.$$

Therefore by Lemma 6.1, we get that 2 is a quadratic non-residue modulo $\mathfrak{q}_1$. This gives us a contradiction. □

Proof of Theorem 1.2 By Lemma 5.3, the above equation has solutions mod $\mathfrak{p}^n$ for all $n \geq 1$ and primes not lying above 2, with $\mathfrak{N}(\mathfrak{p}) > 4$ in K. The only other primes are those above 2 and 3. By the remarks after Lemma 5.3, the above equation has solution mod $\mathfrak{p}^n$ for all $n \geq 1$ and $\mathfrak{p}$ above 2 and 3. Since the Artin symbol for 2 with respect to $\mathbb{Q}(\zeta_{31})/\mathbb{Q}$ has order 5, the prime 2 splits in K. By Lemma 6.2, we have for $a = 3$ and $v = 23$, the equation in (3) has no global solutions. □

7 Prime Factors in Homogeneous Markoff Equations of Degree 3

Markoff numbers are positive integers that appear in the integral solutions of the Diophantine equation $\mathcal{M} = \mathcal{M}_{3,0} : x^2 + y^2 + z^2 = 3xyz$. They form a sequence starting with $1, 2, 5, 13, 29, 34 \ldots$. A triple of integers (x, y, z) that satisfy $\mathcal{M}$ is called a Markoff triple. We say a triple (x, y, z) is normalized if its coordinates satisfy $0 \leq x \leq y \leq z$. We note that all Markoff triples can be easily retrieved from the set of normalized Markoff triples using permutations and sign changes. The Markoff tree $\mathfrak{M}$ with normalized Markoff triples is constructed by the initial triple $(1, 1, 1)$ and subsequent group action of Γ, the group of affine morphisms of $\mathbb{A}^3$ generated by the permutations and the two Vieta involutions ϕ_1 and ϕ_2 from (4). A nice description of the tree can be found as Figure 2 in [21]. There are two (maximal) branches of the tree that have $m \in \mathfrak{M} \backslash \{1, 2\}$. In fact, the unicity conjecture implies that the tree has exactly two (maximal) branches with a fixed smallest coordinate m: when $m \in \mathfrak{M} \backslash \{1, 2\}$ appears as a triple (a, m, b) in the tree. One branch can be obtained by applying ϕ_1 twice, and ϕ_2 infinitely many times, and the other branch can be obtained by first using ϕ_1, and then ϕ_2 infinitely many times. Hence, for a fixed Markoff number m, we have two branches of the Markoff tree that have m as the lowest term for the triples. When $m = 5$, we have two branches that have 5 as the lowest term of each triple, which are

$$(5, 13, 194) - (5, 194, 2897) - (5, 2897, 43261) - (5, 43261, 646018) - \cdots$$

and

$$(5, 29, 433) - (5, 433, 6466) - (5, 6466, 96557) - (5, 96557, 1441889) - \cdots .$$

We summarize the properties of branches of the Markoff tree with fixed smallest coordinates in the following lemma.

Lemma 7.1 *For a fixed Markoff number $m \geq 5$, the unicity conjecture implies that there are exactly two maximal branches of the Markoff tree that have m as the lowest term for their triples*

$$\mathfrak{L}^{(j)} : (m, \ell_1^{(j)}, \ell_2^{(j)}) - (m, \ell_2^{(j)}, \ell_3^{(j)}) - \cdots - (m, \ell_n^{(j)}, \ell_{n+1}^{(j)}) - \cdots ,$$

for $j = 1, 2$, and then the following are true:

(1) For $i = 1, 2$, it is possible to extend the sequence $\{\ell_n^{(i)}\}_{n\geq 0}$ to sequences $\{L_n^{(i)}\}_{n\geq 0}$ that satisfy the linear recurrence

$$L_{n+1}^{(i)} = 3mL_n^{(i)} - L_{n-1}^{(i)}, \quad \text{for } n \geq 1,$$

such that $L_0^{(1)} = L_0^{(2)} = 1$. Without loss of generality, we assume $L_1^{(1)} \geq L_1^{(2)}$.

(2) The terms of the sequences $\{L_n^{(1)}\}_{n\geq 0}$ and $\{L_n^{(2)}\}_{n\geq 0}$ can be represented as linear combinations of terms of the Lucas sequence $\{\mathcal{L}_n\}_{n\geq 0}$ that is defined by

$$\mathcal{L}_0 = 0,\ \mathcal{L}_1 = 1,\ \mathcal{L}_{n+1} = 3m\mathcal{L}_n - \mathcal{L}_{n-1} \quad \text{for } n \geq 0. \tag{17}$$

More precisely, for $n \geq 0$ and $j = 1, 2$, we can write

$$L_n^{(j)} = L_1^{(j)} \left(\frac{\alpha^n - \beta^n}{\alpha - \beta}\right) - L_0^{(j)} \left(\frac{\alpha^{n-1} - \beta^{n-1}}{\alpha - \beta}\right) = L_1^{(j)}\mathcal{L}_n - L_0^{(j)}\mathcal{L}_{n-1}, \tag{18}$$

where α and β are numbers that generate the sequence $\{\mathcal{L}_n\}_{n\geq 0}$.

(3) We have

$$L_n^{(1)} - L_n^{(2)} = (L_1^{(1)} - L_1^{(2)})\mathcal{L}_n, \quad n \geq 0$$

$$L_n^{(1)} + L_n^{(2)} = L_0^{(1)}(\mathcal{L}_{n+1} - \mathcal{L}_{n-1}) = L_0^{(2)}(\mathcal{L}_{n+1} - \mathcal{L}_{n-1}), \quad n \geq 1.$$

Proof The first claim naturally follows by backtracking Vieta involutions. To prove (2), we use properties of the Lucas sequences, and we have

$$\begin{aligned} L_n^{(j)} &= (L_1^{(j)} - 3mL_0^{(j)})\mathcal{L}_n + L_0^{(j)}\mathcal{L}_{n+1} = L_1^{(j)}\mathcal{L}_n + (-3m\mathcal{L}_n + \mathcal{L}_{n+1})L_0^{(j)} \\ &= L_1^{(j)}\mathcal{L}_n - L_0^{(j)}\mathcal{L}_{n-1}, \end{aligned}$$

for $j = 1, 2$. The first equality in (3) naturally follows from (2). The second equality can be proved by induction. □

Lemma 7.1, along with Theorem 3.9 in [6], implies the following.

Theorem 7.2 *Following the notation in Lemma 7.1, let $\{L_n^{(1)}\}_{n\geq 0}$ and $\{L_n^{(2)}\}_{n\geq 0}$ be two linear recurrence sequences in the Markoff tree following the recurrence relation*

$$L_{n+1}^{(i)} = 3mL_n^{(i)} - L_{n-1}^{(i)}, \quad \textit{for } n \geq 1 \textit{ and } i = 1, 2.$$

Then for sufficiently large integers n, the greatest prime factor of the terms $L_n^{(1)}$ and $L_n^{(2)}$ satisfies

$$P(L_n^{(i)}) \gg \log n \frac{\log\log n}{\log\log\log n}, \quad \textit{for } i = 1, 2,$$

where the implied constant depends on the choice of m, the smallest coordinate of triples that contain the terms in $\{L_n^{(1)}\}_{n\geq 0}$ and $\{L_n^{(2)}\}_{n\geq 0}$.

Acknowledgments We thank the organizers of WIN5, Renate Scheidler, Alina Bucur, and Wei Ho and express our deep gratitude to Elisa Bellah, Elena Fuchs, and Lynnelle Ye, and in particular Elena for initiating this WIN group. Moreover, we express our gratitude to the two anonymous referees for their very careful reading of an earlier draft of this article and numerous helpful suggestions.

References

1. M. Aigner, *Markov's Theorem and 100 Years of the Uniqueness Conjecture* (Springer, Cham, 2013)
2. A. Baragar, Asymptotic growth of Markoff-Hurwitz numbers. Comp. Math. **94**(1), 1–18 (1994)
3. A. Baragar, The Markoff-Hurwitz equations over number fields. Rocky Mount. J. Math. **35**(3), 695–712 (2005)
4. E. Bombieri, Continued fractions and the Markoff tree. Expo. Math. **25**(3), 187–213 (2007).
5. J. Bourgain, A. Gamburd, P. Sarnak. Markoff triples and strong approximation. C. R. Math. Acad. Sci. Paris **354**(2), 131–135 (2016)

6. Y. Bugeaud. *Linear Forms in Logarithms and Applications*, volume 28 of *IRMA Lectures in Mathematics and Theoretical Physics* (European Mathematical Society (EMS), Zürich, 2018)
7. L. Carlitz, Certain special equations in a finite field. Monatsh. Math. **58**, 5–12 (1954)
8. L. Carlitz, The number of points on certain cubic surfaces over a finite field. Boll. Un. Mat. Ital. (3) **12**, 19–21 (1957)
9. W. Chen, Nonabelian level structures, Nielsen equivalence, and Markoff triples. Accepted for publication in Annals of Mathematics, arXiv:2011.12940
10. J.-L. Colliot-Thélène, D. Wei, F. Xu, Brauer-Manin obstruction for Markoff surfaces. Ann. Sc. Norm. Super. Pisa Cl. Sci. (5) **21**, 1257–1313 (2020)
11. B. Fine, G. Kern-Isberner, A.I.S. Moldenhauer, G. Rosenberger, On the generalized Hurwitz equation and the Baragar-Umeda equation. Results Math. **69**(1–2), 69–92 (2016)
12. A. Gamburd, M. Magee, R. Ronan, An asymptotic formula for integer points on Markoff-Hurwitz varieties. Ann. Math. (2) **190**(3), 751–809 (2019)
13. A. Ghosh, P. Sarnak, Integral points on Markoff type cubic surfaces. Invent. Math. **229**(2), 689–749 (2022)
14. C. Gurwood, Diophantine approximation and the Markoff chain. Ph.D. Thesis, New York University, 1976
15. N.P. Herzberg, On a problem of Hurwitz. Pac. J. Math. **50**, 485–493 (1974)
16. A. Hurwitz, Ueber eine Aufgabe der unbestimmten Analysis. Arch. Math. Phys. **3**, 185–196 (1907)
17. D. Loughran, V. Mitankin, Integral Hasse principle and strong approximation for Markoff surfaces. Int. Math. Res. Not. IMRN (18), 14086–14122 (2021)
18. A. Markoff, Sur les formes quadratiques binaires indéfinies. Math. Ann. **17**(3), 379–399 (1880)
19. M. Mirzakhani, Counting mapping class group orbits on hyperbolic surfaces (2016). https://doi.org/10.48550/arXiv.1601.03342
20. J.H. Silverman, The Markoff equation $X^2+Y^2+Z^2 = aXYZ$ over quadratic imaginary fields. J. Number Theory **35**(1), 72–104 (1990)
21. D. Zagier, On the number of Markoff numbers below a given bound. Math. Comp. **39**(160), 709–723 (1982)

p-Adic Measures for Reciprocals of L-Functions of Totally Real Number Fields

Razan Taha

1 Introduction

Let K be a totally real number field of finite degree over $\mathbb{Q}$ and let $K(p^{\infty})$ be the maximal abelian unramified-outside-p extension of K. The goal of this chapter is to construct a p-adic measure λ on the Galois group $G(K(p^{\infty})/K)$ from the non-constant terms of the Fourier expansion of holomorphic Eisenstein series on $SL_2(\mathcal{O})$, where $\mathcal{O}$ is the ring of integers of K. Moreover, the Mellin transform of λ turns out to be the reciprocal of the p-adic L-function of the totally real field K. This L-function is the p-adic analog of the classical complex-valued Hecke L-function and has several different constructions which we will briefly describe.

The study of p-adic L-functions starts with the fact that the values of the Riemann zeta function at negative integers are rational and can be expressed in terms of Bernoulli numbers as

$$\zeta(1-k) = (-1)^{k+1}\frac{B_k}{k}, \quad \forall k > 0.$$

Kummer [13] proved that these values satisfy special congruences modulo powers of a prime number p: given two even integers k and k' which are not divisible by $p-1$ and for all $d > 0$, if $k \equiv k' \bmod p^{d-1}(p-1)$, then

$$(1-p^{k-1})\zeta(1-k) \equiv (1-p^{k'-1})\zeta(1-k') \bmod p^d. \tag{1}$$

Kubota and Leopoldt [11] were the first to interpret the Kummer congruences as a p-adic function which interpolates the values of the Riemann zeta function at negative

R. Taha (✉)
Department of Mathematics, University of Michigan, Ann Arbor, MI, USA
e-mail: tahar@umich.edu

A. Bucur et al. (eds.), *Research Directions in Number Theory*, Association for Women in Mathematics Series 33, https://doi.org/10.1007/978-3-031-51677-1_10

integers. We recall the definition of the Kubota–Leopoldt p-adic L-function. Let

$$q = \begin{cases} 4 & \text{if } p = 2 \\ p & \text{otherwise,} \end{cases} \tag{2}$$

and let $\phi(q)$ denote the Euler totient function of q. Let χ be a Dirichlet character and let ω denote the Teichmuller character defined for $x \in \mathbb{Z}_p^*$ by

$$\omega(x) = \begin{cases} \pm 1 & \text{if } p = 2 \text{ and } x \equiv \pm 1 \mod 4\mathbb{Z}_2 \\ \lim\limits_{n\to\infty} x^{p^n} & \text{otherwise.} \end{cases} \tag{3}$$

Kubota and Leopoldt proved that there exists a continuous function $L_p : \mathbb{Z}/\phi(q)\mathbb{Z} \times \mathbb{Z}_p \to \mathbb{C}_p$, except when χ is trivial, such that, for all $k \in \mathbb{Z}_{\geq 0}$ (which embeds diagonally into $\mathbb{Z}/\phi(q)\mathbb{Z} \times \mathbb{Z}_p$),

$$L_p(\chi, 1-k) = L(\chi\omega^{-k}, 1-k)(1 - \chi\omega^{-k}(p)p^{k-1}). \tag{4}$$

The right-hand side of (4) is the value of the complex Dirichlet L-function with the Euler factor at p removed, while the left-hand side of (4) is the value of the p-adic L-function.

For a totally real field K, the Hecke L-function attached to a character χ with conductor $\mathfrak{n}$ is given by

$$L(s, \chi) = \sum_{(\mathfrak{a},\mathfrak{n})=1} \frac{\chi(\mathfrak{a})}{N(\mathfrak{a})^s} = \sum_{[\mathfrak{a}]} \chi([\mathfrak{a}]) \sum_{\mathfrak{b}\in[\mathfrak{a}]} \frac{1}{N(\mathfrak{b})^s}, \tag{5}$$

where $[\mathfrak{a}]$ denotes the class of the ideal $\mathfrak{a}$. Siegel [16] proved that the values of the partial L-function $\sum_{\mathfrak{b}\in[\mathfrak{a}]} N(\mathfrak{b})^{k-1}$ are rational, which led to several different constructions of the p-adic L-function for a totally real field. Serre [14] constructed p-adic modular forms and obtained the p-adic L-function as the constant term of a p-adic Eisenstein series. Shintani [15] proved an elementary formula for this rational number in terms of certain generalized Bernoulli numbers, which Cassou–Noguès [2] then used to define the p-adic L-function in terms of generalized Bernoulli numbers. Coates and Sinnott [3] equivalently proved the generalized Kummer congruences for real quadratic fields, and Deligne and Ribet [6] proved them in general for any totally real field.

One long-term goal of this project is to study p-adic L-functions using methods inspired by the Langlands program. In the complex setting, the Langlands–Shahidi method establishes the analytic continuations and functional equations of many L-functions by studying the non-constant terms of the Fourier expansion of Eisenstein series on reductive groups. In particular, the denominators of these Fourier coefficients contain suitable L-functions. This inspired Gelbart, Miller, Panchishkin, and Shahidi to introduce an analog of parts of the Langlands–Shahidi method to the

p-adic setting in [8]. They proved the existence of a p-adic measure μ^* which can be expressed in terms of the non-zero Fourier coefficients of classical holomorphic Eisenstein series such that its Mellin transform is the reciprocal of the p-adic Riemann ζ-function.

In this chapter, we generalize the result of [8] to the case of a totally real number field K with strict (or narrow) class number h^+ (see Eq. (7) for the definition of the strict class group). We now state our main theorem:

Theorem 1.1 *Let $p \in \mathbb{Q}$ be an odd prime number, $\mathfrak{p}$ be a prime ideal of $\mathcal{O}$ lying above p, and h^+ be the strict class number of K. Assume that*

(1) p does not divide $[K : \mathbb{Q}]$.
(2) p does not divide the class number of $K(e^{2\pi i/p})$.
(3) No prime $\wp$ of the field $K(e^{2\pi i/p} + e^{-2\pi i/p})$ lying above p splits in $K(e^{2\pi i/p})$.

Then

(1) There exists a p-adic measure λ on $G(K(p^\infty)/K)$ whose Mellin transform is the reciprocal of the p-adic L-function of the totally real field K

$$\int_{G(K(p^\infty)/K)} N(x)^{k-1}\, d\lambda = h^+(1 - N(\mathfrak{p})^{k-1})^{-1}\zeta(1-k)^{-1}$$

for all even positive integers k.
(2) For any non-trivial Hecke character of finite type $\chi \mod \mathfrak{p}^m$ with $m > 0$, we have

$$\int_{G(K(p^\infty)/K)} \chi(x) N(x)^{k-1}\, d\lambda = h^+ L(1-k,\chi)^{-1}$$

for any positive integer k satisfying the parity condition $\chi(-1) = (-1)^k$.

Moreover, the measure λ can be expressed in terms of Fourier coefficients of the Eisenstein series defined by χ on the Hilbert upper half plane.

To prove Theorem 1.1, we construct the distribution λ by considering a specific non-constant term in the Fourier expansion of the partial Eisenstein series

$$(\alpha_3 A^{-1} z + \alpha_4)^k \sum_{\substack{\{c,d\}\in\mathfrak{a}\times\mathfrak{a} \\ \gcd\left(\frac{c}{\mathfrak{a}},\frac{d}{\mathfrak{a}}\right)=1 \\ d\equiv a_2 \bmod \mathfrak{p}^m\mathfrak{a}}} \frac{N(\mathfrak{a})^k}{(\mathfrak{p}^m c A^{-1} z + d)^k}. \tag{6}$$

When the character χ is trivial, the values of λ are unbounded, so we modify the definition by adding a multiple of the normalized Haar measure on $\mathcal{O}_p$. Using the p-regularity assumptions, we show that λ is a p-adic unit, and thus it can be formally inverted. Moreover, we show that its reciprocal is Deligne–Ribet's p-adic L-function.

Remark 1.2 Greenberg mentions in the introduction to [9] that the condition $p \nmid [K : \mathbb{Q}]$ may be relaxed. However, the other conditions on p are the regularity conditions and cannot be removed. In fact, we expect the distribution to be unbounded when p is irregular, and we will study this case in a future work.

The chapter is organized as follows: In Sect. 2, we give a brief overview of Hilbert modular forms in the classical setting, and we define holomorphic Hilbert Eisenstein series. In Sect. 3, we review the Fourier expansion of holomorphic Eisenstein series of weight k and level $\mathfrak{n}$. In Sect. 4, we use the non-constant terms of this Fourier expansion to construct a p-adic distribution λ. This is the main construction of the chapter. Moreover, we show that λ satisfies the properties given in Theorem 1.1. In Sect. 5, we prove that λ is in fact a p-adic measure.

2 Hilbert Modular Forms

In this section, we set up the main definitions and results that we need from the theory of Hilbert modular forms. A thorough treatment of the subject can be found in [7], for example.

2.1 The Field K

Let K be a number field of degree r over $\mathbb{Q}$. Let $\mathcal{O}$ denote the ring of integers of K, $\mathfrak{d}$ its different ideal, D its discriminant, h its class number, and $\mathbb{A}_K$ its ring of adeles. Let $N : K \to \mathbb{Q}$ denote the norm map on K.

We fix an ordering of the r embeddings $K \hookrightarrow \mathbb{R}$. For any $\alpha \in K$, we denote by $\alpha^{(1)}, \cdots, \alpha^{(r)}$ the conjugates of α. We say α is totally positive, denoted $\alpha \gg 0$, if $\alpha^{(i)} > 0$ for all $1 \le i \le r$. We will assume throughout that K is a totally real field, i.e., that $\alpha \gg 0$ for all $\alpha \in K$.

Let $\mathfrak{n}$ be a non-zero integral ideal of $\mathcal{O}$. The strict (or narrow) ideal class group of K modulo $\mathfrak{n}$, denoted $G_{\mathfrak{n}}$, is defined by

$$\{\text{prime-to-}\mathfrak{n}\text{ fractional ideals of } K\}/\left\{\langle\alpha\rangle \mid \alpha \gg 0 \text{ and } \alpha \in 1 + \mathfrak{n}\mathfrak{b}^{-1} \text{ for some } \mathfrak{b}\right\}. \tag{7}$$

We denote the strict (or narrow) class number of K modulo $\mathfrak{n}$ by $h^{+}(\mathfrak{n})$.

A Grossencharacter χ is said to be of finite type with conductor $\mathfrak{n}$, denoted χ mod $\mathfrak{n}$, whenever χ is a character $(\mathcal{O}/\mathfrak{n})^* \to \mathbb{C}^*$ and χ is not induced from any other such character.

2.2 Structure of the Hilbert Modular Group

The Hilbert modular group $SL_2(\mathcal{O})$ acts discontinuously on the product of r copies of the upper half plane $\mathfrak{h}$ by Mobius transformations componentwise: For any $A = \left(\begin{smallmatrix} a & b \\ c & d \end{smallmatrix}\right) \in SL_2(\mathcal{O})$ and any $z = (z_1, \cdots, z_r) \in \mathfrak{h}^r$,

$$A \cdot z = \left(\frac{a^{(1)}z_1 + b^{(1)}}{c^{(1)}z_1 + d^{(1)}}, \cdots, \frac{a^{(r)}z_r + b^{(r)}}{c^{(r)}z_r + d^{(r)}} \right). \tag{8}$$

This produces an infinite number of cusps but only a finite number of inequivalent cusps.

Theorem 2.1 ([7], p. 37, Corollary 3.5_1) *The equivalence classes of cusps κ under the action of $SL_2(\mathcal{O})$ are in one-to-one correspondence with the ideal classes of K. In particular, the number of inequivalent cusps is h.*

The correspondence is easy to describe: The cusp represented by $(a : b) \in \mathbb{P}^1(K)$ is mapped to the class of the ideal $\langle a, b \rangle$. This implies that the cusp $\kappa = \infty$ which is represented by $(1 : 0) \in \mathbb{P}^1(K)$ corresponds to the trivial ideal class consisting of all principal ideals of K.

2.3 Modular Forms on Congruence Subgroups

For $k = (k_1, \cdots, k_r) \in \mathbb{Z}^r$, $z = (z_1, \cdots, z_r) \in \mathbb{C}^r$, and $c, d \in K$, we write

$$Tr(k) = \sum_{i=1}^{r} k_i \qquad Tr(z) = \sum_{i=1}^{r} z_i \tag{9}$$

$$(cz + d)^k = \prod_{i=1}^{r} (c^{(i)} z_i + d^{(i)})^{k_i}. \tag{10}$$

The slash operator $|_k$ is defined for a complex-valued function f on $\mathfrak{h}^r$ and an element $\left(\begin{smallmatrix} * & * \\ c & d \end{smallmatrix}\right) \in SL_2(\mathcal{O})$ by

$$(f\, |_k \left(\begin{smallmatrix} * & * \\ c & d \end{smallmatrix}\right))(z) = (cz + d)^{-k} f(\left(\begin{smallmatrix} * & * \\ c & d \end{smallmatrix}\right) \cdot z). \tag{11}$$

Given a congruence subgroup $\Gamma \subset SL_2(\mathcal{O})$, we say that f is a Hilbert modular form of weight $k = (k_1, \cdots, k_r)$ and level Γ if $f\, |_k\, \gamma = f$ for all $\gamma \in \Gamma$. We will only be interested in the case when $k_1 = k_2 = \cdots = k_r = k$, and we will say the Hilbert modular form is of (parallel) weight k. One basic example of Hilbert modular forms is the Eisenstein series attached to a character of K^*. We denote the space of all the Hilbert modular forms of weight k and level Γ by $\mathcal{M}_k(\Gamma)$.

We will be most interested in the congruence subgroups

$$\Gamma_0(\mathfrak{n}) = \left\{ \gamma = \begin{pmatrix} a & b \\ c & d \end{pmatrix} \in SL_2(\mathcal{O}) \,\middle|\, \gamma \equiv \begin{pmatrix} * & * \\ 0 & * \end{pmatrix} \text{ mod } \mathfrak{n} \right\} \tag{12}$$

$$\Gamma_1(\mathfrak{n}) = \left\{ \gamma = \begin{pmatrix} a & b \\ c & d \end{pmatrix} \in SL_2(\mathcal{O}) \,\middle|\, \gamma \equiv \begin{pmatrix} 1 & * \\ 0 & 1 \end{pmatrix} \text{ mod } \mathfrak{n} \right\} \tag{13}$$

of $SL_2(\mathcal{O})$, where $\mathfrak{n}$ is an integral ideal of $\mathcal{O}$.

If $f \in \mathcal{M}_k(\Gamma_0(\mathfrak{n}))$ or $f \in \mathcal{M}_k(\Gamma_1(\mathfrak{n}))$, it has a Fourier expansion at the cusp ∞ of the form

$$f(z) = \sum_{\substack{\nu \equiv 0 \text{ mod } (\mathfrak{d}\mathfrak{n})^{-1} \\ \nu \gg 0}} a(\nu) \exp(\nu z), \tag{14}$$

where $\exp(\nu z) = e^{2\pi i Tr(\nu z)}$.

For any general cusp κ, there is a matrix $A = \left(\begin{smallmatrix} \alpha_1 & \alpha_2 \\ \alpha_3 & \alpha_4 \end{smallmatrix}\right)$ of determinant 1 such that $A\kappa = \infty$ ([12], p. 181). Moreover, A can be chosen so that

$$\kappa = -\frac{\alpha_4}{\alpha_3}, \quad \mathfrak{b} = \langle \alpha_3, \alpha_4 \rangle, \quad \mathfrak{b}^{-1} = \langle \alpha_1, \alpha_2 \rangle, \quad \gcd(\mathfrak{n}, \mathfrak{b}) = 1. \tag{15}$$

In this case, f admits a Fourier expansion at $\kappa = A^{-1}\infty$ of the form

$$(\alpha_3 A^{-1} z + \alpha_4)^k f(A^{-1} z) = \sum_{\substack{\nu \equiv 0 \text{ mod } \mathfrak{b}^2 (\mathfrak{d}\mathfrak{n})^{-1} \\ \nu \gg 0}} a_A(\nu) \exp(\nu z). \tag{16}$$

2.4 Eisenstein Series

Recall that $\mathfrak{n}$ is an integral ideal of $\mathcal{O}$. Let Γ be a congruence subgroup of $SL_2(\mathcal{O})$. Given a Grossencharacter χ mod $\mathfrak{n}$ of finite type, an integer $k \geq 3$, and a cusp κ of Γ, the weight k holomorphic Eisenstein series for Γ at κ is given by

$$E^{\Gamma}_{k,\mathfrak{n}}(\chi, z, \kappa) = \sum_{\gamma \in \Gamma_\kappa \backslash \Gamma} \text{Im}(\gamma z)^k \tag{17}$$

for any $z \in \mathfrak{h}^r$. Here, Γ_κ denotes the stabilizer of κ inside Γ. When $\Gamma = \Gamma_0(\mathfrak{n})$, we will suppress the congruence subgroup from the notation of the Eisenstein series.

Let κ_1 be a cusp of $\Gamma_1(\mathfrak{n})$ lying above a cusp κ of $\Gamma_0(\mathfrak{n})$. Then we have

$$E_{k,\mathfrak{n}}(\chi, z, \kappa) = \sum_{\gamma=\begin{pmatrix} a & b \\ c & d \end{pmatrix} \in \Gamma_1(\mathfrak{n}) \backslash \Gamma_0(\mathfrak{n})} \chi(\langle d \rangle) E_{k,\mathfrak{n}}^{\Gamma_1(\mathfrak{n})}(\chi, z, \kappa_1). \tag{18}$$

We can describe the Eisenstein series explicitly at the cusp $\kappa = \infty$. This cusp corresponds to the trivial ideal class, so all the ideals are principal. Two matrices in $SL_2(\mathcal{O})$ are left-equivalent under Γ_∞ if and only if they have the same bottom row $(c, d) \in \mathcal{O} \times \mathcal{O}$. Fix a principal ideal $\mathfrak{a}$ which is not necessarily integral. A tuple (c, d) is the bottom row of a matrix in $\Gamma_0(\mathfrak{n})$ or $\Gamma_1(\mathfrak{n})$ precisely when $c, d \in \mathfrak{a}$, $\gcd\left(\frac{c}{\mathfrak{a}}, \frac{d}{\mathfrak{a}}\right) = 1$, and the tuple satisfies the congruence condition (12) or (13) of the respective subgroup ([1], p. 706, Equation (46)). Since $\mathfrak{a} = \langle \alpha \rangle$ is principal, every element $a \in \mathfrak{a}$ can be written as $a = a^*\alpha$, and we use the notation $\frac{a}{\mathfrak{a}}$ to denote a^*. Choose a set of representatives $a_2 \in \mathfrak{a}$, which runs over all the congruence classes modulo $\mathfrak{n}\mathfrak{a}$. Now the sum in Eq. (17) at the cusp $\kappa = \infty$ can be written as

$$E_{k,\mathfrak{n}}(\chi, z) := E_{k,\mathfrak{n}}(\chi, z, \infty) = \sum_{a_2 \in (\mathfrak{a}/\mathfrak{n}\mathfrak{a})^\times} \chi(\langle a_2 \rangle) \sum_{\substack{\{c,d\} \in \mathfrak{a} \times \mathfrak{a} \\ \gcd\left(\frac{c}{\mathfrak{a}}, \frac{d}{\mathfrak{a}}\right)=1 \\ \mathfrak{n} | c \\ d \equiv a_2 \bmod \mathfrak{n}\mathfrak{a}}} \frac{N(\mathfrak{a})^k}{(cz + d)^k} \tag{19}$$

$$= \sum_{a_2 \in (\mathfrak{a}/\mathfrak{n}\mathfrak{a})^\times} \chi(\langle a_2 \rangle) \sum_{\substack{\{c,d\} \in \mathfrak{a} \times \mathfrak{a} \\ \gcd\left(\frac{c}{\mathfrak{a}}, \frac{d}{\mathfrak{a}}\right)=1 \\ d \equiv a_2 \bmod \mathfrak{n}\mathfrak{a}}} \frac{N(\mathfrak{a})^k}{(\mathfrak{n}cz + d)^k}. \tag{20}$$

The notation $\{c, d\}$ means that two tuples (c, d) and (c', d') are identified together if there exists $\varepsilon \equiv 1 \bmod \mathfrak{n}$ such that $c = \varepsilon c'$ and $d = \varepsilon d'$.

This definition does not depend on the ideal $\mathfrak{a}$ that was chosen. Given another ideal $\mathfrak{a}'$ in the same ideal class as $\mathfrak{a}$, there exists an element $\alpha \in K^*$ such that $\mathfrak{a}' = \alpha\mathfrak{a}$. This implies that the numerator and the denominator of the inner sum in (19) would both be multiplied by $N(\alpha)^k$, and hence the expression is invariant ([5], p. 662).

We now want to describe the Eisenstein series at a cusp $\kappa \neq \infty$, which corresponds to a (fixed) non-trivial ideal class. The idea is to translate the cusp κ to ∞ where an explicit expression of the Eisenstein series is easy to describe.

By Eq. (16), the Fourier expansion of $E_{k,\mathfrak{n}}(\chi, z, \kappa)$ at κ is equal to the Fourier expansion of $E_{k,\mathfrak{n}}(\chi, z)|_k A^{-1} = (-\alpha_3 z + \alpha_1)^{-k} E_{k,\mathfrak{n}}(\chi, A^{-1}z)$ at ∞. Since $\det A = 1$, a direct computation shows that $(-\alpha_3 z + \alpha_1) = (\alpha_3 A^{-1} z + \alpha_4)^{-1}$. Hence, we have

$$E_{k,\mathfrak{n}}(\chi, z, \kappa) = (\alpha_3 A^{-1} z + \alpha_4)^k \sum_{a_2 \in (\mathfrak{a}/\mathfrak{n}\mathfrak{a})^\times} \chi(\langle a_2 \rangle) \sum_{\substack{\{c,d\} \in \mathfrak{a} \times \mathfrak{a} \\ \gcd\left(\frac{c}{\mathfrak{a}}, \frac{d}{\mathfrak{a}}\right)=1 \\ d \equiv a_2 \bmod \mathfrak{n}\mathfrak{a}}} \frac{N(\mathfrak{a})^k}{(\mathfrak{n} c A^{-1} z + d)^k}. \tag{21}$$

3 Fourier Coefficients of Certain Eisenstein Series

We are interested in calculating the non-constant Fourier coefficients of (21). Indeed, this will be the main calculation needed to construct our p-adic measure λ. We start by defining a more general Eisenstein series by

$$G_k(z, a_1, a_2, \mathfrak{n}, \mathfrak{a}) = \sum_{\substack{\{c,d\}^+ \in \mathfrak{a} \times \mathfrak{a} \\ (c,d) \equiv (a_1, a_2) \bmod \mathfrak{n}\mathfrak{a}}} \frac{N(\mathfrak{a})^k}{(cz + d)^k}. \tag{22}$$

The notation $\{c, d\}^+$ means that two tuples (c, d) and (c', d') are identified together if there exists a totally positive $\varepsilon \equiv 1 \bmod \mathfrak{n}$ such that $c = \varepsilon c'$ and $d = \varepsilon d'$.

Lemma 3.1 *We have*

$$\sum_{\substack{\{c,d\} \in \mathfrak{a} \times \mathfrak{a} \\ \gcd\left(\frac{c}{\mathfrak{a}}, \frac{d}{\mathfrak{a}}\right)=1 \\ d \equiv a_2 \bmod \mathfrak{n}\mathfrak{a}}} \frac{N(\mathfrak{a})^k}{(cz + d)^k} = \sum_{a_1 \in \mathfrak{a}/\mathfrak{a}\mathfrak{n}} \sum_{i=1}^{h(\mathfrak{n})} \sum_{\mathfrak{t} \in C_i} \frac{\mu(\mathfrak{t})}{N(\mathfrak{t})^k} G_k(z, \tau a_1, \tau a_2, \mathfrak{n}, \mathfrak{a}\mathfrak{t}),$$

where $\mu(\mathfrak{m})$ denotes the Mobius function for any integral ideal $\mathfrak{m}$ given by

$$\mu(\mathfrak{m}) = \begin{cases} 1 & \textit{if } \mathfrak{m} = \mathcal{O} \\ (-1)^r & \textit{if } \mathfrak{m} = \mathfrak{p}_1 \cdots \mathfrak{p}_r \\ 0 & \textit{if } \mathfrak{p}^2 | \mathfrak{m} \textit{ for some prime ideal } \mathfrak{p}, \end{cases} \tag{23}$$

and $h(\mathfrak{n})$ denotes the strict ideal class number of K mod $\mathfrak{n}$.

Proof Recall the following property of the Mobius function:

$$\sum_{\mathfrak{t} | \mathfrak{n}} \mu(\mathfrak{t}) = \begin{cases} 1 & \text{if } \mathfrak{n} = \mathcal{O} \\ 0 & \text{otherwise.} \end{cases} \tag{24}$$

We can use this property to remove the coprimeness condition. Recall that $\frac{c}{\mathfrak{a}}, \frac{d}{\mathfrak{a}} \in \mathcal{O}$, which implies that the ideal gcd $\left(\frac{c}{\mathfrak{a}}, \frac{d}{\mathfrak{a}}\right)$ is an integral ideal. We now have

$$\sum_{\substack{\{c,d\}\in\mathfrak{a}\times\mathfrak{a} \\ \gcd\left(\frac{c}{\mathfrak{a}},\frac{d}{\mathfrak{a}}\right)=1 \\ (c,d)\equiv(a_1,a_2) \bmod \mathfrak{n}\mathfrak{a}}} \frac{N(\mathfrak{a})^k}{(cz+d)^k} = \sum_{\substack{\{c,d\}\in\mathfrak{a}\times\mathfrak{a} \\ (c,d)\equiv(a_1,a_2) \bmod \mathfrak{n}\mathfrak{a}}} \frac{N(\mathfrak{a})^k}{(cz+d)^k} \sum_{\mathfrak{t} \mid \gcd\left(\frac{c}{\mathfrak{a}},\frac{d}{\mathfrak{a}}\right)} \mu(\mathfrak{t}) \tag{25}$$

$$= \sum_{\gcd(\mathfrak{t},\mathfrak{n})=1} \mu(\mathfrak{t}) \sum_{\substack{\{c,d\}\in\mathfrak{a}\times\mathfrak{a} \\ (c,d)\equiv(a_1,a_2) \bmod \mathfrak{n}\mathfrak{a} \\ (c,d)\equiv(0,0) \bmod \mathfrak{a}\mathfrak{t}}} \frac{N(\mathfrak{a})^k}{(cz+d)^k}. \tag{26}$$

Let $C_1, \cdots, C_{h(\mathfrak{n})}$ denote the strict ideal classes of K modulo $\mathfrak{n}$. Choose integral ideals $\mathfrak{t}_1, \cdots, \mathfrak{t}_{h(\mathfrak{n})}$ such that $\mathfrak{t}_i \in C_i^{-1}$ for all $i \in \{1, \cdots h(\mathfrak{n})\}$. For each $\mathfrak{t} \in C_i$, let $(\tau) = \mathfrak{t}\mathfrak{t}_i$, where $\tau \equiv 1 \bmod \mathfrak{n}$. Then the previous sum is equal to

$$\sum_{i=1}^{h(\mathfrak{n})} \sum_{\substack{\mathfrak{t}\in C_i \\ \gcd(\mathfrak{t},\mathfrak{n})=1}} \mu(\mathfrak{t}) \sum_{\substack{\{c,d\}\in\mathfrak{a}\times\mathfrak{a} \\ (c,d)\equiv(\tau a_1,\tau a_2) \bmod \mathfrak{n}\mathfrak{a}\mathfrak{t}}} \frac{N(\mathfrak{a})^k}{(cz+d)^k} \tag{27}$$

$$= \sum_{i=1}^{h(\mathfrak{n})} \sum_{\mathfrak{t}\in C_i} \frac{\mu(\mathfrak{t})}{N(\mathfrak{t})^k} G_k(z, \tau a_1, \tau a_2, \mathfrak{n}, \mathfrak{a}\mathfrak{t}). \tag{28}$$

Summing up over the equivalence classes of a_1 modulo $\mathfrak{n}\mathfrak{a}$, we get the desired result. □

The Fourier expansion of the Eisenstein series in Eq. (22) has been computed by Klingen ([12], pp. 181 and 182) as follows:

Theorem 3.2 *The Fourier expansion of $G_k(z, a_1, a_2, \mathfrak{n}, \mathfrak{a})$ at a general cusp κ is given by*

$$\delta\left(\frac{(a_1^*)}{\mathfrak{n}\mathfrak{a}\mathfrak{b}}\right) N(\mathfrak{a})^k \sum_{\substack{d\equiv a_2^* \bmod \mathfrak{n}\mathfrak{a}\mathfrak{b}^{-1} \\ \{d\}^+}} \frac{sgn^k N(d)}{|N(d)|^k}$$
$$+ \frac{(-2\pi i)^{kr} N(\mathfrak{a})^{k-1} N(\mathfrak{b})}{(k-1)!^r N(\mathfrak{n})|D|^{1/2}} \sum_{\substack{c\equiv a_1^* \bmod \mathfrak{n}\mathfrak{a}\mathfrak{b} \\ \nu\equiv 0 \bmod \mathfrak{b}/\mathfrak{n}\mathfrak{a}\mathfrak{d} \\ c\nu\gg 0, \{c\}^+}} sgnN(\nu)N(\nu)^{k-1} \exp(a_2^*\nu + c\nu z), \tag{29}$$

where the following notation is used:

$$\delta(\mathfrak{m}) = \begin{cases} 1 \text{ if } \mathfrak{m} \subset \mathcal{O} \\ 0 \text{ otherwise} \end{cases}$$

$$a_1^* = \alpha_4 a_1 - \alpha_1 a_2 \in \mathfrak{a}\mathfrak{b}$$

$$a_2^* = \alpha_1 a_2 - \alpha_2 a_1 \in \mathfrak{a}\mathfrak{b}^{-1}$$

$$\{m\}^+ = m\{\varepsilon\mathcal{O}^* : \varepsilon \equiv 1 \bmod \mathfrak{n}, \varepsilon \gg 0\}.$$

We use this Fourier expansion to prove the following:

Theorem 3.3 *The Fourier expansion of the partial Eisenstein series*

$$(\alpha_3 A^{-1}z + \alpha_4)^k \sum_{\substack{\{c,d\}\in\mathfrak{a}\times\mathfrak{a} \\ \gcd\left(\frac{c}{\mathfrak{a}},\frac{d}{\mathfrak{a}}\right)=1 \\ d\equiv a_2 \bmod \mathfrak{n}\mathfrak{a}}} \frac{N(\mathfrak{a})^k}{(\mathfrak{n}cA^{-1}z + d)^k}$$

at the cusp ∞ *is given by*

$$\sum_{l\equiv a_2^* \bmod \mathfrak{n}\mathfrak{a}\mathfrak{b}^{-1}} \frac{N(\mathfrak{a})^k}{N(l)^k} \sum_{\substack{\mathfrak{t}\in K \\ \mathfrak{t}|\langle l\rangle \\ \mathfrak{n}\nmid\mathfrak{t}}} \mu(\mathfrak{t})$$

$$+ \frac{(-2\pi i)^{kr} N(\mathfrak{b})^k}{(k-1)!^r N(\mathfrak{n})^k |D|^{\frac{2k-1}{2}}} \sum_{i=1}^{h(\mathfrak{n})} \sum_{\mathfrak{t}\in C_i} \frac{\mu(\mathfrak{t})}{N(\mathfrak{t})^k} \sum_{\substack{\nu\in\mathcal{O} \\ c\nu\gg 0,\{c\}^+}} sgnN(\nu)N(\nu)^{k-1}$$

$$\times \exp\left(\frac{\mathfrak{t}_i a_2' \nu\mathfrak{b}}{\mathfrak{n}\mathfrak{d}} + \frac{c\nu\mathfrak{b}^2 z}{\mathfrak{d}}\right).$$

Proof Lemma 3.1 implies that we need to calculate the Fourier expansion of

$$(\alpha_3 A^{-1}z + \alpha_4)^k \sum_{a_1\in\mathfrak{a}/\mathfrak{a}\mathfrak{n}} \sum_{i=1}^{h(\mathfrak{n})} \sum_{\mathfrak{t}\in C_i} \frac{\mu(\mathfrak{t})}{N(\mathfrak{t})^k} G_k(\mathfrak{n}A^{-1}z, \tau a_1, \tau a_2, \mathfrak{n}, \mathfrak{a}\mathfrak{t}). \tag{30}$$

Theorem 3.2 along with Eq. (16) now shows that the constant term of (30) is nonzero only when $a_1^* \equiv 0 \bmod \mathfrak{n}\mathfrak{a}\mathfrak{t}\mathfrak{b}$. This implies that the contribution from the terms of the outermost sum is all zero except for one equivalence class of a_1. The nonzero term remaining is

$$\sum_{i=1}^{h(\mathfrak{n})}\sum_{\mathfrak{t}\in C_i}\frac{\mu(\mathfrak{t})}{N(\mathfrak{t})^k}N(\mathfrak{a}\mathfrak{t})^k\sum_{\substack{d\equiv\tau a_2^*\bmod \mathfrak{n}\mathfrak{a}\mathfrak{t}\mathfrak{b}^{-1}\\ \{d\}^+}}\frac{\operatorname{sgn}^k N(d)}{|N(d)|^k}. \tag{31}$$

The ideals $\mathfrak{n}$ and $\mathfrak{t}$ being coprime, we can rewrite $d\equiv\tau a_2^* \bmod \mathfrak{n}\mathfrak{a}\mathfrak{t}\mathfrak{b}^{-1}$ as two equivalences modulo $\mathfrak{n}\mathfrak{a}\mathfrak{b}^{-1}$ and $\mathfrak{a}\mathfrak{t}\mathfrak{b}^{-1}$. We now obtain the constant term to be

$$\sum_{i=1}^{h(\mathfrak{n})}\sum_{\mathfrak{t}\in C_i}\frac{\mu(\mathfrak{t})}{N(\mathfrak{t})^k}\sum_{\substack{d\equiv a_2^*\bmod \mathfrak{n}\mathfrak{a}\mathfrak{b}^{-1}\\ d\equiv 0\bmod \mathfrak{a}\mathfrak{b}^{-1}\mathfrak{t}\\ \{d\}^+}}N(\mathfrak{a}\mathfrak{t})^k\frac{\operatorname{sgn}^k N(d)}{|N(d)|^k}. \tag{32}$$

Switching the order of summation and writing $\langle l\rangle$ for the ideal generated by l, the constant term becomes

$$\sum_{\substack{\mathfrak{t}\in K\\ \mathfrak{p}\nmid\mathfrak{t}}}\frac{\mu(\mathfrak{t})}{N(\mathfrak{t})^k}\sum_{\substack{td\equiv a_2^*\bmod \mathfrak{n}\mathfrak{a}\mathfrak{b}^{-1}\\ \{d\}^+}}N(\mathfrak{a})^k\frac{1}{N(d)^k} \tag{33}$$

$$=\sum_{l\equiv a_2^*\bmod \mathfrak{n}\mathfrak{a}\mathfrak{b}^{-1}}\frac{N(\mathfrak{a})^k}{N(l)^k}\sum_{\substack{\mathfrak{t}\in K\\ \mathfrak{t}|\langle l\rangle\\ \mathfrak{n}\nmid\mathfrak{t}}}\mu(\mathfrak{t}). \tag{34}$$

We consider next the non-constant terms. Note that as a_1 varies over the equivalence classes of $\mathfrak{a}/\mathfrak{a}\mathfrak{n}$, a_1^* varies over the equivalence classes of $\mathfrak{a}\mathfrak{b}/\mathfrak{a}\mathfrak{b}\mathfrak{n}$. It is clear then that the non-constant terms are

$$\frac{(-2\pi i)^{kr}N(\mathfrak{b})}{(k-1)!^r N(\mathfrak{n})|D|^{1/2}}\sum_{a_1^*\in\mathfrak{a}\mathfrak{b}/\mathfrak{a}\mathfrak{b}\mathfrak{n}}\sum_{i=1}^{h(\mathfrak{n})}\sum_{\mathfrak{t}\in C_i}\frac{\mu(\mathfrak{t})}{N(\mathfrak{t})^k}N(\mathfrak{a}\mathfrak{t})^{k-1}$$

$$\sum_{\substack{c\equiv\tau a_1^*\bmod \mathfrak{n}\mathfrak{a}\mathfrak{b}\mathfrak{t}\\ \nu\equiv 0\bmod \mathfrak{b}/\mathfrak{n}\mathfrak{a}\mathfrak{t}\mathfrak{d}\\ c\nu\gg 0,\{c\}^+}}\operatorname{sgn}N(\upsilon)N(\nu)^{k-1}\exp(\tau a_2^*\nu+c\nu\mathfrak{n}z). \tag{35}$$

As before, we can split the equivalence modulo $\mathfrak{n}\mathfrak{a}\mathfrak{b}\mathfrak{t}$ into two equivalences modulo $\mathfrak{n}\mathfrak{a}\mathfrak{b}$ and $\mathfrak{a}\mathfrak{b}\mathfrak{t}$. Moreover, we make a change of variables $\nu\to\frac{\nu\mathfrak{b}}{\mathfrak{n}\mathfrak{a}\mathfrak{d}\mathfrak{t}}$ and $c\to\mathfrak{a}\mathfrak{b}\mathfrak{t}c$. Since $N(\mathfrak{d})=|D|$, we get

$$\frac{(-2\pi i)^{kr} N(\mathfrak{b})^k}{(k-1)!^r N(\mathfrak{n})^k |D|^{\frac{2k-1}{2}}} \sum_{i=1}^{h(\mathfrak{n})} \sum_{\mathfrak{t} \in C_i} \frac{\mu(\mathfrak{t})}{N(\mathfrak{t})^k} \sum_{\substack{\nu \in \mathcal{O} \\ c\nu \gg 0, \{c\}^+}} \operatorname{sgn} N(\nu) N(\nu)^{k-1}$$

$$\times \exp\left(\frac{\mathfrak{t}_i a_2' \nu \mathfrak{b}}{\mathfrak{n}\mathfrak{d}} + \frac{c\nu \mathfrak{b}^2 z}{\mathfrak{d}}\right). \tag{36}$$

□

4 Constructing a *p*-Adic Measure

4.1 p-Adic Measures on Galois Groups

Let $\mathfrak{b}_1, \cdots, \mathfrak{b}_h$ represent the strict ideal classes of K, where each $\mathfrak{b}_j$ is the ideal generated by the lower row of the matrix A_j associated with the distinct cusps κ_j. Fix a set of elements $b_j \in \mathbb{A}_K^{\times,(\infty)}$ such that the $\mathcal{O}$-ideal associated with b_j is $\mathfrak{b}_j$. We will abuse notation and identify b_j and $\mathfrak{b}_j$. By construction, all the ideals $\mathfrak{b}_j$ are relatively prime to $\mathfrak{p}$.

Let $K(p^\infty)$ denote the strict ray class field of modulus p^∞. We can decompose $G = \mathrm{G}(K(p^\infty)/K)$ as $\sqcup_j \mathfrak{b}_j^{-1}(\mathcal{O}_\mathfrak{p}^*/U)$, where U is the closure in $\mathcal{O}_\mathfrak{p}^*$ of the subgroup of all totally positive units in $\mathcal{O}^*$ ([10], Section 3.9). Then for any continuous function ϕ on G, we can define a continuous function ϕ_j on $\mathcal{O}_\mathfrak{p}$ by

$$\phi_j(x) = \begin{cases} \phi(\mathfrak{b}_j^{-1}x) & \text{if } x \in \mathcal{O}_\mathfrak{p}^* \\ 0 & \text{otherwise.} \end{cases} \tag{37}$$

We then define p-adic measures on G by setting

$$\int_G \phi \, d\mu = \sum_j \int_{\mathcal{O}_\mathfrak{p}^*/U} \phi_j d\mu_j. \tag{38}$$

4.2 The Main Construction

The structure of the Galois group G allows us to define a distribution λ on G by defining it on each piece $\mathfrak{b}_j^{-1}(\mathcal{O}_\mathfrak{p}^*/U)$, which we identify naturally with $\mathcal{O}_\mathfrak{p}^*/U$. We will denote the restriction of λ to each $\mathfrak{b}_j^{-1}(\mathcal{O}_\mathfrak{p}^*/U)$ by λ_j.

Fix a strict ideal class $\mathfrak{b}_j$. Define a map

$$\varepsilon_{k,\mathfrak{p}^m,\mathfrak{b}_j}(\mathfrak{a}) : \mathcal{O}/\mathfrak{p}^m\mathcal{O} \to \mathcal{M}_k(\Gamma_1(\mathfrak{p}^m)) \tag{39}$$

by

$$\varepsilon_{k,\mathfrak{p}^m,\mathfrak{b}_j}(\mathfrak{a}) := (\alpha_3 A^{-1}z + \alpha_4)^k \sum_{\substack{\{c,d\}\in\mathfrak{a}\times\mathfrak{a} \\ \gcd\left(\frac{c}{\mathfrak{a}},\frac{d}{\mathfrak{a}}\right)=1 \\ d\equiv a_2 \bmod \mathfrak{p}^m\mathfrak{a}}} \frac{N(\mathfrak{a})^k}{(\mathfrak{p}^m c A^{-1}z + d)^k}. \tag{40}$$

Let C_n denote the map which sends a holomorphic Hilbert modular form to the coefficient of $e^{2\pi i Tr(nz)}$ in its Fourier expansion. Theorem 3.3 states that

$$C_{\mathfrak{p}^m}(\varepsilon_{k,\mathfrak{p}^m,\mathfrak{b}_j}) = \frac{(-2\pi i)^{kr} N(\mathfrak{b}_j)^k}{(k-1)!^r |D|^{\frac{2k-1}{2}}} \sum_{u=0}^{m} N(\mathfrak{p})^{u(k-1)-mk} \sum_{i=1}^{h(\mathfrak{p}^m)} \sum_{\mathfrak{t}\in C_i} \frac{\mu(\mathfrak{t})}{N(\mathfrak{t})^k} \exp\left(\frac{\mathfrak{t}_i a_2' \mathfrak{p}^u \mathfrak{b}_j}{\mathfrak{p}^m \mathfrak{d}}\right). \tag{41}$$

Define a distribution on $\mathfrak{b}_j^{-1}(\mathcal{O}_{\mathfrak{p}}^*/U)$ by

$$\lambda_j(\mathfrak{a} + \mathfrak{p}^m \mathcal{O}_{\mathfrak{p}}) = \frac{1}{2^r N(\mathfrak{b}_j)^k} C_{\mathfrak{p}^m}(\varepsilon_{k,\mathfrak{p}^m,\mathfrak{b}_j}) + \gamma(k)\lambda_{Haar}, \tag{42}$$

where

$$\gamma(k) = \begin{cases} \frac{N(\mathfrak{p})^{2k-1}}{N(\mathfrak{p})^k - 1} \frac{(1-N(\mathfrak{p})^{k-1})^{-1}}{\zeta(1-k)} & \text{for k even} \\ 0 & \text{otherwise,} \end{cases} \tag{43}$$

and λ_{Haar} is the unique normalized Haar measure on $\mathcal{O}_p$ such that $\lambda_{Haar}(\mathcal{O}_p) = 1$.

Note that since $a_2' \in \mathfrak{a}\mathfrak{b}_j^{-1}$, the term inside the exponential function, and hence the definition of λ, does not depend on the ideals $\mathfrak{b}_j$. Henceforth we will write $\mathfrak{a}$ instead of $a_2'\mathfrak{b}_j$.

The following lemma implies that λ_j is a distribution.

Lemma 4.1 *Suppose that*

$$\nu(\mathfrak{a} + \mathfrak{p}^m \mathcal{O}_p) = N(\mathfrak{p})^{m(k-1)} \sum_{i=1}^{h(\mathfrak{p}^m)} \sum_{t\in C_i} a(\mathfrak{t}) N(\mathfrak{t})^{-k} \exp\left(\frac{\mathfrak{t}\mathfrak{a}}{\mathfrak{p}^m}\right)$$

for some prime ideal $\mathfrak{p}$ *and integers* $k \geq 2$ *and* $m \geq 0$. *Suppose furthermore that* $a(\mathfrak{t}\mathfrak{p}) = a(\mathfrak{t})$. *If all the sums converge absolutely, then*

$$\nu(\mathfrak{a} + \mathfrak{p}^m \mathcal{O}_p) = \sum_{\mathfrak{b}\in\mathfrak{p}^m\mathcal{O}_p/\mathfrak{p}^{m+1}\mathcal{O}_p} \nu(\mathfrak{a} + \mathfrak{b}\mathfrak{p}^m + \mathfrak{p}^{m+1}\mathcal{O}_p).$$

Proof We have

$$\sum_{\mathfrak{b}\in\mathfrak{p}^m\mathcal{O}_p/\mathfrak{p}^{m+1}\mathcal{O}_p} \nu(\mathfrak{a}+\mathfrak{b}\mathfrak{p}^m+\mathfrak{p}^{m+1}\mathcal{O}_p)$$

$$= N(\mathfrak{p})^{(m+1)(k-1)}\sum_{i=1}^{h(\mathfrak{p}^m)}\sum_{t\in C_i} a(\mathfrak{t})N(\mathfrak{t})^{-k}\sum_{\mathfrak{b}\in\mathfrak{p}^m\mathcal{O}_p/\mathfrak{p}^{m+1}\mathcal{O}_p}\exp\left(\frac{\mathfrak{t}\mathfrak{a}}{\mathfrak{p}^{m+1}}+\frac{\mathfrak{t}\mathfrak{b}}{\mathfrak{p}}\right).$$

Note that

$$\sum_{\mathfrak{b}\in\mathfrak{p}^m\mathcal{O}_p/\mathfrak{p}^{m+1}\mathcal{O}_p}\exp\left(\frac{\mathfrak{t}\mathfrak{b}}{\mathfrak{p}}\right)=\begin{cases}N(\mathfrak{p}) & \text{if } \mathfrak{p}\mid\mathfrak{t}\\ 0 & \text{otherwise.}\end{cases} \tag{44}$$

Hence, the last sum becomes

$$N(\mathfrak{p})^{(m+1)(k-1)}N(\mathfrak{p})^{1-k}\sum_{i=1}^{h(\mathfrak{p}^m)}\sum_{t\in C_i} a(\mathfrak{t}\mathfrak{p})N(\mathfrak{t})^{-k}\exp\left(\frac{\mathfrak{t}\mathfrak{a}}{\mathfrak{p}^m}\right). \tag{45}$$

Knowing that $a(\mathfrak{t}\mathfrak{p})=a(\mathfrak{t})$, the last sum is equal to $\nu(\mathfrak{a}+\mathfrak{p}^m\mathcal{O}_p)$. □

We now calculate the Mellin transform of λ.

Theorem 4.2 *Let χ be a Hecke character of finite type with conductor $\mathfrak{p}^m$ for some $m\geq 1$. Let k be an integer of the same parity as χ. Then we have*

$$\int_G \chi^{-1}(x)\,d\lambda(x)=\frac{h^+\left(1-\chi(\mathfrak{p})N(\mathfrak{p})^{k-1}\right)^{-1}}{L(1-k,\chi)}, \tag{46}$$

where h^+ denotes the strict class number of K.

Proof First, we consider the case when χ is the trivial character and k is even. By finite additivity, we have

$$\lambda_j(\mathcal{O}_\mathfrak{p}^*)=\sum_{\mathfrak{a}\in\mathcal{O}/\mathfrak{p}\mathcal{O}}\lambda_j(\mathfrak{a}+\mathfrak{p}\mathcal{O}_\mathfrak{p}) \tag{47}$$

$$=\sum_{\mathfrak{a}\in\mathcal{O}/\mathfrak{p}\mathcal{O}}\frac{1}{2^r}C_\mathfrak{p}(\varepsilon_{k,\mathfrak{p},\mathfrak{b}_j})+\sum_{\mathfrak{a}\in\mathcal{O}/\mathfrak{p}\mathcal{O}}\gamma(k)\lambda_{Haar}. \tag{48}$$

We calculate each sum separately and start with the first sum. Since the sum of roots of unity modulo $\mathfrak{p}$ is 0, we have

$$\sum_{\mathfrak{a} \in \mathcal{O}/\mathfrak{p}\mathcal{O}} \exp\left(\frac{\mathfrak{t}_i \mathfrak{a}}{\mathfrak{p}\mathfrak{d}}\right) = -1. \tag{49}$$

Moreover, $\exp\left(\frac{\mathfrak{t}_i \mathfrak{a}}{\mathfrak{d}}\right) = e^{2\pi i Tr(\mathfrak{t}_i \mathfrak{a}/\mathfrak{d})} = 1$ for our choice of $\mathfrak{t}_i$ and $\mathfrak{a}$, so

$$\sum_{\mathfrak{a} \in \mathcal{O}/\mathfrak{p}\mathcal{O}} \exp\left(\frac{\mathfrak{t}_i \mathfrak{a}}{\mathfrak{d}}\right) = N(\mathfrak{p}) - 1. \tag{50}$$

These comments directly imply that the first sum is equal to

$$\frac{(-2\pi i)^{kr}}{2^r (k-1)!^r |D|^{\frac{2k-1}{2}}} \sum_{\substack{\mathfrak{t} \in K \\ \mathfrak{p} \nmid \mathfrak{t}}} \frac{\mu(\mathfrak{t})}{N(\mathfrak{t})^k} \sum_{u=0}^{1} N(\mathfrak{p})^{u(k-1)-k} \sum_{\mathfrak{a} \in \mathcal{O}/\mathfrak{p}\mathcal{O}} \exp\left(\frac{\mathfrak{t}_i \mathfrak{p}^u \mathfrak{a}}{\mathfrak{p}\mathfrak{d}}\right) \tag{51}$$

$$= \frac{(-2\pi i)^{kr}}{2^r (k-1)!^r |D|^{\frac{2k-1}{2}}} \sum_{\substack{\mathfrak{t} \in K \\ \mathfrak{p} \nmid \mathfrak{t}}} \frac{\mu(\mathfrak{t})}{N(\mathfrak{t})^k} \left(-N(\mathfrak{p})^{-k} + \frac{N(\mathfrak{p}) - 1}{N(\mathfrak{p})}\right). \tag{52}$$

The term $\sum_{\substack{\mathfrak{t} \in K \\ \mathfrak{p} \nmid \mathfrak{t}}} \mu(\mathfrak{t}) N(\mathfrak{t})^{-k}$ is the reciprocal of the Dedekind zeta function with the Euler factor at $\mathfrak{p}$ removed, so the previous sum is equal to

$$\frac{(-2\pi i)^{kr}}{2^r (k-1)!^r |D|^{\frac{2k-1}{2}}} \frac{(1 - N(\mathfrak{p})^{-k})^{-1}}{\zeta(k)} \left(-N(\mathfrak{p})^{-k} + \frac{N(\mathfrak{p}) - 1}{N(\mathfrak{p})}\right). \tag{53}$$

The functional equation of the Dedekind zeta function is given by

$$\frac{1}{\zeta(1-k)} = \frac{(-2\pi i)^{kr}}{2^r (k-1)!^r |D|^{\frac{2k-1}{2}}} \frac{1}{\zeta(k)}. \tag{54}$$

It directly implies the first sum is equal to

$$\frac{(1 - N(\mathfrak{p})^{-k})^{-1}}{\zeta(1-k)} \left(\frac{N(\mathfrak{p})^k - N(\mathfrak{p})^{k-1} - 1}{N(\mathfrak{p})^k}\right). \tag{55}$$

The second sum is

$$\frac{N(\mathfrak{p})^{2k-1}}{N(\mathfrak{p})^k - 1} \frac{(1 - N(\mathfrak{p})^{k-1})^{-1}}{\zeta(1-k)} \frac{N(\mathfrak{p}) - 1}{N(\mathfrak{p})}. \tag{56}$$

Adding Eqs. (55) and (56) gives

$$\lambda_j(\mathcal{O}_\mathfrak{p}^*) = \frac{(1 - N(\mathfrak{p})^{-k})^{-1}}{\zeta(1-k)}. \tag{57}$$

Summing up over j equivalence classes, we get the required result.

Next we consider the case when χ is a non-trivial Hecke character of finite type with conductor $\mathfrak{p}^m$ and k is an integer of the same parity as χ. Since χ is orthogonal to Haar measure, the integral of χ against λ_{Haar} is zero. The integral of χ against λ_j is

$$\int_{\mathcal{O}_\mathfrak{p}^*} \chi^{-1}(x)\, d\lambda_j(x) = \frac{(-2\pi i)^{kr}}{(k-1)!^r |D|^{\frac{2k-1}{2}}} \sum_{u=0}^{m} N(\mathfrak{p})^{u(k-1)-mk}$$
$$\sum_{i=1}^{h(\mathfrak{p}^m)} \sum_{\mathfrak{t} \in C_i} \frac{\mu(\mathfrak{t})}{N(\mathfrak{t})^k} \sum_{\mathfrak{a} \in (\mathcal{O}/\mathfrak{p}^m)^\times} \chi(\mathfrak{a}) \exp\left(\frac{\mathfrak{t}_i \bar{\mathfrak{a}} \mathfrak{p}^u}{\mathfrak{p}^m \mathfrak{d}}\right). \tag{58}$$

We define the Gauss sum

$$\tau(\chi) = \sum_{\mathfrak{x} \in \mathcal{O}/\mathfrak{p}^m} \chi(\mathfrak{x}) e^{2\pi i Tr(\mathfrak{x}/\mathfrak{p}^m \mathfrak{d})}. \tag{59}$$

If χ is primitive and $x \in \mathcal{O}$, then

$$\sum_{\mathfrak{x} \in \mathcal{O}/\mathfrak{p}^m} \chi(\mathfrak{x}) e^{2\pi i Tr(\mathfrak{x}x/\mathfrak{d})} = \begin{cases} \overline{\chi}(x)\tau(\chi) & \text{if } \gcd(x, \mathfrak{p}^m) = 1 \\ 0 & \text{otherwise.} \end{cases} \tag{60}$$

Then the term with $u = 0$ is the only non-zero term in Eq. (58), so we have

$$\frac{(-2\pi i)^{kr}}{2^r (k-1)!^r N(\mathfrak{p}^m)^k |D|^{\frac{2k-1}{2}}} \sum_{i=1}^{h(\mathfrak{p}^m)} \sum_{\mathfrak{t} \in C_i} \frac{\mu(\mathfrak{t})}{N(\mathfrak{t})^k} \sum_{\mathfrak{a} \in (\mathcal{O}/\mathfrak{p}^m)^\times} \chi(\mathfrak{a}) \exp\left(\frac{\mathfrak{t}_i \bar{\mathfrak{a}}}{\mathfrak{p}^m \mathfrak{d}}\right) \tag{61}$$

$$= \frac{(-2\pi i)^{kr}}{2^r (k-1)!^r N(\mathfrak{p}^m)^k |D|^{\frac{2k-1}{2}}} \sum_{\substack{\mathfrak{t} \in K \\ \mathfrak{p} \nmid \mathfrak{t}}} \frac{\mu(\mathfrak{t})}{N(\mathfrak{t})^k} \chi(\mathfrak{t})^{-1} \tau(\overline{\chi}) \tag{62}$$

$$= \frac{(-2\pi i)^{kr}}{2^r (k-1)!^r N(\mathfrak{p}^m)^k |D|^{\frac{2k-1}{2}}} \tau(\overline{\chi}) \frac{1}{L(k, \overline{\chi})}. \tag{63}$$

The functional equation of the Hecke L-function of a totally real field

$$L(k, \chi) = \frac{(2\pi i)^{kr}}{2^r (k-1)!^r |D|^{\frac{2k-1}{2}} N(\mathfrak{p}^m)^{k-1} \tau(\chi^{-1})} L(1-k, \chi^{-1}) \tag{64}$$

along with the relation $\tau(\chi)\tau(\overline{\chi}) = \chi(-1)N(\mathfrak{p}^m)$ gives the integral against λ_j to be $L(1-k,\chi)^{-1}$. Summing over the strict ideal classes of K proves the result. □

5 p-Adic Measures of Totally Real Fields

It remains to show that the distribution λ is bounded. It is desirable to prove the boundedness directly by properties of the Eisenstein series. Instead, we show that the p-adic L-function is invertible as an element of the Iwasawa algebra. We first introduce the p-adic L-function of totally real fields and follow the construction of Deligne and Ribet [6].

Let $\mathfrak{n}$ be a non-zero integral ideal of K and let $G_{\mathfrak{n}}$ denote the strict ideal class group of K modulo $\mathfrak{n}$. Let $K(\mathfrak{n}p^\infty)$ denote the strict ray class field of modulus $\mathfrak{n}p^\infty$. By Class Field Theory, we can identify $\varprojlim_n G_{\mathfrak{n}p^n}$ with the Galois group $\mathrm{Gal}\,(K(\mathfrak{n}p^\infty)/K)$ of the largest abelian extension of K which is unramified at all the finite places of K prime to $\mathfrak{n}p$.

Let χ be any $\mathbb{C}$-valued locally constant character of $G_{\mathfrak{n}}$. We extend the norm map $N : K \to \mathbb{Q}$ to the unique continuous norm map $G \to \mathbb{Z}_p^*$ which we again denote by N. Siegel [16] showed that the values of $L(s,\chi) = \sum_{\gcd(\mathfrak{a},\mathfrak{n})=1} \chi(\mathfrak{a})N(\mathfrak{a})^{-s}$ at negative integers are rational whenever the values of χ are rational.

We now assume that all the primes lying above p divide $\mathfrak{n}$. Any locally constant function $\chi : G \to \mathbb{Q}_p$ factorizes through $G_{\mathfrak{n}'}$ for some $\mathfrak{n} \mid \mathfrak{n}'$, so we can define the values $L(1-k,\chi) \in \mathbb{Q}_p$ unambiguously. In particular, for any $c \in G$, $k \geq 1$, and $\chi : G_{\mathfrak{n}} \to \mathbb{Z}_p$, the values

$$\Delta_c(1-k,\chi) = L(1-k,\chi) - N(c)^k L(1-k,\chi_c), \tag{65}$$

where χ_c is the twist of χ by $c \in G_{\mathfrak{n}}$, are well defined independently of the choice of $\mathfrak{n}$ and define a p-adic measure. Explicitly, the following p-adic L-function exists:

Theorem 5.1 ([6] Theorem 8.2) *Let $\mathfrak{c}$ be an ideal of K which is relatively prime to p. Then there exists a $\mathbb{Z}_p$-valued measure $\mu_{1,\mathfrak{c}}$ on G such that for every integer $k \geq 0$ and every locally constant character χ, we have*

$$\int_G \chi(x)N(x)^{k-1}\, d\mu_{1,\mathfrak{c}} = (1-\chi(\mathfrak{c})N(\mathfrak{c})^k))L(1-k,\chi). \tag{66}$$

5.1 *Invertibility of Measures in the Iwasawa Algebra*

We begin by describing the p-adic measures on G as elements of the Iwasawa algebra $\Lambda(G) \cong \mathbb{Z}_p[[G]]$. We refer the reader to [4] for more details.

Let μ_p denote the group of all pth roots of unity and let K^{cyc} be the unique $\mathbb{Z}_p$ extension of K in $K(p^\infty)$. The group $G = Gal(K(p^\infty)/K)$ decomposes as $\Delta \times \Gamma$, where $\Delta = Gal(K(\mu_p)/K)$ and $\Gamma = Gal(K^{cyc}/K) \cong \mathbb{Z}_p$. Let $\delta : G \twoheadrightarrow \Delta$ be the projection map. Then, any element $x \in G$ can be written uniquely as $\delta(x) \cdot \delta(x)^{-1}x$ with $\delta(x) \in \Delta$ and $\delta(x)^{-1}x \in \Gamma$.

Assume that $p > 2$. Let $\omega : \mathbb{Z}_p^\times \to (\mathbb{Z}/p\mathbb{Z})^\times$ be a character and $N : K \to \mathbb{Q}$ be the norm map. We then have a character $\omega \circ N : G \to (\mathbb{Z}/p\mathbb{Z})^\times \subset \mathbb{Z}_p^\times$, which we will identify with a character $\omega : G \to \Delta$.

Any measure ν on Γ can be extended to the rest of G by the relation

$$\nu(aU) = \nu(U) \tag{67}$$

for any $a \in \Delta$ and compact open $U \subset \Gamma$. Let $\omega^i \nu$ be the product of the measure ν by ω^i, which is defined by

$$\int_G f\, d(\omega^i \nu) = \int_G f \omega^i\, d(\nu) \tag{68}$$

for any continuous function f on G. Since each ω^i is a continuous function on G and since the measures on G, endowed with the convolution product, form an algebra over the ring of continuous functions on G, the product $(\omega^i \nu)(aU)$ is a measure on G satisfying

$$(\omega^i \nu)(aU) = a^i (\omega^i \nu)(U). \tag{69}$$

On the other hand, every measure μ on G can be decomposed as a sum of measures

$$\mu = \frac{1}{q} \sum_{i=1}^{q} \mu_i \tag{70}$$

where

$$\mu_i(U) = \sum_{a \in \Delta} \omega^{-i}(a) \mu(aU). \tag{71}$$

Lemma 5.2 *There exists a unique measure ν_i on G which is induced from a measure $\widetilde{\nu_i}$ on Γ such that $\mu_i = \omega^i \nu_i$.*

Proof Let $\mathbf{1}_\Gamma$ denote the characteristic function of the finite index subgroup $\Gamma \subset G$. Set

$$\widetilde{\nu_i} = \mathbf{1}_\Gamma \mu_i. \tag{72}$$

Then $\widetilde{\nu_i}$ is a measure on Γ. By Eq. (67), we extend $\widetilde{\nu_i}$ to a measure ν_i on G such that $\nu_i(aU) = \nu_i(U)$ for any open $U \subset \Gamma$.

Define the measure $\omega^i \nu_i$ to be the product of the measures ω^i and ν_i on G. We need to show that $\omega^i \nu_i = \mu_i$. Since $\Gamma \subset G$ is a finite index subgroup and G is equipped with the profinite topology, then any open set $U \in G$ can be written as a finite, disjoint union of open sets

$$U = \bigcup_{\substack{b \in G \\ V \subset \Gamma \\ V \text{ open}}} bV. \tag{73}$$

Thus, it is enough to show that $\omega^i \nu_i(bV) = \mu_i(bV)$ for any $b \in G$ and any open set $V \subset \Gamma$ to prove the existence of ν_i. To this end, we have

$$\begin{aligned}\mu_i(bV) &\overset{(71)}{=} \sum_{a \in \Delta} \omega^{-i}(a)\mu(abV) = \omega^i(b) \sum_{a \in \Delta} \omega^{-i}(ab)\mu(abV) \\ &= \omega^i(b) \sum_{a \in \Delta} \omega^{-i}(a)\mu(aV) \overset{(71)}{=} \omega^i(b)\mu_i(V) \overset{(72)}{=} \omega^i(b)\widetilde{\nu_i}(V) = \omega^i(b)\nu_i(V).\end{aligned} \tag{74}$$

It remains to show that $(\omega^i \nu_i)(bV) = \omega^i(b)\nu_i(V)$. Let $\mathbf{1}_{bV}$ denote the characteristic function of the open set $bV \subset G$. Then we have

$$(\omega^i \nu_i)(bV) = \int_G \mathbf{1}_{bV}\, d(\omega^i \nu_i) \overset{(68)}{=} \int_G \omega^i \mathbf{1}_{bV}\, d\nu_i = \int_{bV} \omega^i\, d\nu_i. \tag{75}$$

Since $\omega|_\Gamma = 1$ and $V \subset \Gamma$, we have $\omega|_V = 1$. Hence,

$$(\omega^i \nu_i)(bV) = \int_{bV} \omega^i\, d\nu_i = \omega^i(b)\nu_i(bV) = \omega^i(b)\nu_i(V). \tag{76}$$

This shows that there exists a measure ν_i on G which is induced from a measure $\widetilde{\nu_i}$ on Γ such that $\mu_i = \omega^i \nu_i$. The uniqueness of ν_i follows from the fact that

$$\mathbf{1}_\Gamma \nu_i = \widetilde{\nu_i} = \mathbf{1}_\Gamma \mu_i. \tag{77}$$

□

For any continuous homomorphism $\chi : G \to \mathbb{C}_p^*$, let $\chi|_\Gamma$ denote the restriction of χ to Γ. Recall that $\delta : G = \Delta \times \Gamma \twoheadrightarrow \Delta$ is the projection map. The map $\chi \circ \delta : G \to \mathbb{C}_p^*$ factors through $\chi|_\Delta : \Delta \to \mathbb{C}_p^*$. Since Δ is a cyclic group of order q, so is its character group. Hence, the character group of Δ is generated by any character of order q. Note that $\omega : \Delta \to \mathbb{Z}/p\mathbb{Z}$ is such a character. This implies that $\chi(\delta(x)) = \omega^{-j}(\delta(x))$ for some $j \bmod q$.

Since χ transforms under Δ by ω^{-j}, the integral of χ against μ only involves the term for $i = j$ and equals

$$\int_G \chi(x)\,d\mu = \frac{1}{q}\int_G \chi(x)\,d(\omega^j \nu_j) = \int_\Gamma \chi|_\Gamma\,d\nu_j, \tag{78}$$

reducing the integration of the character χ on G to that of $\chi|_\Gamma$ on Γ.

Now assume that $\nu_i \equiv 0$ whenever i is even and that $\nu_i|_\Gamma$ is a unit in the Iwasawa algebra $\Lambda(G)$ whenever i is odd. Since $\Lambda(G)_{\omega^i} \cong \Lambda(\Gamma) \cong \mathcal{O}[[T]]$ ([2] Section 4), the invertibility condition is equivalent to the constant term

$$(\omega^i \nu_i)(\Gamma) = \sum_{a\in\Delta} \omega(a)^{-i}\mu(a\Gamma) = \int_G \omega(x)^{-i}\,d\mu \tag{79}$$

being a p-adic unit. In this case, for each odd value of i mod q, there exists an inverse measure ν_i^{-1} such that $\nu_i * \nu_i^{-1}$ is equal to the δ-distribution δ_1 at the identity by Iwasawa's isomorphism theorem, or in other words

$$\int_\Gamma f(x)\,d(\nu_i * \nu_i^{-1})(x) = \int_\Gamma\int_\Gamma f(xy)\,d\nu_i(x)\,d\nu_i^{-1}(y) = f(1) \tag{80}$$

for any continuous $\mathbb{C}_p$-valued function on Γ. In particular, when i is odd,

$$\int_\Gamma \chi|_\Gamma\,d\nu_i^{-1} = \left(\int_\Gamma \chi|_\Gamma\,d\nu_i\right)^{-1}. \tag{81}$$

We can then define an inverse measure μ^{-1} on G by

$$\mu^{-1} = \frac{1}{q}\sum_{\substack{1\le i\le q\\ i \text{ odd}}} \omega^i \nu_i^{-1}. \tag{82}$$

Moreover, evaluating the inverse measure μ^{-1} against odd characters χ,

$$\int_G \chi(x)\,d\mu^{-1} = \int_\Gamma \chi|_\Gamma\,d\nu_j^{-1} = \left(\int_\Gamma \chi|_\Gamma\,d\nu_j\right)^{-1} = \left(\int_G \chi(x)\,d\mu\right)^{-1}, \tag{83}$$

we see that the integrals of μ and μ^{-1} are reciprocals of each other.

5.2 Boundedness of $\boldsymbol{\lambda}$

We will prove that λ is a p-adic measure, that is, a bounded distribution, by showing that the p-adic measure on totally real fields $\mu_{1,\mathfrak{c}}$ is a unit in the Iwasawa algebra and that the regularization μ^* of its inverse $\mu_{1,\mathfrak{c}}^{-1}$ is equal as a distribution to a multiple of λ.

We start by considering the p-adic measure on totally real fields $\mu_{1,\mathfrak{c}}$. By Eq. (79), ν_i is invertible whenever the integral

$$\int_G \omega^{-i}(x)\, d\mu_{1,\mathfrak{c}} \tag{84}$$

is a p-adic unit. Using the property

$$\int_G \chi(x) N(x)^{n-1}\, d\mu_{1,\mathfrak{c}}(x) = \left(1 - \chi(\mathfrak{c}) N(\mathfrak{c})^n\right) L(1-n, \chi), \tag{85}$$

the integral in (84) is equal to $\left(1 - \omega^{-i}(\mathfrak{c}) N(\mathfrak{c})\right) L(0, \omega^{-i})$.

Lemma 5.3 *Let w be the number of roots of unity in $K(\mu_p)$. We can choose an ideal $\mathfrak{c} \in \mathcal{O}$ with $(\mathfrak{c}, p) = 1$ such that $1 - \omega^{-1}(\mathfrak{c}) N(\mathfrak{c}) \in w\mathbb{Z}_p^\times$.*

Proof Let $\chi : Gal(\mathbb{Q}(\mu_{p^\infty})/\mathbb{Q}) \xrightarrow{\sim} \mathbb{Z}_p^\times$ denote the cyclotomic character defined by

$$\sigma(\zeta) = \zeta^{\chi(\sigma)} \qquad \forall \sigma \in Gal(\mathbb{Q}(\mu_{p^\infty})/\mathbb{Q}),\ \zeta \in \mu_{p^\infty}. \tag{86}$$

Let χ_K denote $\chi|_{Gal(K(\mu_{p^\infty})/K)} : Gal(K(\mu_{p^\infty})/K) \to \mathbb{Z}_p^\times$.

Since $\chi_K|_{Gal(K(\mu_p)/K)} = \omega$ and

$$Gal(K(\mu_{p^\infty})/K) = Gal(K(\mu_{p^\infty})/K(\mu_p)) \times Gal(K(\mu_p)/K), \tag{87}$$

we have

$$\chi_K \omega^{-1} : Gal(K(\mu_{p^\infty})/K(\mu_p)) \hookrightarrow 1 + p\mathbb{Z}_p. \tag{88}$$

Now, Galois theory shows that

$$\begin{aligned} Gal(K(\mu_{p^\infty})/K(\mu_p)) &\xrightarrow{\sim} Gal(\mathbb{Q}(\mu_{p^\infty})/K \cap \mathbb{Q}(\mu_{p^\infty})) \\ &= Gal(\mathbb{Q}(\mu_{p^\infty})/\mathbb{Q}(\mu_{p^{ord_p(w)}})) \\ \sigma &\to \sigma|_{\mathbb{Q}(\mu_{p^\infty})}, \end{aligned} \tag{89}$$

where ord_p denotes the p-adic valuation with $ord_p(p) = 1$. Thus,

$$\chi_K \omega^{-1} : Gal(K(\mu_{p^\infty})/K(\mu_p)) \xrightarrow{\sim} 1 + w\mathbb{Z}_p. \tag{90}$$

Now, $N\omega^{-1} : \mathbb{A}_K^\times / K^\times \to \overline{\mathbb{Q}}_p^\times$ is the p-adic Hecke character over K which corresponds to $\chi_K \omega^{-1}$, and since the global Artin map is surjective, we get that the image of $N\omega^{-1}$ is $1 + w\mathbb{Z}_p$, which proves the claim. □

Theorem 5.4 *With $\mathfrak{c}$ chosen as in Lemma (5.3),*

$$p \nmid \left(1 - \omega^{-i}(\mathfrak{c})N(\mathfrak{c})\right) L(0, \omega^{-i}) \textit{ for all odd } 1 \leq i \leq q.$$

Proof By Lemma 5.3, we have that

$$\left(1 - \omega^{-i}(\mathfrak{c})N(\mathfrak{c})\right) L(0, \omega^{-i}) = p^{e_i} L(0, \omega^{-i}) u_i \tag{91}$$

for every $1 \leq i \leq q$, where $u_i \in \mathbb{Z}_p^{\times}$ and

$$e_i = \begin{cases} ord_p(w) & \text{if } i = 1 \\ 0 & \text{otherwise.} \end{cases} \tag{92}$$

Hence, it is enough to show that

$$p \nmid p^{e_i} L(0, \omega^{-i}) \quad \text{for every } 1 \leq i \leq q. \tag{93}$$

Greenberg's class number formula ([9], p. 250) states that

$$a \cdot h^{-} = \left(\frac{w}{2^{e+1}}\right) \prod_{\substack{i=1 \\ i \text{ odd}}}^{q} L^{(p)}(0, \omega^{i}), \tag{94}$$

where a is a power of 2, h^{-} is the class number of K divided by the class number of $K(e^{2\pi i/p} + e^{-2\pi i/p})$, w is the number of roots of unity in $K(\mu_p)$, and $L^{(p)}$ is the L-function with the Euler factor at p removed. By assumption, p is regular, which implies that p does not divide the left-hand side of Eq. (94). Hence,

$$p \nmid \left(\frac{w}{2^{e+1}}\right) \prod_{\substack{i=1 \\ i \text{ odd}}}^{q} L^{(p)}(0, \omega^{-i}). \tag{95}$$

Since ω has conductor p, all the Euler factors of the L-function at p are equal to 1, which implies that

$$p \nmid p^{e_i} \prod_{\substack{i=1 \\ i \text{ odd}}}^{q} L(0, \omega^{-i}). \tag{96}$$

This implies that each factor in the product on the right-hand side of Eq. (96) is not divisible by p. Therefore, Eq. (93) holds true. □

We conclude from the above discussion that the measure $\mu_{1,\mathfrak{c}}$ is a p-adic unit. Moreover, $\mu_{1,\mathfrak{c}}^{-1}$ is a measure which satisfies the property

$$\int_G \chi(x)N(x)^{k-1}\, d\mu_{1,\mathfrak{c}}^{-1} = (1-\chi(\mathfrak{c})N(\mathfrak{c})^k)^{-1}L(1-k,\chi)^{-1} \tag{97}$$

for any Hecke character χ and non-negative integer k of the same parity. By direct computation, the regularization

$$\mu^*(U) = \mu_{1,\mathfrak{c}}^{-1}(U) - N(\mathfrak{c})\mu_{1,\mathfrak{c}}^{-1}(\mathfrak{c}U) \tag{98}$$

satisfies the equation

$$\int_G \chi(x)N(x)^{k-1}\, d\mu^* = L(1-k,\chi)^{-1}. \tag{99}$$

We are now ready to prove that λ is a measure.

Theorem 5.5 *The distribution λ defined in Eq.* (42) *is a p-adic measure.*

Proof Define a distribution $\overline{\lambda}$ on G by $\overline{\lambda}(U) = \lambda(U^{-1})$. The integral in Eq. (46) is then equal to $\int_G \chi(x)\, d\overline{\lambda}$. Moreover, since μ^* is a measure, so is the product $N(x)^{k-1}\mu^*$ of the measure μ^* with the continuous function $N(x)^{k-1}$ for any k. Comparing Eqs. (46) and (99), we see that $\overline{\lambda} - h^+N(x)^{k-1}\mu^*$ vanishes when integrated against any Hecke character of finite type whenever $k \geq 3$. Since these characters span the space of locally constant functions on G, the distribution $\overline{\lambda} - h^+N(x)^{k-1}\mu^*$ is identically zero. This implies that $\overline{\lambda}$ is equal to $h^+N(x)^{k-1}\mu^*$ as distributions and so it is bounded. This implies that λ is a p-adic measure. □

Acknowledgments The author would like to thank Andrei Jorza, Ellen Eischen, and A. Raghuram for their helpful conversations, Romyar Sharifi for his explanation of the structure of the Iwasawa algebra, and the anonymous reviewer for their insightful comments and suggestions. This work is partially supported by Freydoon Shahidi's NSF grants DMS-1500759 and DMS-1801273 as a part of the thesis written under him at Purdue University.

References

1. R.W. Bruggeman, R.J. Miatello, I. Pacharoni, Density results for automorphic forms on Hilbert modular groups. Geom. Funct. Anal. **13**(4), 681–719 (2003)
2. P. Cassou-Noguès, Valeurs aux entiers négatifs des fonctions zêta et fonctions zêta p-adiques. Invent. Math. **51**(1), 29–59 (1979)
3. J. Coates, W. Sinnott, On p-adic L-functions over real quadratic fields. Invent. Math. **25**, 253–279 (1974)
4. J. Coates, R. Sujatha, *Cyclotomic Fields and Zeta Values*. Springer Monographs in Mathematics (Springer, Berlin, 2006)

5. H. Cohen, F. Strömberg, *Modular Forms*. Graduate Studies in Mathematics, vol. 179 (American Mathematical Society, Providence, 2017). A classical approach
6. P. Deligne and K.A. Ribet, Values of abelian L-functions at negative integers over totally real fields. Invent. Math. **59**(3), 227–286 (1980)
7. E. Freitag, *Hilbert Modular Forms* (Springer, Berlin, 1990)
8. S. Gelbart, S.D. Miller, A. Panchishkin, F. Shahidi, *A p-Adic Integral for the Reciprocal of L-Functions*. Automorphic Forms and Related Geometry: Assessing the Legacy of I. I. Piatetski-Shapiro, Contemporary Mathematics, vol. 614 (American Mathematical Society, Providence, 2014), pp. 53–68
9. R. Greenberg, A generalization of Kummer's criterion. Invent. Math. **21**, 247–254 (1973)
10. H. Hida, *Elementary Theory of L-Functions and Eisenstein Series*. London Mathematical Society Student Texts, vol. 26 (Cambridge University Press, Cambridge, 1993)
11. T. Kubota, H.-W. Leopoldt, Eine p-adische Theorie der Zetawerte. I. Einführung der p-adischen Dirichletschen L-Funktionen. J. Reine Angew. Math. **214**(215), 328–339 (1964)
12. H. Klingen, Über den arithmetischen Charakter der Fourierkoeffizienten von Modulformen. Math. Ann. **147**, 176–188 (1962)
13. E.E. Kummer, Über eine allgemeine Eigenschaft der rationalen Entwickelungscoefficienten einer bestimmten Gattung analytischer Functionen. J. Reine Angew. Math. **41**, 368–372 (1851)
14. J.-P. Serre, *Formes modulaires et fonctions zêta p-adiques*. Modular Functions of one Variable, III (Proceedings of International Summer School, Univ. Antwerp, 1972). Lecture Notes in Mathematics, vol. 350 (1973), pp. 191–268
15. T. Shintani, On evaluation of zeta functions of totally real algebraic number fields at non-positive integers. J. Fac. Sci. Univ. Tokyo Sect. IA Math. **23**(2), 393–417 (1976)
16. C.L. Siegel, Über die Fourierschen Koeffizienten von Modulformen. Nachr. Akad. Wiss. Göttingen Math.-Phys. Kl. II **1970**, 15–56 (1970)

Printed in the USA
CPSIA information can be obtained
at www.ICGtesting.com
CBHW072128290724
12402CB00003B/11

9 783031 51676